AF722929

WITHDRAWN

Methods of Inference from Animal to Human Behaviour

Maison des Sciences de l'homme, Paris
Publications 3

MOUTON · THE HAGUE · PARIS

Methods of Inference from Animal to Human Behaviour

Edited by

MARIO VON CRANACH

Conference on the Logic of Inference from Animal to Human Behavior

ALDINE · CHICAGO / MOUTON · THE HAGUE · PARIS

QL
751
.C65
1973

Proceedings of the Conference on "The Logic of Inference from Animal to Human Behaviour", held in Murten, Switzerland, in March 1973, sponsored by the Maison des Sciences de l'Homme, Paris, with the aid of the Fonds National Suisse de la Recherche Scientifique.

Copyright © 1976 by Mouton & Co., The Hague, and Maison des Sciences de l'Homme, Paris. No part of this publication may be reproduced, stored in a retrieval system or transmitted in any form or by any means, electronic, mechanical, photocopying, recording or otherwise, without the written permission of Mouton Publishers, The Hague.

Published in the United States of America and Canada by Aldine Publishing Company, Chicago, Illinois.

ISBN 90-279-7763-1 (Mouton)
0-202-02020-7 (Aldine)

Printed in France

Two saints were crossing a flooded river by walking upon the water, when a pilgrim arrived at the shore and asked them: "How can you walk upon the water, why don't you sink?" "It is faith that enables us to walk upon the water," answered the first saint, and when the pilgrim hesitated, the second added: "Just be strong in your faith and follow us." The pilgrim prayed shortly, turned his gaze towards heaven, set his feet on the water and immediately disappeared. The first saint turned to the second one and asked: "Should we have told him about the stepping stones?"

Reported at the conference by
MICHAEL CHANCE

Preface

This book represents the proceedings of the conference on "The logic of inference from animal to human behaviour", which I organized in March, 1973, in Murten, Switzerland, with the generous help of the Maison des Sciences de l'Homme, Paris, for which I am deeply grateful. Thanks are also due to the Schweizerische Nationalfonds for financial support, and to the Hochschulstiftung der Universität Bern who offered aid in case the other sources would not suffice. The book would never have been finished without the aid of my wife, Agnes von Cranach, who did all the technical and a good deal of the scientific editing. Since I could not persuade her to sign as a co-editor, I can only say that her name should appear on the book's front page; it is her work as well as mine.

During the conference, each presentation was followed by a discussion; we decided to edit these discussions since they seemed to contain ideas that are crucial for the solution of the meeting's central problem. In the editing, we sometimes changed the temporal sequence of the remarks for the sake of contextual continuity; for the same reason, and for shortness' sake, we had to omit many interesting remarks, mainly on details of animal behaviour. Since A. J. Cain could not send us a paper, we included the summary he provided for the conference. M.R.A. Chance, F. Bresson and K. Foppa included the gist of the ensuing discussions into their papers. Unfortunately our English is far from perfect; we beg indulgence from our readers for any mistakes we have made. Anthony Ambrose came to Bern and worked hard for two days answering our questions on language problems, we owe him special thanks; the fact that he helped us to avoid numerous mistakes does not make him responsible for those we have made, though.

Furthermore we want to express our thanks to our Bernese staff, Christiane Mauderli and Charlotte Apostolou who did the typing, and to Anne-Marie Allemann who helped in the organization of the conference; to Anne Balandier, Paris, for her coordinative efforts and to Maryann

Macdonald who worked over the manuscripts. This list of acknowledgments would be sadly incomplete, however, if it did not include the name of Clemens Heller, Administrateur Adjoint of the Maison des Sciences de l'Homme, whose encouragement and help made the enterprise possible.

Bern, May 1975

MARIO VON CRANACH

Contents

List of Participants

Anthony Ambrose
Dundon House
Minster Lovell
Oxford

François Bresson
Laboratoire de Psychologie
École des Hautes Études en Sciences Sociales
Paris

Arthur J. Cain
Department of Zoology
University of Liverpool

Michael R. A. Chance
Sub-Department of Ethology
University of Birmingham

Mario von Cranach
Psychologisches Institut
Universität Bern

John H. Crook
Department of Psychology
University of Bristol

Klaus Foppa
Psychologisches Institut
Universität Bern

Siegfried Frey
Psychologisches Institut
Universität Bern

Alan Goodall
Department of Zoology
University of Liverpool

Henry Hécaen
Laboratoire de Pathologie du Langage
École des Hautes Études en Sciences Sociales
Paris

Jan A.R.A.M. van Hooff
Laboratorium voor Vergelijkende Fysiologie
Rijksuniversiteit Utrecht

Alfred Lang
Psychologisches Institut
Universität Bern

Paul Leyhausen
Arbeitsgruppe Wuppertal
Max-Planck-Institut für Verhaltensphysiologie

Roger D. Masters
Department of Government
Dartmouth College
Hanover, New Hampshire

Thomas Pitcairn
Arbeitsgruppe für Humanethologie
Max-Planck-Institut für Verhaltensphysiologie
Percha/Starnberg

Detlev Ploog
Max-Planck-Institut für Psychiatrie
Munich

O. E. Rössler
Abteilung für theoretische Chemie
Universität Tübingen

Scania de Schonen
Laboratoire de Psychologie
École des Hautes Études en Sciences Sociales
Paris

Uta Seibt
Max-Planck-Institut für Verhaltensphysiologie
Seewiesen bei Starnberg

David G. Senn
Zoologische Anstalt
Universität Basel

Wolfgang Wickler
Max-Planck-Institut für Verhaltensphysiologie
Seewiesen bei Starnberg

MARIO VON CRANACH

Introduction
The Order of Topics in this Book

This outline is essentially identical with the introductory paper I gave to the participants at the beginning of the conference. Although, compared to my present understanding of the comparison problem, the views expressed here seem somewhat naive, it may still be of value for the reader to read this short exposition, since it introduces the topic and presents the considerations that led me to structure the conference and thus the realm of problems in this particular way.

For some time, I have been confronted again and again with various questions that were relevant for inferences from animal to human behaviour; these questions suggested a logical order that seemed to represent a possible methodology of the field. Accordingly, I tried to find specialists for these questions, and arranged their contributions in a corresponding order. Of course, it soon became obvious that this previewed order could not be maintained entirely. Neither was it possible to find contributors for all the topics (especially social scientists proved to be not particularly interested in the problem), nor did all of the participants consider the pre-arranged order appropriate; and in the course of the conference, aspects which had not been preconsidered became important. Nevertheless, this introduction may help the reader to better understand the course of the proceedings and to get acquainted with the ideas involved.

Aim of the Conference

I started from the observation that explicit comparison of the behaviour of animals and that of man became more and more frequent in literature (in psychology, human ethology, anthropology, sociology, political science, etc.); this sensitized me to the fact that in many exacting and method-observant disciplines (like physiology, pharmacology or learning theory in experimental psychology) implicit comparisons of the same kind are quite common, and pass undebated.

As I saw it, these comparative inferences obviously imply methodological assumptions and decisions, but these are rarely discussed. Instead, arguments of a more polemic nature can frequently be found. On the other hand, some authors seem to be convinced that efforts to clarify the methodology involved were bound to fail (*e.g.* H.P. Harlow *et al.,* 1972).

The aim of this conference was a rational discussion of the problems involved in the comparison of animal and human behaviour, including basic questions of comparative methodology as well as the clarification of the most important methodological steps and their proper order, illustrated by pertinent examples. It was further hoped that the discussion would lead to the establishment of a set of inferential rules, and that the most important cases of the realm would be treated.

Basic Types of Reasoning

In our conference, we had to discuss inferences from animal to man: and occasionally, it could be helpful to consider inferences from man to animal; both kinds of inference could be considered special cases of comparison of systems. Three basic types of reasoning seem to be involved (von Cranach, 1970):

— All of these comparisons are based on judgments of similarity/dissimilarity of whole systems, or parts of systems (subsystems).
— If the comparison somehow, as a presupposition or as a goal, comprises the notion of relationship, the assumption that structures of both systems are based on information that has been passed from one system to the other, or from a common ancestor to both, similarity between structures is considered based on *homology*.
— If two systems have something in common without being related, conclusions are drawn on the basis of *analogy*.

Thus, both homology and analogy are based on the notion of *similarity*. The establishment of similarity would be essential in all further discussions. (See S. Frey, paper and discussion.) In each case it has to be stated what should be judged to be similar to what in what respect; in other words, we have to specify the level on which a comparison is to be made. Some further problems arose: which are the methods for the assessment of similarity, how do they work, and what are their advantages and disadvantages?

Similarity is, in most cases, assessed on the basis of global perceptual judgments, that is to say in a subjective way. It was interesting to see what objective methods (*e.g.* multidimensional analysis) exist for this purpose.

Since the concepts of homology and analogy were central for our thinking, particular emphasis was placed on their definitions.[1] (See A.J. Cain, paper and discussion.) For both of them, differentiating criteria had been developed that had to be re-examined. Intermediate forms between homology and analogy had also been distinguished, and it was to be discussed whether they constituted their own methodology, and whether they were important for our problem.

The Assessment of Behavioural Homology

Since the method of homologizing originated from comparative morphology, it seemed important to understand how it was used in this discipline. Next, we had to understand how the original method was applied to behavioural structures. (See also D.G. Senn and W. Wickler, papers.) It seemed very probable that not all of the participants would consider this application justified.

Various problems were expected; here are some examples: in morphology, the homology method is used to determine genetic relationship. In ethology, genetic relationship of the compared organisms is presupposed, and the method is used to establish the identity of single behaviours.

Whatever the function of relationship may be, just what is relationship? On what circumstances does it depend, what degree of relationship is required? (Man–chimpanzee? Man–butterfly?). What is considered as "related", "identical", or "different" seems to depend on various factors that are, up to now, not well described. (Different individuals of the same species may be treated as *related* or *identical, e.g.* in statistical designs, or as *different, e.g.* in differential psychology or behaviour genetics. The same individual may be treated as relatively unrelated to itself in developmental studies, considering different stages of ontogenetic development, or in successive steps of an experiment.)

Analogical Comparison

Although comparative conclusions based on analogy are quite common, I felt that their basic logical nature did not emerge. For this reason, two papers were provided. (See R.D. Masters and O.E. Rössler.) [2] What I hoped for was the solution of questions like the following:

- What are the logical and factural presuppositions of analogy-conclusions (especially with regard to similarities between systems, or system–environment relationships)?
- Is the use of a "model" one form of analogy?
- Are there general factors or circumstances from which the usefulness of a particular analogy can be predicted?

— Can an unjustified homology always be considered an analogy (as some people seem to believe)?

We had the feeling that analogical conclusions would perhaps turn out to be the most important form of reasoning in our realm of problems.

Brain Structures

We could, however, not stop with the comparison of behaviour. Especially in relation to the problem of homology, further questions arose. If, in ethology, the method was applied to movements instead of morphological structures, the implicit justification would be that these movements were visible manifestations of invisible neurophysiological structures. If so, we should have to consider the function of similarities/dissimilarities of the central nervous systems of related species.[3] (See H. Hécaen and D. Ploog, papers.)

On the other hand, it could be argued that movements were not only considered as visible manifestations of central nervous organizations, but as structures with their own value, especially since they are directly subject to selective mechanisms.

Complex Behaviour

If we assumed that we did understand the operation of homologous and analogous conclusions, and that their methodological and factual requirements were spelled out in detail, we had to reconsider the matter in regard to different levels of complexity of behaviour. Facial expressions in man and sub-human primates constitute a relatively simple example although difficult enough. (See J.A.R.A.M. van Hooff, paper.) The comparison of processes on the receptor side seems even more difficult. Cognitive schemata of various kinds, from "taxis" to "innate releasing mechanisms", postulate an interaction between a stimulus and a neurophysiological mechanism. What method of comparison would be applied to these? Could we state that all comparisons involving system–environment interactions have the status of functional analogies? This, then, would also be true for all kinds of "interactive behaviour" that may be analysed, in the simplest case, as mutually stimulated behaviour.

Unfortunately, we could not provide a special paper for these problems. Another important case can be found in instincts and motivations, that are generalizations somehow summing up events of single behaviours. (Psychologists would talk about "intervening variables" or "constructs".) The decision whether or not these concepts should be considered to represent neurophysiological mechanisms will influence their comparative treatment. In any case, homological conclusions appeared difficult. (See P. Leyhausen, paper.)

The most complex case for comparison was perhaps that of collective behaviour and of social systems. Obviously, we had to rely on analogical reasoning. The usefulness of animal–man comparisons in this area seemed to depend on the impact of cognitive and cultural factors and their integration into behavioural structures. Comparison of formal aspects on a rather high level of analysis could eventually become accessible.[4] (See M.R.A. Chance and J.H. Crook, papers.)

Finally, in the context of complex behaviour the animal-to-man inferences in learning theory and in the comparison of learning achievement also have to be considered. (See K. Foppa, paper.)

Comparative Studies of Development

Problems similar to those occurring in the comparison of complex behaviour arose when we looked at comparative studies of development. In the comparison of different developmental states of one system, homological conclusions (*e.g.* in tracing the ontogenetic development of one particular behaviour, let us say the "smiling response") as well as analogical conclusions (in comparing different states of system–environment relationships) could occur. The comparison of the development of different species, namely animal–man, was even more complicated.[5] (See J.A. Ambrose, and F. Bresson, this vol.)

Advice to the Reader

From this order and intent of the papers, some advice for the reader can be derived. All the single papers can stand for themselves and yield information about their specific topic. Achievement of full understanding of the comparison problem, however, requires the study of the papers in their order. This is, of course, even more true for the discussions, which can only partially be understood without reference to the whole conference. The final conclusions are my summary of the comparison problem, but not a summary of the book.

NOTES

1. A.J. Cain's paper and the following discussion deal with the use of these concepts in evolutionary theory, but definitional problems are also treated in other papers and discussions.
2. R.D. Masters discusses the most relevant form of analogy, *functional analogy*. O.E. Rössler's paper clarifies the issue by discussing a more formalized example from theoretical chemistry.

3. H. Hécaen's paper discusses hemispheric specialization as a pertinent example; D. Ploog's paper treats the problem under the special aspect of cerebral evolution and dissolution.
4. M.R.A. Chance and J.H. Crook have discussed the relevance of primates in respect to human social organization in their papers. Unfortunately no correspondent could be recruited to represent the social sciences.
5. A. Ambrose's paper gives a treatment of methodological problems in the comparison of development. Unfortunately we received Mr. Bresson's paper too late to consider it in our conclusions.

REFERENCES

Cranach, M. von. 1970. "Ethology and Human Behaviour", pp. 370-392 in: J. Israel/H. Tajfel (eds.) *The Context of Social Psychology*. London/New York, Academic Press.

Harlow, H.F./Gluck, H.P./Suomi, Stephen J. 1972. "Generalization of Behavioral Data between Nonhuman and Human Animals", *American Psychologist* (August) 709-716.

SIEGFRIED FREY

The Assessment of Similarity

It appears from the program of this conference that during this week we will spend a great deal of time discussing whether behaviours of different organisms show a homologous or analogous relationship. The problem I am concerned with in my talk precedes these questions. Similarities among behaviours serve as empirical references for both homological and analogical reasoning. Actually it seems that there must be agreement as to the similarity of phenomena before any controversy about homology or analogy can arise. In this presentation I shall try to show that the concept of similarity, as well as data on similarity, constitute a potential source of controversy, too. In reviewing the literature, it becomes obvious that there are marked differences as to what we mean if we talk about similarity and that there are marked differences as to what we do in order to arrive at a statement about similarity. The latter question concerns the crucial methodological problem, but before we go into this, I shall first try to characterize some conceptual problems of similarity.

CONCEPTUAL PROBLEMS OF SIMILARITY

Generally speaking, a statement about similarity reflects an experimenter's claim that investigated phenomena have common characteristics. The term similarity also indicates that the experimenter has observed characteristics in which these phenomena differed. In classifying phenomena as being similar, however, the experimenter implies that he considers the differences he observed as negligible, possibly because they are so few in number or possibly because he considers them unimportant.

From this it becomes clear that similarity data do not give a description of phenomena themselves, but provide an evaluation of descriptions made prior to the statement about similarity. Therefore if we want to understand adequately the experimenter's statement about the similarity among the behaviours he has investigated, we must have information about two things: 1) The rules used by the experimenter to transfer observations into raw data. Technically speaking, we must know what kind of scale provided the reference for the experimenter's differentiation of behaviours. 2) The rule by which the experimenter derives statements about similarity from raw data.

As von Cranach and Frenz (1969) pointed out several years ago, the experimenter decides by the choice of his scale which differences he allows to occur in an empirical representation of actual behaviour. If a scale consists of few large intervals, von Cranach and Frenz expect that "behaviours of considerably different kinds and considerably different meanings" might be attributed to the same interval. This, of course, has severe implications with respect to similarity since, as Thomas warned already in 1932, behaviours once attributed to the same classificatory interval will be considered as identical in any further analysis. In general we can expect that the extent to which a given scale can discriminate among phenomena is inversely proportional to the extent to which it will identify similarities among phenomena. We shall go into these problems again when we discuss methodological strategies. In order to avoid misconception, however, I hasten to add that the problems arising from the choice of the scale can be considered as being largely independent of the problem of objective and reliable assignment of phenoma to a given scale. While by using scales we preorganize behavioural phenomena with respect to their similarities, there is always the possibility of examining the descriptive data for further similarities by stipulating rules for such evaluations.

Similarity must be considered a quality that has no dimension; thus it can be employed for comparisons among any phenomena. In view of this flexibility it is not surprising that quite a number of scientific disciplines refer to the similarity of the phenomena they investigate. Disciplines vary widely, however, in the extent to which they make use of this term. In geometry and physics only particular sets of data are evaluated in terms of similarity; psychologists and human ethologists seem to have a general interest in evaluating data in terms of similarity. Accordingly, the criteria given for the assessment of similarity are either applicable only to particular types of data, or they are applicable to a great variety of phenomena. To my knowledge, in geometry the term similarity is employed only with data that reflect the shape of geometrical figures, so that the criteria used for the assessment of similarity refer to relations among the points of a geometrical figure, or to angles constituted by lines that ultimately constitute the geometrical figure. In physics, the use of the term similarity seems to be restricted largely to the comparison of currents measured in fluids; the criteria for the assessment of similarity are defined with respect to these data.

With the advent of psychology as a science, an incalculable amount of data has been collected in order to describe an incalculable number of behavioural properties. In the permanent attempt to create order among the huge body of data, psychologists made every effort to find possibilities for the assessment of similarity. Their decision to look for similarities among sets of data, however, has led to serious restrictions as to the possible ways in which criteria for the assessment of similarity could be defined. Of course, if we want to make sure that *any* set of data will provide the information necessary for the assessment of similarity, we have to chose criteria that are non-specific, at least with respect to the label given to behavioural properties. This actually is the case with regard to the criteria employed by psychologists for the assessment of similarity. The rules for the assessment of similarity refer to the scale quality of the basic data rather than to the label given to behavioural properties. However, as a consequence of the fact that psychologists employ several types of scales in their investigations, several sets of rules for the assessment of similarity had to be developed. Along with the assumptions about the scale quality of the raw data, different strategies are recommended to be used for sets of data that constitute either nominal, ordinal or interval scales. The most prominent strategies used for the evaluation of sets of data with respect to their similarities are correlation techniques. They are employed in most of the cases in which the experimenter intends to answer the question of whether the data he gathered about two or more behavioural properties show a similar distribution. The prescriptions for the calculation of correlation values are defined in such a way that a perfect linear correspondence in the order of the different sets of data yields a correlation value of either + 1 or − 1. If the sets of data are in a completely random order with respect to each other, the correlation results in the value "zero". As to the way psychologists actually interpret correlation values in terms of the similarities among the sets of data, there appears to be an important difference between two groups of correlation techniques: the bivariate and the multivariate correlation techniques. If bivariate correlation techniques are applied, inferences about similarity are usually made with reference to a significance test: there is similarity among the sets of data if the correlation value is found to be significant, while there is no similarity if the correlation value is insignificant. Clearly, if we base our decision about similarity on the criterion of statistical significance, we no longer conceive similarity as a quantitative variable. Consequently, correlation values of, for example, .3 and .6 would not reflect any difference in similarity if both of these values are significant or if both of these values are insignificant.

In the first case, when both coefficients are significant, we could only conclude that there was some similarity among the sets of data. More precisely, we could only conclude that the sets of data are not completely unordered with respect to each other. In the second case, where both coefficients are shown to be not significant, we could only conclude that we did not find any sign of similarity among the sets of data. In particular, in the case of a significant correlation as well as in the case of an insignificant correlation the inference rules do not permit the conclusion that the .6 coefficient indicates a higher degree of similarity between the sets of data than does the .3 coefficient. These principles refer, of course, only to the orthodox approach, and in fact psychologists seem to violate these inference rules all the time, but that is another problem.

In multivariate correlational techniques such as factor analysis or cluster analysis, similarity is conceived in a different way, namely, as a quantitive variable. According to the factor analytic concept, correlation coefficients reflect degrees of similarity. Using our previous example again, the .6 correlation would reflect a higher degree of similarity than would the .3 coefficient, no matter whether the correlation coefficients are significantly or insignificantly different from a zero coefficient. In factor analysis, we arrive at statements about the similarity of variables by correlating each set of data with each other set of data and by subsequent grouping of the results into groups of variables that correlate highly with each other. The different groups of highly correlated variables are represented by factors that are considered to reflect the similarities among a fairly large number of variables.

Of course, if we take the conservative point of view that correlation coefficients can only tell whether there is no zero correlation, and do not yield reliable information about the magnitude of similarity, we can hardly approve of the factor analysts' rules for the assessment of similarity. On the other hand, the inference strategy employed in classical bivariate techniques seems to be unsatisfactory, too, because we ought to be skeptical about the value of a similarity concept that assumes similarity whenever sets of data are found to be not completely zero-correlated. A quick glance at the significance tables for correlation coefficients shows that even correlation coefficients close to zero will nevertheless be classified as significant if a large number of data have been analysed. Since, on a mere descriptive level, a correlation coefficient that is close to zero indicates that variables are differently distributed with respect to each other, we might ponder the legitimacy of such statements of similarity.

In any case, the conceptual inconsistencies as they exist with respect to similarity between bivariate and multivariate correlation strategies certainly deserve continued attention. For the purpose of this presentation, however, it may suffice merely to state them.

No matter how much we agree or disagree with any particular concept of similarity, as far as the correlational techniques are concerned, they are all formalized to the degree that allows an objective and reliable use of the provisions given to assess similarity.

In human ethology, formalized models for the assessment of similarity have, as far as I know, not yet been developed. The assessment of similarity seems to be integrated into the strategies used for the differentiation of behavioural phenomena. Thus we might postpone the question about how human ethologists arrive at statements about similarity and turn our attention to problems classified earlier as methodological.

THE CHOICE OF A MEASUREMENT SCALE AND ITS IMPLICATIONS FOR THE ASSESSMENT OF SIMILARITY

At the beginning of my talk I mentioned that we introduce a systematic bias with respect to similarity at the moment we choose the scale with which we discriminate among behavioural phenomena. Which bias we introduce depends on the particular features of the chosen scale (especially on the number and lenght of its intervals). However, there is a kind of bias with respect to similarity that is characteristic for types of scales. If nominal scales (or their special cases, the dichotomous and the indicator scales) are being used for the differentiation of behaviours, we preorganize behavioural phenomena with respect to a similarity of a very special kind, namely identity. If we discriminate behaviours on the basis of ordinal interval or ratio scales, the particular order of the intervals constitutes the bias for similarity. A simple example might serve as an illustration: suppose we are asked to describe a subject's behaviour in terms of walking, sitting, kneeling, standing, lying, crawling, jumping and running. If we decide to assess such behaviours on the basis of a nominal scale, we can start right away with the assessment of such behaviours. Of course, the different ways of walking, sitting, kneeling, standing and so on would be assigned to their respective intervals and thus be considered as being identical in any further analysis. If we decide to employ an ordinal scale for this coding task, we should first have to establish an order among our phenomena; we should have to define a scale. We might list phenomena in this order: lying, sitting, kneeling, standing, crawling, walking, jumping and running. Or we might see reason to establish the list in this order: lying, kneeling, crawling, sitting, jumping, standing, walking, running. Of course, depending upon the order given in our scale, kneeling would in the first listing be considered most similar to sitting and standing, in the second listing it would be similar to lying and crawling. This example shows that in order to arrive at objective and reliable statements about the similarity of behaviours, it is necessary to make stipulations about the particular type of scale serving as a reference system. But once we have made the necessary stipulations about which scale we use in order to discriminate among behaviours, the problem arises as to whether we assign similar behaviours to the intervals in an objective and reliable way. Inasmuch as behavioural properties have to be assessed via observation, psychologists frequently use "behavioural checklists" or "rating scales" as reference systems for their observations. Thus, depending on what scale is being used, the observer might be required to assess phenomena like *stereotyped, nervous, dominant, agitated, aggressive, appeasing* or *threatening* behaviour.

Of course, such data suffer from the fact that insofar as no agreement has been reached about the definition of concepts such as stereotypy, nervousness, dominance or aggressiveness, the observational criteria for sterotypical, nervous, dominant or agressive behaviours cannot meet methodological requirements for objectivity and validity. Whether they satisfy the postulate of reliability depends on whether the item definition exactly specifies the coding operations. This does not seem to be the case, however, for the great majority of complexly defined behavioural properties. The assessment of these data most often follows the methodological principle termed by Campbell (1958) "reductive coding". In this procedure, the observer decides during actual observation whether and to what degree the behaviour of the observed subject represents the property in question. Since these data result from the registration of an unknown number of behavioural details computed in an informal weighing process, their reliability is doubtful.

Thus, if we are dealing with data generated by reductive coding techniques, we can assume that von Cranach and Frenz (1969) are right in expecting that data attributed to the same scale interval constitute a pool of data considerably different in kind and meaning.

BEHAVIOURAL ANALYSIS BY MEANS OF METHODS THAT PERMIT THE SEPARATION OF THE DESCRIPTIVE AND THE EVALUATIVE ASPECTS OF THE MEASUREMENT PRECEDURE

In the course of our research, von Cranach and I have developed a strategy by which we hope to be able to cope with the methodological problems involved in the assessment of data about complexly defined behavioural parameters. If we characterize complex behavioural parameters via reductive coding techniques, the descriptive and evaluative activities of the experimenter are confounded in the statement about the behavioural parameters. In contrast to the reductive coding approach, we recommend a strategy that strictly separates the descriptive and the evaluative activities that give rise to the statement about a behavioural parameter. Of course, if the data are to be generated by such a procedure, the experimenter is required to develop a definition of the parameter that exactly specifies *a)* which behavioural details are to be recorded and *b)* which computing processes are to be applied to the data recorded. Depending upon the complexity of the definition of a parameter, a large or a small number of behavioural details have to be recorded. Naturally, this strategy requires much more work and much more technical equipment than the reductive coding procedures, since the operational specifications usually imply the use of videotape recorders, time counters, event counters and, in some cases, small computers. If one follows such a strategy, however, the improvements in the reliability of the data are striking. Since the evaluative process that gives rise to a statement about the parameter in question follows formalized prescriptions, the reliability of the statement is reduced only by errors that occur in the assessment of the relevant behavioural details. If we take advantage of the technical properties of videotape recorders and event counters, the influence of these errors can be kept minimal.

With respect to the economic aspect of such a procedure, it might be worthwhile to note that different behavioural parameters usually require the assessment of different behavioural details. If the investigator knows already at the beginning of his experiment which behavioural parameters are to be evaluated, it might be economical to restrict the description of behaviour to the assessment of those details that are required for an evaluation of the parameters in question. On the other hand, if the experimenter gives a complete specimen record of the behaviour occurring, the resulting data allow an evaluation of an almost infinite number of behavioural parameters.

There seems to be strong agreement in the literature that a complete specimen record that shows isomorphism with behaviour cannot be achieved. Although we do not doubt the validity of this opinion, it is clear that we can describe behaviour much better than we commonly believe. Behaviour certainly occurs in the form of a continuous process. For practical purposes, however, we can consider behaviour as a sequence of discrete states. Movie cameras register visible behaviour in the form of 25 discrete positional localizations per second, videotape recorders, in the form of 50 or 60 discrete positional localizations per second. As we all know, movie cameras can give a quite satisfactory record of visible behaviour on the basis of 25 discrete positional states per second. Thus, if we could code the discrete positional states that are registered on movie films, the coded record would constitute a quite satisfactory protocol of behaviour, too. In order to be able to evaluate these assessments of behaviour within fairly complex parameters, we must specify, on the basis of the definition given for the behavioural parameters, which behavioural details are to be selected and how they are to be evaluated in order to give rise to a quantitative statement about the parameters in question. If the codings are being put on punch cards, we can economically evaluate the coded behaviours with respect to all parameters that require information about visible behaviour.

The development of a coding system that obtains a sufficient resolution of the discrete positional states appears to be more difficult than it actually is. In the course of our work at the University of California, Jeremy Pool and I developed a coding scheme for the assessment of positional states as they occur in situations where people are communicating while sitting in a chair. The operational specifications of this scheme require the investigator to code off of videotape at each given point in time, a set of 8 separate positions which refer to the head, trunk, right hand, left hand, right thigh, left thigh, right foot and left foot.

Studies on re-test and inter-coder reliability brought satisfactory results. Percentages of agreement on the assessment of positional states varied from 83 to 97. The median value of agreement for both re-test and inter-coder reliability is 94 per cent. In order to test the validity of positional codings, we performed a study in which we tried to reconstruct the original positions of subjects from our data. For this purpose, the head, trunk, hands, thighs and feet of a human model were placed in specific positions, read back from the codings obtained with the original subjects. Pictures taken of this model were then compared with the pictures showing the subject's original position. In this way, the degree of isomorphism between original behaviour and the codings could be assessed. Thus we were able to asses the degree of accuracy with which the codings represented the subject's actual positions.

As illustrated in the photographs, the results of this study can be considered to be very satisfactory with regard to the power of accurate discrimination between the different positions provided by our coding system. Only recently have we begun to concentrate on the development of the specifications necessary for a multidimensional evaluation of the behavioural records, but this work is proceeding rapidly. We have already defined the formalized prescriptions for the computerized evaluation of about 40 behavioural parameters. We started, of course, with the definition of parameters of relatively low conceptual complexity like *variability of body movement, taxis of head movement, spatial expansion, openness of the behaviour of hands and feet, amounts of displacement, rate of movement, etc.* Our next step will be the definition of more conceptually complex parameters like friendliness, dominance, shyness, excitement.

This approach will be strictly empirical. In order to acquire the information necessary for the definition of such parameters, we will try to assess the behavioural phenomena which discriminate among people who have been consistently labelled as showing, for example, friendly, dominant, shy behaviour. At present, we use our coding system to analyse videotape records of persons whose visually observable behaviour has been consistently labelled as either friendly, shy, dominant or excited. After completion of the description, we will analyse the data in order to identify the behaviours that discriminate reliably among people whose behaviour has been labelled as friendly–unfriendly, dominant–submissive, etc. Careful experimental design, combined with our high resolution method of description should ultimately provide the information about the behavioural details that represent the different complex parameters in question. Once we know the behavioural details that constitute a conceptually complex parameter, it is possible to construct a formalized prescription for their assessment. *

It should be clear from my explanation that a methodological strategy strictly separating descriptive and evaluative activities in the measurement of behavioural parameters guarantees that all relevant behavioural details are selected and evaluated in accordance with the definition given for a particular parameter. This, of course, has important implications for the assessment of similarity among behaviours. If we arrive at behavioural data via formalized procedures, we can safely claim that the similarities in our data reflect similarities in the behaviours that have been described. In contrast, if we arrive at behavioural data via reductive coding procedures, the similarities in our data might reflect little more than the fact the behaviours observed have been given a common label. Lest the picture of the value of the obviously much respected reductive coding techniques look too black, I should mention that reductive coding techniques can well be of value in pioneering studies in which the data gathered do not serve for the testing of an hypothesis but for the construction of one. Thus, reductive coding data can be quite helpful if they are considered as heuristic.

It seems to me that the observational strategies used by ethologists often follow the reductive coding principle. Ethologists have so far, however, mainly concentrated on behavioural patterns that occur repeatedly and show little variation as to the elements involved. Their way of analysis has led to the discovery of some very important behavioural phenomena, such as the fixed action patterns. Due to the small actual variation that occurs in such patterns, the assessment of their occurrence as well as data about their similarity are perhaps not problematic from a methodological point of view. On the other hand, if ethologists turn to the investigation of behaviours that show a lower degree of rigid organization, the use of methods that allow the separation of the descriptive and evaluative operations leading to the behavioural measurement, will prove necessary.

In general, we can say that the need for a descriptive method of high resolution increases with increasing complexity of the parameters to be assessed. As to similarity, it is quite obvious that with growing complexity of a behavioural pattern, the liklihood also increases that any different behavioural patterns have at least partially common features. If the method we employ for the description of complex patterns cannot account for the full range of differences between these patterns, we have to be afraid that data reporting similarity reflect the observer's weakness in identifying actual differences rather than a factual similarity between the patterns compared. Similarity can be established artificially simply by ignoring the differences between patterns, and this possibility should be considered a serious one. Among other investigators, Lorenz (1959) pointed out that our perceptual apparatus is quite capable of ignoring or even falsifying differences that do not meet our expectations. In this conference we will certainly focus on the question of whether similarity of behaviours has an analogous or a homologous basis. However, we should be aware of the problem that if we accept too readily reports indicating similarity between different phenomena, we risk basing the discussion on misconceptions. Such misconceptions may prove to be insignificant; on the other hand, they may have consequences as alarming as in the case of those famous greylags which, on the basis of a very modest similarity, mistook Professor Lorenz (the human experimenter) for their mother.

DISCUSSION

Elements versus Patterns

Wickler: A fixed action pattern is an easily recognizable pattern of behaviour of a given species; comparison starts in most cases when fixed action patterns of two or more related species are considered. Why do you think that the assessment of similarities should be less problematic in this case than with more variable and less sharply defined behaviour patterns within one species?

Frey: What I had in mind was not so much the comparison of patterns between species, but the problem of recognizing patterns in one species.

Crook: While Frey is thinking about the diversity of components of one particular sequence of behaviour, and consequently thinking in terms of flexibility, Wickler is concerned with the presence, absence or variability of components within fixed action patterns. One of the main parts of the ethological problem of the comparison of species is the enumeration, frequency and number of components, of behaviour elements. The kind of analysis proposed by Frey may be useful in the study of mammalian behaviour, which shows more flexibility and less stereotypy than does the behaviour of, for example, fishes or birds.

Frey: I doubt that the human perceptual apparatus is capable of adequately analysing behaviour occuring in a multitude of unorganized elements. If there is organization, identification of elements may be possible.

Wickler: Before establishing similarities one must necessarily establish the units to be compared. Trying to compare fixed action patterns of different species implies the assumption that these action patterns can be found in several species. The actual comparison thus deals with the variability of one fixed action pattern in different species. It does not actually matter, however, whether you compare fixed action patterns between or within species, provided the unit of comparison is correctly defined. Biologists have for decades been trying to develop methods of finding and classifying similarities of characters of different species, starting with Gestalt perception and ending up with statistics and computing data. All these methods have their advantages and disadvantages, they have to be chosen according to the particular problems being investigated.

Methods of Assessment of Similarities

Foppa: As far as I understand Dr. Frey and the discussion following his paper, similarity has been treated as an absolute attribute of objects, or rather of two objects. But are two structures, or two elements, ever similar except in respect to a third one? To assess similarity, you have to define a criterion to which you refer. But I have absolutely no idea how to define a criterion which makes absolute comparison possible; bivariate or multivariate methods of measurement in themselves are certainly no solution.

Frey: Certainly not, they are just procedures, methods of organizing data in order to arrive at statements of similarities. To achieve this, three units are needed (for statements of identity two would suffice). This third unit in a comparison can be the reference system or scale. If one wants to compare two elements, one can arrive at a statement of similarity by assigning them to an ordinal. In this case the scale serves as the necessary third unit because the distance on the scale defines the degree of similarity.

Crook: Where are the limits of statistical analysis in the study, for example, of time sequences? I am thinking not so much in terms of sequences of behaviour emitted by one individual, but about the difficulties which may arise when you are trying to compare interactions between two individuals of species A with interactions between two individuals of species B. In this situation, one must consider the entire interactive patterns: It is well known that when you attempted mathematical analysis of processes going on in large groups, problems become so vast that it is extremely difficult to handle them with modern techniques. I would like to know on what level of analysis—moving from the individual sequence to the analysis of interaction between individuals—one encounters difficulties that may prove unsolvable by means of multivariate analysis.

Frey: Multivariate analysis can only be useful to the extent that it gives rise to statements of underlying similarities between all phenomena related to each other ; the concept of a factor would give rise to such statements. Other ways of assessment of similarities may be developed to meet the criteria of other aspects of similarity.

Gestalt Perception of Similarity

Leyhausen: The main problem in behaviour studies is not so much to identify clear-cut, small elements, but to identify an interaction of a number of elements in space and time. This is, of course, a configuration, if you like, and for that reason, because it has so many properties that ought to be described in their proper setting, sequence and placing, it is so difficult to give a really appropriate and satisfactory description of such a complex element or unit. That is why in so many cases Gestalt perception is superior to even the best analysis or description from film or anything else. Even when you are analysing from film you need it, because the complexity you have to consider is so overwhelming that it is impossible by any piece-by-piece description to do justice to them all, and especially to their combination.

von Cranach: It is not that the elements must be clear-cut, but that they must be identified to be compared. Foppa said that Frey's method stipulates absolute attributes, but this may be not quite correct, because once you have decided on a scale of measurement, you have agreed on the relevant dimensions of the behaviour to be measured. That is why I am interested in the problems one has to overcome before one can start multivariate, or any, analysis, for example, the problem of finding out which elements are relevant for a comparison. Beginning, as usual, with a global perceptual judgement that "this one is similar to that one", one is always in a way dependent on the good will of any discussant. Can one get away from that "good will" by a preliminary agreement? Can one define conditions under which similarity can be agreed upon beforehand?

Leyhausen: What actually happens when you start an investigation of that kind is that *first* the similarity strikes you, and only then do you start to think about it, to investigate into the mechanisms of causal network that produce it or the illusion of it—which latter, of course, is possible. But what strikes you first is the impression of similarity, it is so evident that you immediately take it for granted that there is something you ought to investigate. I think we have tried to put the cart before the horse.

Chance: I am concerned with the question of how it comes about that similarity strikes you and also how it happened that ethology, in its basic procedures, developed understanding of behaviour. I think it is because it has again raised the question of repeatability in science, which in the experienced sciences like chemistry or physics is partly resolved by control experiments. Repeatability exists in at least two ways in ethology: from the common experience of two people seeing the same piece of behaviour—this is the first form of communicability, and the basis of all understanding; the next step is the linguistic description of this agreement. This is where the origins of the behavioural sciences really confuse the basis of understanding in some respects, because behaviour has been in some way with us since man began to understand himself, and has been traditionally incapsulated in various forms of describing it. The whole of our language has been used to describe it complexly, in literature and so on, and many of these uses have been transferred without our awareness of the problem of describing behaviour in fact. Thus, the process of reducing experience to a single entity that is recognizable as the same is a question of bringing linguistics into focus with common experience.

Leyhausen: We have to keep in mind that the range of phenomena to which our concepts can be applied is limited. The concept of homology, for example, may still be useful on the level of biochemistry, but as soon as one gets down to single chemical elements or ions it becomes absolutely useless. What one can compare within the framework of the concepts of homology and analogy is only a range of phenomena of medium complexity.

Masters: I am struck by the extent to which this discussion so far has repeated the history of modern epistemology. This is playing the same game that has gone on in western science for the last two or three centuries, and maybe we should not worry too much. Modern science has consistently attempted to get behind Gestalt impressions of similarity by means of methodology, but has always run into difficulties because they forgot something in the Gestalt which cannot be coded. This seems to be a kind of dialectic impass: in the process of elaborating a new approach to behaviour, which is what we are all doing, we may discover that we solved the problem by doing it.

Crook: I think I have a point against that, and I think we should worry. It seems to be deciding on scales and units which is crucial. Very often two people do not see the same things, and this may cause great difficulties. In the study of any complex sequence of behaviour, or in field studies, the actual decision of what is to be described is a very real difficulty, which can be overcome either by a pannel of established agreements or by automatic recording. The latter method is to be preferred, because it permits repetition of observation and examinations from different points of view at different times.

Von Cranach: In my experience it has always been possible to work out an exact quantitative method once one knows in what respect similarities should be assessed and what degree of similarity should be accepted as a criterion. If these things are unknown, however, one must fall back on Gestalt perception; for this reason, one has to rely on Gestalt perception at the beginning of most studies. Frey has been talking about the assessing of similarity referring to one single element on one dimension. This leads to all the correlational considerations he mentioned. The present discussion, however, is more concerned with the position of one element within a structure, whatever structure that may be. This problem could be solved by dissolving the structure, by breaking it up into elements and then measuring these and their relations, as one does in morphology. As to similarities of behaviour, every step of the procedure is more problematic. How do you actually start on a problem of that kind?

Behaviour and Morphology

Wickler: I do not actually see a difference between behavioural and morphological structures, because one normally tries to transform behavioural structures into morphological ones by means of recording them on film or

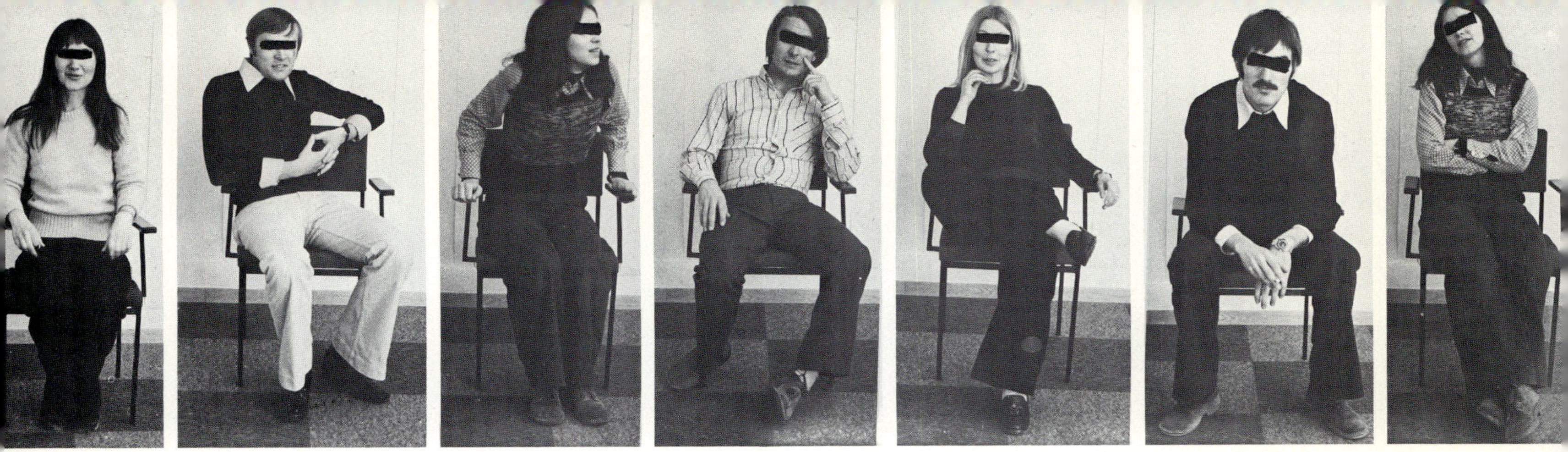

Positions of original subjects

Positions of model, read back from the codings

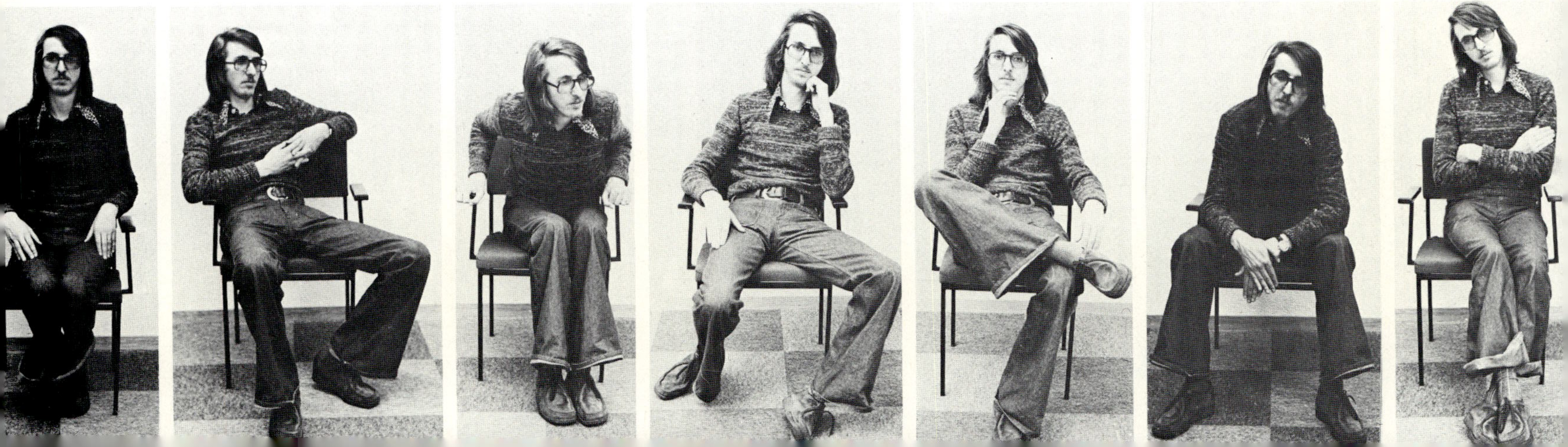

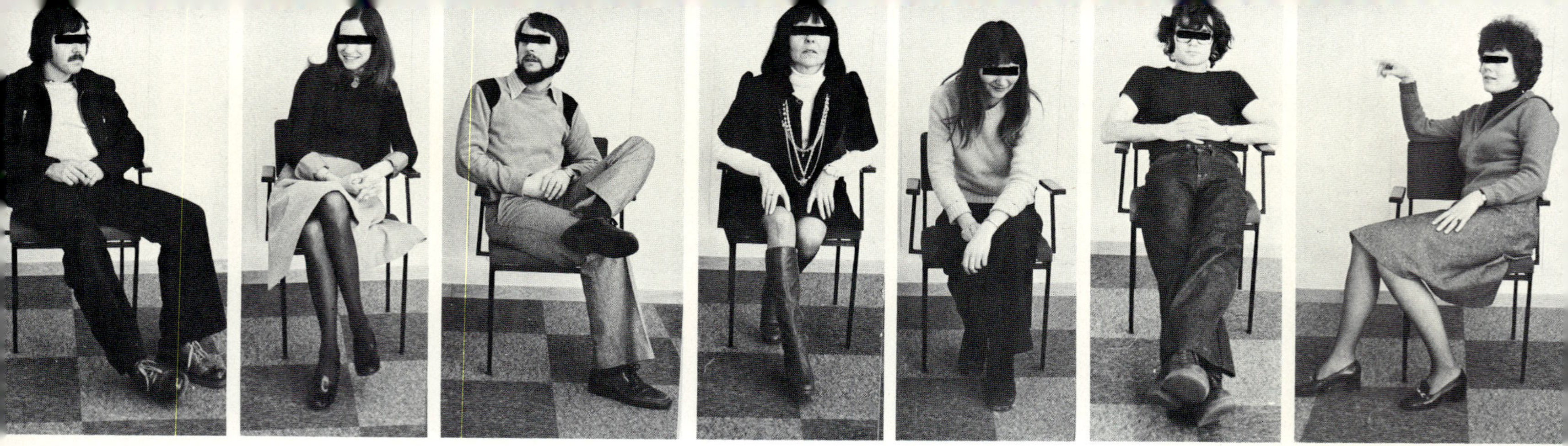

Positions of original subjects

Positions of model, read back from the codings

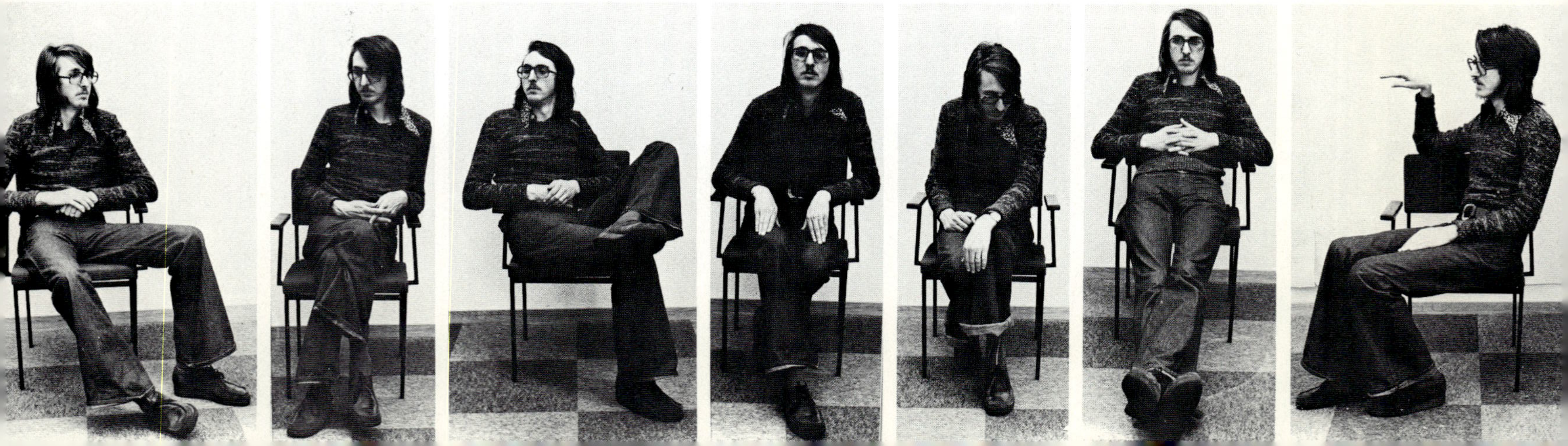

tape. Behavioural structures may be more complex and more variable; the difficult part is not finding the pattern, but describing it. One possible way would be to record on film, and to rely on that film as a morphologist does on an organ. One could say that to describe correctly means to break up the structure into elements. What one *cannot* do, however, is to take one element out of the structure, because that would infer the similarity as given.

von Cranach: There is always the danger of endless regress, of finding elements of elements of elements.

Ploog: The comparison of animal vocalizations gives a very good example of what Wickler just said, of the transformation of a fixed action pattern into a morphological structure by using sound spectrography. The definition of an element of a single call can be based either on description, on an agreement on certain morphological criteria, or on the question "for what is it good", what is the function of that element? There may be cases where that element, that fraction of a fixed action pattern, is the one that carries the function or elicits a specific behaviour. One cannot really decide whether it is an element or not except in relation to its function. If a visually discernible element of a call has no function, it can be combined with other elements until function emerges. To find out about function as well as about the function of variability, one has to experiment.

Wickler: For practical reasons, one should doubtless distinguish between similarity of structure, or morphology, and similarity of function. One can easily say that there is similarity in so far as both units of a comparison have the same colour or the like; it is something different, however, to find out whether they both serve the same function. Therefore, one should, in the basic assessment of similarity, try to omit function. Of course, this again depends on the respective definition of function, but the same problems are inherent in an analysis of interaction of units which may be dependent on presenting something like an imput–output relationship in which the output part again is closely connected with function. These same units may also influence each other in turn. Quite a number of sciences avoid these complications for practical reasons.

Three Forms of Behaviour Genesis

Leyhausen: I think that similarity is certainly a starting point, but of course the matter does not rest with similarities. With homology and analogy, however, the similarity is the *problem*. We have to find out how it came about, what mechanisms produced it. Structural similarities that may develop in an individual, or in a number of individuals, are *there* to be compared. As to behaviour, we are, in a way, one step removed from this. What develops is a latent mechanism which only on occasion produces the phenomena which we want to compare. In behaviour, we have to consider three different kinds of genesis which in turn contribute to these phenomena. The basis is phylogenesis, which provides mechanisms of ontogenesis, which in turn provides mechanisms which touch off the behaviour. So what we really should compare are not the overt similarities, which are the superficial appearance of our phenomenon, but the mechanisms which bring them about. We should inquire into the phylogenetic and ontogenetic mechanisms which produce such readiness. We are faced with phylogenesis, ontogenesis and actogenesis—if it is permissible thus to translate the German term *Aktualgenese*. The causal relationship between them is a very loose one, and they form more or less closed systems which have to be considered each in turn. The relationship which closes the circle between actogenesis, the actualized item of behaviour, the elicitation of the actual behaviour, and its phylogenesis is selection, and selection works by selecting phenomena, selecting phenes and not genes. Genes come in here only indirectly.

Homeology

The second thing I wanted to say is that you have been wrestling with the phenomenon for which the proper name is homeology, which is something in between homology and analogy. It has a common basis in the ancestry of the different organisms. When these later get into similar ecological situations, they produce very much the same answers. They develop rather analogical behaviours or structures which, however, have a common homologous root. The problem actually is to decide at which point to forget about homologous roots which, if removed far enough in time, become universal. As to behaviour, there is a special catch in these developments which has been pointed out by Adolf Haas, who called it generic behaviour: he showed that many related organisms have a more or less common repertoire of latent possibilities of behaving in certain situations, but some of these behaviours have such high thresholds of elicitation that they are never shown under normal conditions. If the patterns with lower thresholds of elicitation "run into difficulties", however, this otherwise latent potential of more ancient behaviours reappears.

NOTES

Part of the research underlying this paper was made possible by a fellowship awarded to the author by the Foundations' Fund for Research in Psychiatry (FFRP), New Haven, Connecticut. The author gratefully acknowledges the generous support received from the FFRP.
For a more detailed account of this work see Frey and Pool (in prep.).

REFERENCES

Campbell, D.T. 1958. "Systematic Error on the Part of Human Link in Communication Systems", *Information and Control* 1:334-369.

von Cranach, M./Frenz, H.G. 1969. "Systematische Beobachtung", pp. 269-331 in: C. Graumann (ed.) *Handbuch der Psychologie* vol. 7: *Sozialpsychologie*. Göttingen, Hogrefe.

Frey, S./von Cranach, M. 1971. "Ein Verfahren zur Messung motorischer Aktivität", *Zeitschrift für experimentelle und angewandte Psychologie* 18:392-410.

— 1973. "A Method for the Assessment of Body Movement Variability", pp. 389-418 in: M. von Cranach/J. Vine (eds.) *Social Communication and Movement*. London, Academic Press.

Frey, S./Pool, J. In prep. A New Approach to the Analysis of Visible Behavior.

Lorenz, K. 1959. "Gestaltwahrnehmung als Quelle wissenschaftlicher Erkenntniss", *Zeitschrift für experimentelle und angewandte Psychologie* 6:118-165.

Thomas, D.S. 1932. "An Attempt to Develop Precise Measurement in the Social Behavior Field", *Sociologus* 8:436-456.

ARTHUR J. CAIN

The Use of Homology and Analogy in Evolutionary Theory

Homology and analogy were first defined, in the senses in which comparative anatomists use them, by Richard Owen (1804 – 1892). In his Hunterian Lectures (1844 on), in an address to the British Association in 1846, and especially in *On the Archetype and Homologies of the Vertebrate Skeleton* (1848), and *On the Nature of Limbs* (1849), he worked out their application. His definitions (1848, p. 7) are: *analogue,* a part or organ in one animal which has the same function as another part or organ in a different animal; *homologue,* the same organ in different animals under every variety of form and function.

These are not very satisfactory as they stand, but are amply elucidated by the examples he gives, of which the most celebrated is the vertebrate limb. It was news to English anatomists of Owen's time that the five-fingered limb of man, the wing of the bat, the trowel of the mole, the one-fingered limb of the horse, all showed the same essential structure when dissected; they had the same bony parts (a few absent in some forms) connected in the same sequence to each other, and could be recognized as the same structures, under all modifications of size, proportions, exaggeration of some parts, and reductions of others even to absence. Even the paired fins of fishes could be shown to be homologous with the fore and hind limbs of mammals, although at first sight so different.

Owen made no reference to evolution in his discussions of homology. As an ardent follower of the *Naturphilosophen,* he referred the essential structure of the body of vertebrates, in all its variety, to actualizations of a single archetype, a Platonic Idea in the mind of the Creator. He was quite explicit that minor variations in the mode of development of a structure (*e.g.* a bone from two centres of ossification in one species and from three in another) were not sufficient to disprove homology. Indeed, he rather neglected embryology, and it was left to others to show how valuable the study of early stages could be in elucidating homological relationships of organs (*e.g.*

the Eustachian tube or thyroid) which in the adult state would never be recognized as homologous at all.

On the acceptance of the theory of evolution, a simple explanation could be offered for homologies and analogies. Analogies were the result of response by different organs to the same selective pressures. To use one of Owen's own examples, the "flying" organs of the little lizard *Draco volans* are developed from the ribs, while those of birds are produced from the forelimbs. Homologies, in contrast, are the response of the same organ, inherited from a common ancestor, to different selective pressures. Thus the glandular groove (endostyle) in the pharynx of *Amphioxus* becomes the thyroid of mammals, one pair of gillslits in fishes becomes the Eustachian tube of mammals, and the articulatory bones of the reptilian jaw form two of the auditory ossicles of mammals.

Homology and analogy in such senses are relative to the taxonomic positions of the forms concerned. Thus, as Owen pointed out, the forelimbs of a porpoise and a fish are homologous in their structure, but analogous in their functioning. "So, likewise, the pectoral fin of the flying-fish is analogous to the wing of the Bird, but, unlike the wing of the Dragon, *(i.e. Draco volans)* it is also homologous with it."

In evolutionary terms this means that the same organ has been modified twice independently for the same function. The forelimb of a porpoise and a fish are both true vertebrate limbs, directly related by homology; but the modifications of the general plan or archetype by which they function as swimming organs have arisen quite independently, and are therefore analogous.

The basic requirement, therefore, in determining homology and analogy, is to describe a structure, sufficiently complex to be recognizable even though distorted in various ways, which is recognizably the same whenever it occurs. There seems no reason why a corresponding complexity of behaviour, with definite related features connected in a definite way, should not be recognizable as homologous whenever it appears.

However, there are certain difficulties, especially when an evolutionary interpretation is used. In the first place, a structure may not be so complex that one is certain that it cannot have arisen several times independently. Secondly, the same evolutionary pressure may call out the same response in closely related forms. An example is the development of the same type of horns independently in several closely related lines of early titanotheres. Here, independent origin is undoubted, but surely the resulting structures are homologous. A further example is the possession of haemoglobin. Almost all living things have a cytochrome system, which can provide the (homologous) basis for the evolution of various respiratory pigments, haemoglobin being one. For all I know, haemoglobin may be produced by the same enzyme pathways in all the organisms that have it. But we find it in a few insects, a few snails, almost all the vertebrates, various annelid worms, and even in some root nodule bacteria. These are far too diverse for the haemoglobin to have been inherited from a common ancestor, and so it is not from an evolutionary point of view homologous in these various forms.

Lastly, an attempt has been made to say that organs, structures, behaviour, or anything else mediated by the same genes in the genotype must be homologous. No doubt this is true; but the same locus can mutate to produce the same mutation independently in very different individuals; evolutionarily, even this might not be a useful criterion, always supposing that we could apply it to fossil titanotheres. Nor can we exclude the possibility that the genetic mechanism underlying a particular structure may change, still leaving it the same structure.

Convergence is the production of the same characteristics in different evolutionary lines. It is the more likely the more closely related those lines are provided it is ecologically possible.

DISCUSSION

Homology and Analogy

Wickler: I think I understood that you agree that, whatever Owen defined initially, one should at least define homologous and analogous as mutually exclusive.

Cain: No, that is the difficulty. As he said himself when he defined them, homologue and analogue are not mutually exclusive. A homologue is necessarily an analogue, but an analogue is not necessarily a homologue. The actual definition of homology is what I read to you before, namely the same organ in a different animal in every variety of form and function.

Wickler: The definition is dependent on what is the *same*.

Cain: Right.

Wickler: Homologous organs are the same organs.

Cain: Yes.

Wickler: This is simply changing the term without expanding it.

Cain: Yes, what he did was not to explain, but to illustrate. He used, if I may use the expression, ostensive definition by pointing out examples where he believed everyone would say that this is the same organ, his example being the vertebrate limb.

Wickler: I doubt that this is still feasible today; if you look at the eye of a goat and the eye of a squid you will regard it to be the same organ and nevertheless you will be wrong.

Cain: Darwin, discussing the issue with another anatomist, expressed the view that two organs whose anatomical development is different are not the same organ even if they perform the same function. This would be an instance of two organs being analogous.

Circularity of Homology

Wickler: I think it is not correct to say that homologies are dependent on our knowledge of the phylogenetic tree, because the phylogenetic tree is a reconstruction of the relatedness of animals, whether living or extinct, and our reconstruction of that tree is based on correct identification of homologies.

Cain: Right.

Wickler: So we cannot possibly close the circle now. The titanotheres you mentioned are not labelled as such, but simply identified by the similarities which you think are homologous.

Cain: This is partly but not wholly true. This is what I was aiming at in the last part of my lecture. When comparing all organ systems of the whole animal, you may find that something you previously thought was homologous cannot be so. For example, although the production of haemoglobin is the same throughout the animal kingdom, it is impossible to say that it is homologous if it occurs in a few molluscs and a few arthropods. So, under certain circumstances, if you take into consideration as far as possible the whole plan of the animal, this would enable you to break up the vicious circle with respect to particular organs. On the other hand, where the evidence is insufficient, or where you have a completely dubious case, you are still left uncertain as to whether you are speaking strictly about a homology or about an analogy.

Homeology

Wickler: This is very nice because you always lead me to the next point. I think that this organ is partly homologous and partly analogous, or is one of these homeologies Paul Leyhausen mentioned. I would like to persuade the conference to forget completely about homeologies. I think they only result from not correctly defining the units which you want to compare. If you ask whether the wings of a bat and a bird are homologous or not, you cannot answer the question. Insofar as they are forelimbs of tetrapods, they are homologous, but as far as they are wings they are clearly not. If you say that they are homologous, you simply mix these two levels of comparison.

Cain: Right.

Wickler: If you stick to the monophyletic origin of animals, all your homologues may, via homeologues, end up as analogues, because you keep the whole thing in a continuous flow and never precisely define the units you want to compare.

Cain: I think I don't want to agree with that, because that would imply that evolution, in a strict sense, consists merely of a process of genetic unfolding, where everything is already there. This would imply that there is no novelty in evolution and I just cannot believe that.

Wickler: No, of course not. You can only compare characters which are present, you cannot compare or homologize characters that are not present.

Cain: Right.

Wickler: So you are always bound to the appearance of the character; and then you should define again whether the whole variety of, let us say, legs, are homologues? Are all legs homologues? No, of course not, but legs within vertebrates are. And what I have done now is simply specify what I mean. If I ask whether wings in vertebrates are homologous, I have to specify: homologous as limbs, yes, or as specialized wings, no.

Cain: That is right. Otherwise you find yourself paying immense respect to the vertebrates and very little to the birds. Now I pay equal respect to both: within the group of birds as far as we can see the forelimbs are homologous wings. They are not always used, but they are there. The bat also flies. It is clearly a tetrapod, clearly a vertebrate. Its basic plans are the same; one must therefore be able to compare the forelimbs of a bird and a bat, and as long as you stay within the vertebrates, comparing their special features for flight, they are analogous and not homologous. By thinking in terms of these distinctions I cannot think how you can propose homology.

Wickler: What I wanted was to avoid the homeologue.

Crook: Could you clarify the meaning of the term homeologue?

Wickler: It is actually defined in literature as analogous modifications of homologous organs, the development of a wing out of the forelimb of a tetrapod for example; the same seems to be true for different horns of the extinct titanotheres. If you can really say that these organs evolved at different times independently, you have already said that they are analogous variations, independent variations of homologous bones.

Ploog: Dr. Cain, could you for a moment go back to the question of the haemoglobin molecule? Would you think that it can occur in a totally unsystematic context, just as another invention of the same thing in quite a different context of development, and that although it has absolutely the same structure, it is still not homologous?

Cain: Yes.

Evolutional Sequence

Ploog: So we have not only to consider the question whether something occurs in the same patterning, and whether the structure is the same, but we also have to test the question in time, when and in what context has that something been invented, and if it has been invented in a non-continuum of time, it is not homologous, but is what we call convergent. So we have to include the time scale, the evolutional sequence into our criterion for homology.

Cain: Yes.

Leyhausen: I should think the problem in this case is the phylogenetics of these molecules. If we could find out about the phylogenetics of these molecules in an insect, a mollusc and a bird, or a reptile, we could say there is an actual homologous process, or there is not, but I am afraid biochemistry is not as far advanced as all that, so it is no use arguing.

Homology and Analogy

Cain: I am afraid that applies to the faculty of resurrection, not to biochemistry. In fact, one almost certainly never will be able to apply such criteria—which would be delightful—to totally extinct forms.

von Cranach: I am puzzled. I always thought the question of homology came up because one could not possibly observe the actual process of evolution; in this discussion, however, we say that we must know the evolutionary sequence in order to resolve the problem of homologies. If we knew the complete evolutionary sequence we would not need to discuss the problem of homology.

Wickler: The modern question that is behind the problem of homology is that we would like to know how some of the similarities we observe can be explained. They could be pure chance; our aim is to rule out chance, normally by way of statistics. Let us take the example of rain droplets. They all have exactly the same form. Is that similarity of form homologous? I do not think so, because this form is moulded by one given character, namely, falling down from heaven through the air. It will occur in all droplets by the physical laws acting, and you need no special information store within them to obtain this form. However, you have in all birds or in all vertebrates or in any group of animals particular structures that cannot be explained generally by simple physics; in some cases you know already that there is a special information storage that is handed on from one generation to the next and ensures that under some conditions this pattern occurs and nothing else. Whereas in another species under the same conditions something else will occur. And this is something different from the moulding of rain droplets falling down.

Cain: The only difference is in time scale, in generation scale. Droplets are moulded almost instantaneously. It may take a certain number of generations to mould a species. But if a species has a particular character, it has it because it needs it, not because its ancestors had it.

Leyhausen: No, because its ancestors needed it: it may no longer need it itself.

Information Store

Wickler: The main point seems to be the information stored in the genome (let us take the simplest case). If that information is different from one species to another, this will ensure, if you keep those animals under the same conditions, that different things will come out. If similar things occur, we assume that the same information is stored in the genomes and therefore these animals are normally said to be closely related. Homology criteria as developed by comparative anatomists are meant to ascertain whether the character under study can be classified as containing the said information or not. Close relationship is not a criterion for homology, though it can be its reason. The criteria should allow one to find out whether two characters are homologous, not to find out whether two species are related.

von Cranach: Is this really the question everybody wants to solve by applying homology methods? As far as I remember, Remane's problem was to establish relationships between organisms, and the homology method was used to solve it. I cannot help feeling that your method already implies the relationship.

Wickler: No, it does not. I only imply the relationship of characters. If you need homologous characters only to establish the relationship of species, you need not worry about non-homologous similarities.

Masters: I think part of the problem with reference to homology arises out of the difference between phylogenetic and ontogenetic levels. One could define homology in terms of common ancestors or, I suppose, one could define homology in terms of identical or nearly identical DNA-sequences in the genome (which would, hypothetically but not necessarily, be correlated with phylogenetics).

Cain: I am not sure one could. I purposely refrained from raising the question of genetics at all since I have a strong suspicion that, because of other selection pressures on the genotype, the basis for a particular organ may vary genetically, while still leading to the same organ. The other reason why I did so is that you cannot hope to apply such a concept to extinct animals anyway.

Leyhausen: You cannot imply homology always being demonstrated by identical sequences of DNA. We know examples from population genetics where absolutely identical characters in closely related species have a different genetical basis (Kühn, 1971; Darlington, 1958; Dobzhansky, 1959; Searle, 1968; Williams, 1966). That is to say, while at one point in evolution the selection pressure to maintain some character has been very strong indeed, the genetic basis evidently was not able to cope with some other requirement, and had to be altered, while the superstructure of the said character was maintained. So that for all we know and for all the criteria we could apply, this character in these closely related species is homologous, but its genetical bases are different. That is one reason why we have to keep the concept to a medium range of complexity. And there is another reason why we simply cannot escape the circle about which Wickler talked earlier: because this is a self-supporting system. I think we have to consider the concept of homology in a very pragmatic way. There is a certain limited range within which it seems useful. As soon as we know all about the causal network which led from the first living thing on this earth to the variety we see today, and to all the extinct ones, we shall not need the concept anymore.

Cain: But we never will.

Masters: Yes, I agree to that. What I was trying to say was that at some point of the circle one could *also* look for genetic sequences, if these where very different, that might lead us in some cases to a hypothesis whether we are dealing with analogy or a homology.

Injunctive Concepts

Leyhausen: When Wickler says that when talking about similarities we should rather forget about homologies, and when Dr. Cain says that in Owen's language homology is very nearly included in analogy, and that all homologous organs are analogous, they are in both cases disregarding a fundamental point of biological concepts that Hassenstein defined as injunctive: they are all a little bit blurred, a little bit frayed at the fringes. If you try to define them too sharply, without any doubt as to where the limits are, or where the borderline should be, you run into difficulties. In this particular case I think I can even put a finger on the reason why this must be so: because of the process of population genetics. Everything that evolves and then is selected for later depends on selecting not so much particular genes, but combinations of genes. And so the history of each characteristic is oscillating, and we can never put a finger on some particular point in time and say here this stage was left, and the border of that stage crossed, because *there is no borderline*. Our attempts to draw such fixed borderlines across phenomena that in the course of evolution and development do not have such lines cannot lead to any practical or even theoretical advance. We have to accept the fact that this is impossible, and that we can use these concepts only as far as they can carry us. We have rather to concentrate on the points where these concepts *are* useful. I should think we have to be content with what nature allows us to do.

Homology and Phylogenetic Relationship

Senn: I think we should fix the term homology to the phylogenetic tree despite the danger that a phylogenetic tree might be based on homologies, or the inverse. The problem is how to escape this circle. I think it is important that there are several disciplines in biology working together. The osteologists can really study fossils and somehow establish a phylogenetic tree in a group, and this helps other investigatiors to interpret soft tissues, in terms of homologizing them in actual animals. I think the cooperation of the different disciplines can help us to escape from this circle. A second point that Dr. Cain mentioned may also help: using some idea of a phylogenetic tree, for example, one established by palaeontologists, we must look at the whole organism as a complex system as well as at the position of a single organ.

von Cranach: If you say "based on the phylogenetic tree" does this mean that you first imply relationship and then go on with the method of homology? What does this mean operationally?

Senn: Operationally, it can mean that we first try to establish some idea about the phylogenetic tree on the basis of the available fossils. And it can also mean that without knowledge of evolution, as Owen did it, the interpretation of actual animals comes to some term of homology which later can be compared from the viewpoint of palaeontology. The important point is that one discipline's work need not be done first, but that both advance together.

Crook: Following up what Paul Leyhausen was saying I wish to remind you that evolution theory is always, by necessity, synthetic. Very often we simply do not have sufficient data on a particular group of animals to make any assertions about homology or analogy with complete certainty. Dr. Cain once wrote a paper called *Taxonomists' Judgement of Affinity;* this problem of deciding about affinity is very difficult in certain groups. For example, it was once thought that all reptiles and birds had a common origin, and similarly about the pinnipedia. It is now thought that some pinnipeds come from bears, and some from otters. The implication here is that, given the information available at any one time in the history of a subject, one may make assertions that a certain kind of character is homologous. Later on, one may discover, say by accumulation of evidence, that one is dealing with a bi- or even triphyletic group, and what was considered a homology may then become an analogy. So we have to realize that the ascription of homology is relative to the amount of information available at any one time. This is the pragmatic use of this term; its operational usage is in very close relation to the knowledge available at a certain time, and the theory is of a necessity synthetic. We can not get away from this.

Wickler: Is it true that in the case of a group of animals originally regarded as monophyletic and finally found to be polyphyletic, characters originally regarded as homologous turned out to be analogous? Or is it not rather that those characters still are considered homologous, but that by now more characters have been studied, and it turns out that there must have been a secondary split within the group? Therefore, the main problem is again not that of correctly identifying homologous characters, but of correctly judging the relationship of species from them.

Crook: If one considers the pinniped form and looks at the flippers, I think one would say they are homologous, but suppose you consider the adaptation to deep diving, this probably comes independently from two sources; they would now be called analogies rather than homologies.

Wickler: Yes, you are right, but your judgement of homology is again based on the probability that a character does show such a form. You think it is a homologous character because it is sufficiently complex to permit a judgement that it has the same structure. Finally, you find out that this development is a likely one, that it was within the variability of that character anyway, and you think that it is actually the same.

If you have a character not sufficiently complex, it is very difficult to find out whether it is a true homology. Not because it is difficult to compare (which it is not), but because the problem of judgement as to how this form happened to evolve may be unsolvable. That means, as far as I know, that you cannot answer the question of how much special information has been passed on to ensure that, under those conditions, this form will emerge.

Cain: What Dr. Wickler has said is perfectly right. There are a lot of characters that are thought to be very complex which turn out to be simple.

You could not possibly use that explanation for the conversion of a bear's foot to a seal's fin. In the case mentioned by John Crook, you must think either that the diving adaptation is analogous, or that the fins are analogous; you have to recognize one or the other, even if you cannot say which.

Senn: I just wanted to give an example for Dr. Wickler's important question about the different characters we are using for the phylogenetic tree. At least ten years ago, some palaeontologists came up with the idea that reptiles have polyphyletic origins. They based their arguments on structural differences in the skull and came to the conclusion that a very typical character in reptiles, their horny scales and scutes, developed independently in different groups. But then another discussant presented a different argument: he said that all reptiles have the same type of egg and early embryo. The embryo develops with a yolk-sack, an alantois, and the same kind of amnion; this would indicate that the reptiles originated monophyletically. This is a good example to illustrate that not all characteristics have, let us say, the same weight on the taxonomic tree. Some characters tell more about the basis of a group and others tell more about its branches.

van Hooff: I would like to go back to the levels of comparison. It turns out that, just as in the problems of similarity we were talking about in a more abstract sense, there is always the question of how many things are included in the group of phenomena you are comparing. It takes at least three to make these comparisons, and if you have characters like the fins of the pinnipeds and dolphins and so on, you will find a common origin if you include many, and go back far enough, and then you will use the term homology. If you keep within a more restricted group with your comparisons, you would perhaps regard these characters as analogous. Does it not depend on the group at which you look?

Wickler: That depends on whether you are lucky enough to find that character in the supposed common ancestor. If you do not, you will be in difficulty.

van Hooff: But at least you could probably find a common ancestor that has some propensity which is due to those things?

Wickler: Oh yes, but then you homologize the propensity and not the character.

van Hooff: Sure, but this propensity is in some way a structure.

Wickler: In some way, but not in the same way as in the character you started with. That means another shift to another unit.

van Hooff: That of course raises the question of how narrowly to define a unit.

Wickler: With proceeding knowledge you have to refine your units more and more.

van Hooff: But that would mean that you gradually explain all homologies in term of analogies with increasing knowledge of intermediate steps.

Behaviour and Morphology

Chance: First, I want to say that in my work I look for behaviour structures which hold together, which are dependent upon homologous structures below the surface of the animal, not immediately visible features and, in a negative sense, to agree with Dr. Crook that at any stage of knowledge we have only a certain amount of information relevant to help us decide on appropriate methods. We should at some point be concerned with unearthing the *actual range* of structure, the actual totality of structure of which behaviour is composed. This is the object of ethology.

The second thing I wanted to say is this: I understand that when a nervous system differentiates it does so in a very special way. Neurons form into groups and then portions of these neurons die, they literally do not multiply. This means that there must be a neural anatomical basis for behaviour. This assumption was made by Prof. D.M.S. Watson, and he made it quite clear that the plesiosaurus, ichthyosaurus and the like were only different in their anatomy because their behaviour was different. He said that they had different ways of moving into the sea. One type waddled, the other reduced its limbs, and became a reptile or a fish. Another sank to the bottom and did not do any moving at all except to catch its prey. Now these three were differences in behaviour producing differences in morphology, and yet these differences in behaviour are based on differences in neural morphology. This would be a totally different basis for the comparability of behaviour in the morphology of the central nervous system.

von Cranach: I should like to ask an ethologist or a zoologist whether there is a true definition of behaviour as opposed to changes in morphology?

Crook: I should think that in most cases it would be very difficult to say whether or not behavioural preceded morphological changes. I would say that almost of a necessity the two would always correlate, that they are co-selective, that they represent a syndrome. It seems to me that they would evolve together, that behaviour is an expression of morphology, and morphology is, in a way, predetermining behaviour. But there are some cases in the evolution of displays where the behaviour *has* come first.

REFERENCES

Cain, A.J. 1964. "The Perfection of Animals", pp. 36-63 in: J.D. Carthy/C.L. Doddington (eds.) *Viewpoints in Biology*. London, Butterworth.

— 1968. "The Assessment of New Types of Character in Taxonomy", pp. 229-234 in: J.G. Hawks (ed.) *Chemotaxonomy and Serotaxonomy*. London, Systematics Association.

Darlington, C.D. 1958. *The Evolution of Genetic Systems*. New York, Basic Books.

De Beer, G. 1971. *Homology, an Unsolved Problem*. Oxford Biology Readers 11. London, Oxford University Press.

Dobzhansky, Th. 1959. "Evolution of Genes and Genes in Evolution", *Cold Spring Harbor Symposia on Quantitative Biology* 24:15-30.

Kühn, Alfred. 1971. *Grundriss der Vererbungslehre* 5th ed. (rev. Albrecht Egelhaaf.) Heidelberg, Quelle & Meyer.

Owen, R. 1848. *On the Archetype and Homologies of the Vertebrate Skeleton*. London, for the author.

— 1849. *On the Nature of Limbs. A Discourse Delivered on Friday, February 9, at an Evening Meeting of the Royal Institution of Great Britain.*

Remane, A. 1952. *Die Grundlagen des natürlichen Systems, der vergleichenden Anatomie und der Phylogenetik. Theoretische Morphologie und Systematik I.* Leipzig, Geest & Portig. Reprinted 1971, Koenigstein-Taunus.

Searle, A.G. 1968. *Comparative Genetics of Coat Colour in Mammals*. London, Logos & Academic Presses.

Williams, G.C. 1966. *Adaptation and Natural Selection*. Princeton University Press.

DAVID G. SENN

Methodological Steps in the Use of the Concept of Homology

TERMINOLOGY

Homology as Similarity and Homology as Correspondence

A concept of homology, which uses homology as similarity, was developed by R. Owen (1843). Especially since the development of Darwin's theory of natural selection (1859), the concept of homology appears with increasing frequency within the framework of evolutionary theory. Homologous structures are traced back to common ancestral patterns. Today, many biologists use this concept of homology. Different types of homology which are recognized as similarities are distinguished by Smith (1967) and Northcutt (1969).

— *Serial homology* refers to similarity of structures within an individual organism, primarily to the phenomenon of segmentation. The somites of an early vertebrate embryo look similar to each other. Structures which originate from somites, for instance the vertebrae, elements of the axial skeleton, show such similarities. The third and the fourth thoracic vertebrae in a mammal are almost identical in appearance. Serial homology corresponds to homonomy (Remane, 1952). A series of structures is located one behind the other in a pattern of segmentation.

— *Discrete homology* refers to similarity based on common ancestry of structures in different organisms. The pelvic girdle of birds and the pelvic girdle of mammals are discretely homologous; they are similar structures considered to be derived from a common ancestral pattern.

— *Field homology* refers to the common ancestry of structures in terms of developmental fields. Field homology is important in comparative embryology. The buds of forelimbs in different vertebrate embryos are similar fields. These forelimb buds are considered as a field, regardless of whether later differentiation will produce a wing, a leg or a flipper. The term field homology may also cover problematic examples: the inferior lobe of the lung of a primate is field homologous with the inferior lobe of the lung of a chick. These fields are similar in terms of a division of the lung into lobes. But based on comparative embryology, the homology between avian and mammalian lobes of lungs could also be disproved. Thus, field homology is problematic as a descriptive method in comparative anatomy (see p. 45, *Comparative Embryology*).

— *Patristic homology* refers to similarity among corresponding organs in organisms of which one is directly descendant from the other. Single elements of the skeleton of the paired fins in *Eusthenopteron* (a Crossopterygian fish of the late Devonian) can be homologized with elements of the limb skeleton of land vertebrates. Another example is the patristic homology of the wing of an Archaeopteryx (a bird-like reptile of the Jurrassic) and the wings of modern birds.

— *Cladistic homology* is a similarity between organs of animals which represent divergent ancestral lines. For example, the wing and the human arm can be traced back to an ancestral land vertebrate forelimb.

In the use of these categories of homology, the similarity originally was the major criterion. Under the influence of evolutionary theory, the dimension of historicity increasingly influenced the use of the term homology. Now, the categories are used on the basis of similarities which reflect historical processes.

The term homology can also be established primarily with regard to the evolutionary dimension. In this respect, the similarity is of secondary importance. In other words, homology is a correspondence rather than a similarity, a correspondence directly depending on the evolutionary process. Structures may be homologous, but completely dissimilar in appearance; for example, the hyomandibular of a fish and the stapes of a mammal. Organs of different animals are homologous if they correspond, whether they are similar or not. The use of the term homology in this sense was introduced by J. W. von Goethe; later, it was adopted by most of the classical morphologists, and it has been explicitly stated by Simpson (see below p. 43).

In this respect, it is not necessary to distinguish so many kinds of homology. If structures of two different animals are considered homologous, this includes homology with a structure of a common ancestor. On the other hand, a homology has not to be specified as a patristic one when an ancestral and an actual structure are compared. A homology concerns also other actual descendants of that ancestor. Homology as a correspondence of structure is the same whether two or more actual structures, or actual and ancestral structures, are compared.

Homology and Analogy

If structures in different animals (actual or fossil) correspond, they are homologous. The homology is established whether the structures have the same, similar or different functions. The following examples will demonstrate this: the paired eyes of all vertebrates are homologous, in gross structure and arrangement as well as in many details. The two eyes are connected with the ventral part of the diencephalon by the optic tract, which continues into the diencephalon and in part into the mesencephalon. As to details, the retina contains the same cellular layers in all vertebrates: visual cells, bipolar cells and ganglion cells. Variation among vertebrates concerns only detailed differentiations of these main layers. Thus, among representatives of the group of vertebrates, the paired eyes correspond both in their states as whole organs and in their structural details (Fig. 1, 2).

FIGURE 1. *Examples of paired eyes in vertebrates.*

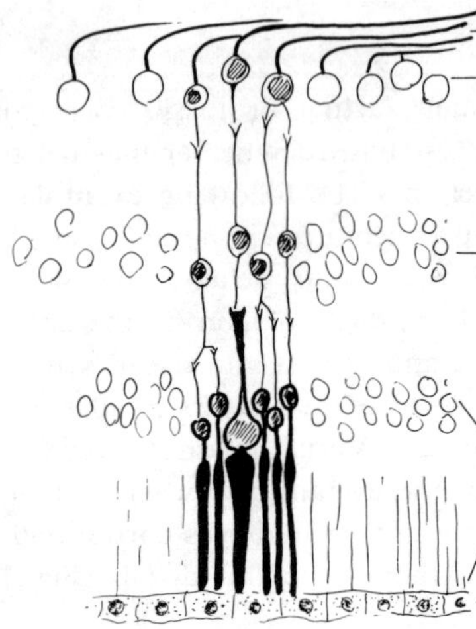

FIGURE 2. *Structure of the retina of vertebrates.*

Another example is the brain of vertebrates. The fundamental division is the same in all vertebrate brains; higher association and integration functions are located in the mesencephalic roof, the telencephalon is mainly olfactory, or the telencephalon differentiates a number of association and integration centres as especially seen in birds and mammals.

Structural correspondence is termed homology; in contrast, a correspondence in function is termed analogy. Structures of different animals are analagous if they have the same function, in spite of having different origins. This can casily be demonstrated in animals which are not systematically related such as *Sepia,* a cephalopod mollusc, and *Rhinecanthus,* a bony fish of the coral reef (Fig. 3), both of which have a similar locomotion apparatus. *Sepia* possesses fins on each lateral side of its mantle; these fins produce undulatory movements which allow the animal to perform slow and precise forward and backward movements and turns. If necessary, however, *Sepia* can switch to fast movement which is not produced by the fins, but by the funnell through which respiratory water is expelled. In comparison, *Rhinecanthus* can also perform slow and precise movements and turns in the coral reef by means of undulatory movements of the fins. If rapid movement is necessary, it is not produced by the fins, but by a general undulatory movement of the whole body. The fins as structures have nothing in common in *Sepia* and the *Rhinecanthus* fishes, but they work the same way; they allow the animals to perform the same kind of precise movement. They are analogous structures.

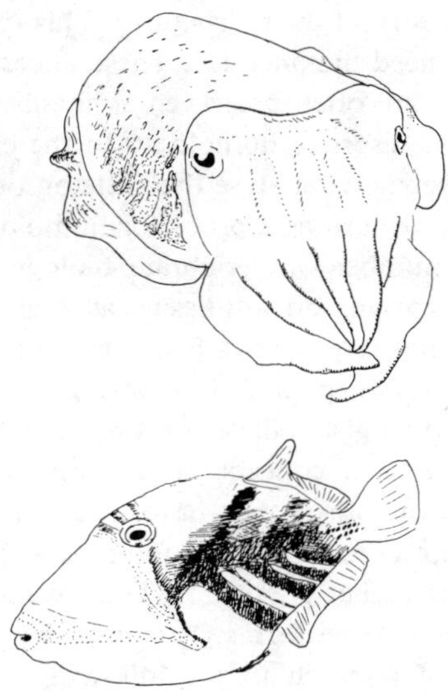

FIGURE 3. Sepia officinalis *(above) and a trigger fish,* Rhinecanthus aculeatus *(below).*

THE CONCEPT OF HOMOLOGY

Simpson's Definition of Homology

Simpson's concept of homology is based on common ancestry. Structures of two or more organisms are said to be homologous if a common ancestor of these organisms also had this structure. Structures are homologous if they can be traced back to the common ancestor.

This concept expresses the essence of the term homology. In this respect homology is included into the evolutionary concept: "For instance the basic forelimb of the land vertebrates, five-toed and adapted for walking, has undergone remarkable radiations among later animals. Among other things, it has become a one-toed, tip-toe running apparatus (in proterotheres and horses), an arm and hand (in you and me), a webbed paddle (in duck bill dinosaurus, otters and others), a flipper (in ichthyosaurus, whales, etc.), a wing (in pterodactyls, birds and bats), and so on." Regardless of the actual function and differentiation of organs, they can be traced back to a common ancestral pattern; they are homologous.

The concept of homology is one problem, its practical use is another. The criterion whether organs of two or more actual animals are homologous or not is the common ancestry of their structures. This means that to check the homology, we should need the ancestor. These ancestors, or at least forms closely related to them, are often preserved as fossils. The fossil record can be analysed to prove or disprove homology. In the case of morphology of vertebrates we can, essentially, analyse the skeleton only. For an osteologist the use of the ancestral pattern as a proof for homology is the most helpful method. But a good number of vertebrate biologists do not work with bones. Morphologists working on soft tissues as well as ethologists studying behaviour patterns cannot rely on any fossil record to find arguments for a given case of homology. Morphologists working on the brain or on blood vessels, as well as ethologists, have to distinguish between Simpson's definition of homology as a concept on the one side, and the practical method in research on the other; they have to use indirect methods. Simpson's theoretical concept is essential, but for practical purposes we have to interpret *actual* animals, recent organisms. Thus, we are looking for vestiges of ancestral characters in actual forms. This must be kept in mind while we deal with the details of research in the following sections of this paper.

Homology and Methods in Comparative Morphology

Topographical Position of Organs within the Organism

The topographical position is an important criterion in establishing a homology. To begin with, the homology of topographical position is often formulated as a guess. There is some feeling about a homology, for instance, that stomachs are homologous in the following animals: dog, cat, turtle, chick, frog, shark and trout. This sounds somehow self-evident. But let us check: the stomach is in every one of these animals part of a complex; it is located between the esophagus and the duodenum; it has a definite position within the intestinal tract. In addition, it is a field which is fixed within the coelom by the same part of the mesenterium. The stomach is homologous in all the animals mentioned, because it always has the same position within an entire complex.

In praxis, one could proceed in the following way: that the stomach is homologous in a bird and a mammal can be affirmed, because it is the same part within a homologous complex in each case. Then, keeping in mind Simpson's theoretical concept, one could ask for a common ancestral pattern of a stomach. The common ancestors of birds and mammals are early reptiles. Actually, the stomachs of those fossils are not preserved and cannot be examined. What we can do, however, is to check other descendants of the early reptiles: the modern (actual) reptiles. If the reptilian stomach is in a comparable position, this helps to establish the homology of avian and mammalian stomachs.

The same course of reasoning can be applied in the case of the paired eyes of vertebrates which have corresponding positions in the head. As a part of the set of the main sensory organs, they are located between the nose and the inner ear. The eyes have a definite connection with a definite part of the brain. Thus, the paired eyes of vertebrates are not only comparable because of their internal structure, but also because of their comparable position within a complex, that is, the head. Similar considerations can apply to a comparison of the human arm with the wing of a bird, or the lung of a land vertebrate with the air bladder of a fish.

Historicity as a main factor in the concept of homology may lead to some special questions, for instance: why is the mammalian central nervous system located dorsally? This question cannot be answered in terms of function. The central nervous system (CNS) as a coordinating organ could be located almost anywhere in the body to serve its function—but it is, in fact, located dorsally. Why? The reason is not functional, but historical. Primary vertebrates had a dorsal CNS, and therefore all their descendants, including the mammals, have it likewise. All vertebrates have a dorsal central nervous system, a character which was conservatively maintained during the evolution of vertebrates. These examples should illustrate the importance of the topographical relations of organs within the organism.

A. Portmann (1959) described another example, namely, the inverse retina of vertebrates. In the vertebrate retina, the receptor cells are not located at the surface directed towards the incoming light, but point towards the inside of the head; in other words, the light coming into the eye has to penetrate all retinal layers before it affects the receptor cells (Fig. 2). From the viewpoint of functional construction, the receptors should rather be located at the frontal surface, and the nerve should connect the inner side and the CNS. The inverse retina can be explained only by historical reasons. The eye develops from a part of the brain tube in the early embryo. The brain tube itself is a specialized dorsal part of the epidermal layer, which becomes invaginated in the process of neurulation. Potential sensory cells are directed towards the ventricular cavity of the brain in this invagination process. When the eye vesicle grows in a lateral direction, the ventricle points towards the inner side, and later developing visual elements point towards the centre of the head.

Comparative Embryology and the Significance of Haeckel's Biogenetic Principle

In 1828, C. E. von Baer, the famous embryologist, wrote the following sentence: "All animals are equal in their first form of appearance, they are nothing but hollow spheres. The farther back we trace their development, the more we find partial congruence among different animals. This leads to the question of whether at the beginning of their development all animals were essentially the same, and whether there is a single ancestral form for all animals." In 1902, Ernst Haeckel developed this idea into the biogenetic principle: "Ontogenesis, the development of the individual, is a short and fast recapitulation of the phylogenesis or the development of the group; in other words, of the ancestors which constitute the generations which lead to this individual."

Of course, we cannot accept Haeckel's biogenetic principle in its widest sense; man, for instance, does not in his ontogenesis repeat all the differentiated developmental stages of primitive vertebrates. On the other hand it is a striking fact that early embryos, for instance, in the vertebrate group, resemble one another very closely (Fig. 4). They can easily be compared. The *Anlagen* of corresponding organs resemble one another; fins, legs and wings of early vertebrate embryos are extremely similar in appearance.

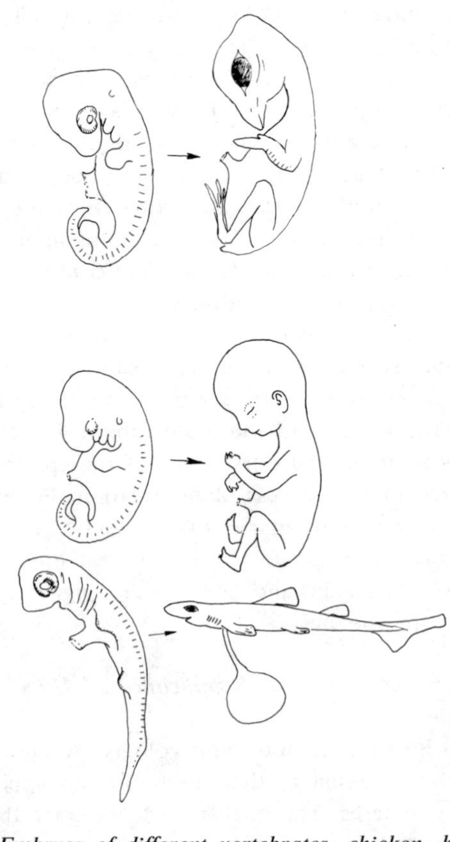

FIGURE 4. *Embryos of different vertebrates: chicken, human, fish.*

Comparative embryology as a method is a very important part of comparative morphology. Organs or structures which evidently correspond in early embryonic stages are considered homologous, regardless of later differentiation. In this general respect, the ideas of von Baer and Haeckel are still essential for actual comparative morphology.

CHARACTERISTICS OF HOMOLOGOUS STRUCTURES

Homology and Bauplan

A *Bauplan* includes all characters which are common to a group of animals; thus, the term is always applied in relation to a definite group. We describe the vertebrate Bauplan, for instance, by including all those characters that are found in all vertebrates. Or we specify Bauplan of birds by including specific characters found in all birds. Characters common to all members of a group which are used to describe its Bauplan are homologous structures. Hence, a Bauplan is a complex set of homologous structures (see p. 44 *Topographical Position of Organs*).

In formulating the Bauplan of vertebrates, certain characters, such as the presence of a notochord covered by the growing vertebral column, are listed. This implies the homology of these structures in all vertebrates. Similar considerations apply to other parts of the vertebrate Bauplan, such as the gross structure of the skull, the main sensory organs, the brain, the segmented musculature, the coelom, etc.

It is interesting that in the description of the vertebrate Bauplan a certain category of characters is not listed, such as the quality of the skin, whether it consists of bony or horny scales, whether feathers or hairs are present or whether the skin is soft or differentiated as a carapace. Not all structures show the same degree of variability throughout the vertebrate group. Some structures remained rather constant during evolution, others have been more affected and, consequently, show a much wider variation throughout the vertebrate group. Such a variable structure system is the externally differentiated skin which may show characters of a subgroup of vertebrates; birds have feathers, mammals can be characterized by hairs, elasmobranch fishes have placoid scales, etc. The entire vertebrate group, however, has to be characterized by rather constant structures. It is a stroke of luck that the skeleton is a relatively constant system, because fossil vertebrates documenting the past consist of skeletons. As far as we know, one of the most constant structures is the central nervous system. The structure of the midbrain roof, the optic tectum, looks almost identical in some turtles and some lizards (Senn, 1970, 1971), even in extreme detail. However, palaeontologists affirm that the line of the turtles and the line of the saurians must have separated at least 230 million years ago. The resemblance of the two midbrain structures makes it almost impossible to assume parallel evolution; thus, the common ancestor must have had this type of *tectum opticum* even 230 million years ago.

Another example demonstrating the constancy of brain structure is that of diencephalic and mesencephalic structures in reptiles and birds. Birds, which originate from early reptiles, show definite differences from reptiles. Palaeontologists have demonstrated that the large group of archosaurians, which gave origin to the dinosaurs, also produced the crocodiles and the birds. In other words, birds are more closely related to crocodiles than they are to any other reptiles. Externally, birds and crocodiles look very different. Anatomically, they have some resemblance, especially in their brains. In actual crocodiles structures which are also characteristic for birds are developed to a certain extent: the subdivision of the dorsal thalamus, the nuclear structure of the telencephalic striatum and the advanced differentiation of the optic tectum are present in both crocodiles and birds. In these respects, crocodiles differ from all other reptiles (Senn, 1969).

In summary, comparative neuroanatomy contributes, on the one hand, additional arguments to the discussion of the relationship of large groups, such as classes, orders and families. It can complement arguments based on osteology. On the other hand, variable structures can be used to characterize small groups, such as genera, species and subspecies. These variable characters (for instance, the surface of the skin) reflect evolutionary facts of shorter periods of time than do other, less variable systems.

Differentiation

It is important to note that homologous structures may show various differentiations. On the one side this concerns phylogeny, on the other, ontogeny. For our present purpose, it is the question of differentiation in the phylogenetic system that is important. In a group of related animals in which homologous structures are present in all representatives, we can also consider a factual (fossil) or hypothetical common ancestral pattern. Each descendant form shows differences when compared with the ancestral form. In other words, structures changed in shape, size, and often also in function. But not all descendants of a common ancestor have undergone the same changes, and not all representatives show the same amount of change. Thus in a group of actual animals related to one another, some remain morphologically closer to the ancestor than others. *Generalized* forms have many qualities in common with the ancestor, *specialized* forms have fewer. A fish is a rather generalized vertebrate, a bird is a specialized one.

It is characteristic of homologous structures in a group of related animals that they appear in a certain variety and that some are more generalized and some more specialized (always in relation to the hypothetical ancestral pattern).

Homology in Closely Related Animals

The relationship of structure and function can be studied in a group of closely related animals, within a genus or two closely related genera. Structures usually can be homologized easily by similarity. Their functions may be more or less similar, too. In different forms within a genus function may vary, reflecting more generalized or more specialized types. A generalized type is more similar to the hypothetical ancestral form of the genus than is a specialized one.

An example is given in Wickler's study on mouth-breeding fish (1970). From the viewpoint of an anatomist, the African cichlids of the genera *Haplochromis* and *Hemichromis* are closely related. Representatives of the genus *Haplochromis* are mouth-breeding fish. In a specialized species, *Haplochromis burtoni,* the female, after laying the eggs, immediately turns and takes them up into her mouth before they can be fertilized by the male. This is a problem. Then, the male presents his anal fin which carries a spectacular pattern: a number of roundish orange spots resembling the species' eggs in shape and colour. The female responds to this signal by snapping at the dummy eggs on the fin, thereby picking up the sperm necessary to fertilize the real eggs in her mouth.

One can assume that the species *Haplochromis burtoni* is not a generalized, but a specialized form within the *Hemichromis-Haplochromis* group. One of Wickler's questions was this: why does the male, after the female has laid and taken up the eggs, present the lateral side of his body, and expand his anal fin to present the dummy eggs? Perhaps this action is not a particularity of *Haplochromis burtoni,* but a modified behaviour which in a general form occurs in other species as well? If so, it must have evolved from a pattern occurring in non-mouth-breeding fish. A representative is *Hemichromis bimaculatus* which deposits the eggs on the bottom. The male of the species often shows an interesting behaviour, the so-called *Rütteln,* part of the courtship behaviour. It consist of a presentation of the lateral side towards the female and has the function of a signal to keep the couple together. This fixed action pattern, a character of these cichlids, was modified in the course of evolution from general substrate-breeding to specialized mouth-breeding species. The function changed from keeping the couple together to a signal for the female to try to pick up the dummy eggs from the male's fin; in doing so, she picks up the sperm to fertilize the eggs in her mouth. Parallel to this evolution of function, the pattern of dummy eggs on the anal fin of the male evolved. A behaviour and a structure evolved together.

This relationship between a structure and a behaviour raises the question whether a behavioural pattern can be treated like a morphological structure. An action pattern, a behaviour can be considered from two different points of view: one is that of its *function,* for example, as a signal. Another aspect of the behaviour pattern is that of a *character* of the animal. As a character, it can be considered in a way similar to that in which structures are used in morphology. A certain animal can be characterized by structures and differentiations of these structures on the one hand and behavioural patterns and their differentiations on the other. As we have seen in Wickler's example of the mouth-breeding fish, the presentation of the dummy eggs by the male, after the female has picked up the unfertilized real eggs has not only the *function* of attracting the female, to have her pick up the sperm, but is also a *character* of the species *Haplochromis burtoni.* As a character, it can be compared with behaviour patterns of related animals. In other words, among related animals characters can be compared regardless of whether they are structures or behaviour patterns.

AN EXAMPLE OF A COMPLEX SITUATION: THE FAMILY OF IGUANID SAURIANS

Iguanids are a family of saurians (lizards) which includes several hundred species. Most of the representatives are distributed in the New World. The members of the iguanid family show common features in terms of morphology; for instance, the differentiation of the optic system in the brain (Senn, 1971). Iguanid saurians are specialized; they differ from other saurians by a greater degree of differentiation in the optic system, particularly of the superficial layers in the optic tectum. This part of the primary optic centre represents a sensory correlation system receiving also non-optic inputs (tactile, temperature, pain and vestibular). But this additional differentiation of the superficial tectal layers is not isolated. In saurians, the superficial part of the tectum is connected with the dorsal thalamus, especially with the nucleus rotundus. The fibre bundle which connects the tectum with the nucleus rotundus and the nucleus itself is of remarkable size in iguanids, compared with other saurians. In all saurians the nucleus rotundus is connected with the dorsal ventricular ridge of the telencephalic hemispheres (Northcutt, 1967). The bundle forming this connection is the dorsal peduncle of the lateral forebrain bundle. It is interesting that, compared to that of other saurians, the dorsal ventricular ridge of iguanids is a huge nuclear mass. The whole chain of optic centres, that is, the superficial optic tectum, the dorsal thalamus of the diencephalon and the dorsal ventricular ridge of the telencephalon are remarkably larger in all iguanids than they are in other saurians (Fig. 5).

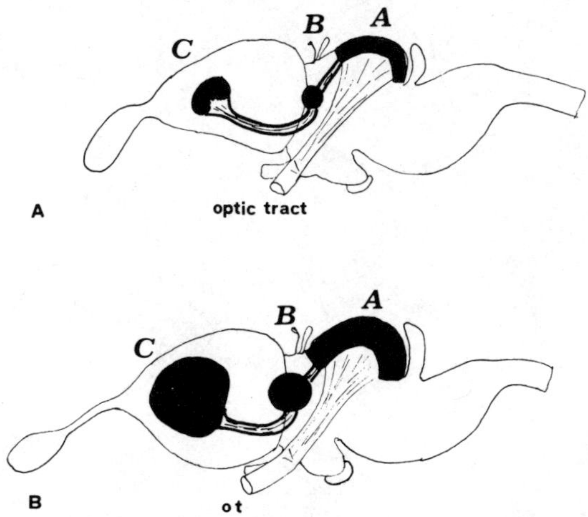

FIGURE 5. *The optic system in the brain.* A. *Generalized saurians.* B: *Iguanid. Centre A, the superficial part of the optic tectum, is reached by the optic tract. This centre is much larger in the Iguanids than in most other saurians. Centre B represents the nucleus rotundus of the dorsal thalamus. This nucleus receives projections from centre A. Centre C, the dorsal ventricular ridge in the telencephalon, receives projections from centre B. Centres A, B and C form a chain of the optic centres. In the Iguanids, all three optic centres synchronously increased in size and differentiation.*

Why do iguanids have such a remarkably differentiated optic system?

All saurians have a well developed optic system (disregarding specialized forms which reduced it). But iguanids have an exclusively advanced optic system. When one examines their behaviour, one discovers that the optic system is closely involved in their social behaviour. Optic signals play an important role in organizing social interaction.

When observing iguanids in the field, one of the first things one notices is their "bobbing behaviour", a vertical shaking movement of the head. The male of *Anolis carolinensis* (Fig. 6) which lives in the swamp forests of south-eastern North America, often paces up and down a tree, especially during courtship season. He often stops, looks around, starts bobbing and expands his orange-coloured throat skin which acts as a moving optic signal when the animal shakes its head. This signal is clearly visible to other individuals, even in the structurally complicated, arboraceous biotope.

FIGURE 6. *The throat skin in* Anolis carolinensis *(above) and* Iguana iguana *(below).*

Another iguanid, the green iguana (*Iguana iguana,* Fig. 6), shows a similar behaviour. However, the signal function is achieved in a slightly different manner: the expanded throat skin does not itself act as a signal, but changes the shape of the animal's head to an imposing silhouette. Similar kinds of bobbing behaviour also occur in other iguanids such as the marine iguana *(Amblyrhynchus)* and the land iguana *(Conolophus)* of the Galapagos Islands.

There is no doubt that the receivers of these signals recognize the bobbing of other individuals; they give special attention to the environment to identify other individuals and to watch their bobbing. In this context, the advanced differentiation of the optic system in the brain of iguanid saurians can be explained. Good vision is important for the individuals to recognize their relation to others and, consequently, for social interaction.

The family of *Iguanidae* can be described by characters. The increased optic system of the brain is one such character, an anatomical one. The bobbing behaviour, also common to all iguanas as a behaviour pattern, is another. One can conclude that an ancestral iguanid form, from which the whole family as a variety originated, had both characters. During the course of time, differentiation reflected the adaptive radiation, including that of bobbing behaviours.

Hunsacker (1962) demonstrated this in the American fence lizards *(Sceloporus),* representatives of the iguanid family. This genus includes a number of species which have very similar bobbing behaviours. Experiments have demonstrated that females are able to select from a sample of males of different species those belonging to their own species. Each species has a characteristic pattern of moving its head in bobbing. Differences between species concern speed, amplitude and number of movement elements in a single bobbing action. The ancestral *Sceloporus* gave rise to all the actual forms; every branch varied the bobbing movement.

Similarly, differences occurred in the whole group of the iguanid family. Every member has the fundamental bobbing behaviour, differentiated to specific appearance. This behavioural "structure" is homologous. *Anolis* and *Iguana,* to give an example, despite specific peculiarities, possess the following homologous characters: the optic system, including the nuclear dorsal ventricular ridge, enlarged dorsal thalamus and specialized optic tectum; the expansile throat skin and ritualized bobbing behaviour.

An interesting aspect might be added. To the question which of the other saurian families is most closely related to the family of *Iguanidae,* a systematic anatomist would answer: the family of *Agamidae*. These animals are found in the Old World in Africa, Asia and also in Australia. Anatomy shows that agamids and iguanids must be very closely related. The agamid brain also shows the increased optic system. Thus, it is not astonishing that agamids have the same bobbing behaviour universal among iguanids.

The concept of homology is linked to the concept of evolution. Homology is possible only in animals which are systematically related. Homology concerns characters—traditionally, structural characters—, but also behaviour patterns. Homology shows the commonality of characters of a group of animals, regardless of the appearance of variation. Homologous elements can vary. Variation is an essential result of evolution. Constancy and homology on the one side, variation and differentiation on the other cannot be separated in the living organism; they are coherent factors: this is what makes biology so exciting and so rich.

DISCUSSION

How To Identify Elements

von Cranach: Could you perhaps work out some of the details? You said, for example, that to begin the homologizing of a structure, you would make a guess. I assume—to keep to the example of stomachs—that after guessing that what you have here is a stomach, you will proceed to find out the locations, relative positions of similar organs in other animals, and this brings us back to the question of distinguishing elements. Then, you said that you must find out about connections to other organs, which is something different from location.

Senn: The term location always refers to relative location; an organ is topographically connected with other organs. In addition, connection includes functional aspects. In brain centres, for example, connections are of functional importance; in some cases, homologies may be established on the basis of fibre connections.

von Cranach: What I would like to know more about is your guess. To guess that an organ is a stomach in a mammal is probably easy, but how about unknown structures? If you come up against structures you have not seen before, what do you actually do?

Chance: Examples would be the pineal body and the neurohypophysis.

Senn: The pineal bodies are a good example. In dissecting the head of a reptile, we will find two pineal bodies on the roof of the diencephalon. Both, the rostral pineal body (parietal eye) and the caudal pineal body (epiphysis) show a characteristic connection with the habenular complex. In other vertebrates, such as mammals and birds, there is only one pineal body. It can be demonstrated that its characteristic connection with the habenular complex resembles that of the caudal body in reptiles; thus, it is the epiphysis.

Ploog: But the pineal gland is part of the epithalamus in an iguana as well as in mammals, so you have the common structure and position requirement which you need to identify a pineal body as you identify other organs.

von Cranach: Somehow this confirms my idea that in all the examples given you already had a fair knowledge about the structures you compared, and were aiming at the single organs. I would like to know whether this is true generally, and whether you have to already know very much about the general structure of an organism in order to work on its single parts.

Senn: Let me give an example: when I began to study the caudal part of the midbrain roof, the *colliculi inferiores,* I found comparisons among reptiles remarkably difficult. The pair of inferior colliculi is found in all reptiles. Then I found some species with a remarkable speciality: they have two pairs of inferior colliculi in the midbrain. These animals *(Dibamus, Typhlops, Anomalepis, Liotyphlops)* show intermediate patterns between saurians and snakes in other anatomical respects. This leads to the question whether the two pairs of inferior colliculi also represent an intermediate pattern between lizards and snakes. The following question was: are the inferior colliculi of most lizards and snakes (which possess one pair) homologous? To answer these questions, I studied the embryonic development of the colliculi in lizards and snakes. In both animals, the general pattern of the early embryonic neural tube which develops an internal cell matrix layer and an external marginal layer is found in the area of the inferior colliculi. In lizards, in the subsequent stages of development, the cell matrix is rearranged. The matrix cells (which remain within the matrix layer) differentiate a nucleus, the inferior colliculus. This oval cell nucleus is in a periventricular position (in the former matrix position). In embryonic snakes, on the other hand, the matrix cells leave the matrix, migrate in peripheral direction to settle within the marginal layer, where they differentiate an oval nucleus, the inferior colliculus. Origin and location of the colliculus are different in lizards and snakes. In addition, the fibre connections are different.

Thus the animals which have two colliculi on each side represent an intermediate stage between lizards and advanced snakes. Their internal colliculus corresponds to that of lizards, including the pattern of fibre connections; the external colliculus corresponds to that of snakes. The evolution of snakes, which originate from lizards (saurians), includes the development of a second external pair of inferior colliculi. Later, the original (internal) pair has been reduced.

Ploog: I would like to know whether you can relate behaviour and morphology in this case with respect to hearing. Is there any homology in the function of the colliculi?

Senn: We do not know yet. There are still technical problems in making small lesions in a reptile head. However, a difference in function is indicated by the different fibre connections. In addition, my colleague, R. G. Northcutt of Ann Arbor, Michigan, found different histochemistry in the saurian and ophidian inferior colliculi.

Levels of Comparison

von Cranach: This morning we learned that it is very difficult to determine a level on which comparisons can be based. I think this example is suited to show what is a level of comparison.

Senn: It is.

von Cranach: Well, what is the level of comparison? How would you define it?

Senn: Structure rather than function.

Comparing Structure and Function

von Cranach: If we keep to structure, there are probably also various levels. It has been pointed out again and again that this is the main problem. And I should like to hear how it is handled in practice.

Wickler: The aspects of function and structure are being mixed now, aren't they? In your answer to the question of Professor Ploog you were talking about homology of function.

Ploog: No. Dr. Senn kept to the level of comparison of structures in various aspects of homology which he listed before. He did what I expected comparative neurologists to do: namely, to compare the structure on the one hand, and try to find out how it is related to the function, that is, to behaviour, on the other.

Wickler: For what reason should this be done? To clarify the homology of these structures?

Ploog: Take the example of the eye and the visual system; if we would stick only to comparisons of the various visual systems without knowing that they are made for seeing, this would not make much sense.

Wickler: Would it not make sense to find out that the auditory ossicles, in spite of having quite a different function, are homologous with the original joint of the jaw? You would never find that out if you concentrated on function, because function is what changed.

van Hooff: If you find correspondence in structure, position, and so on, and then also find corresponding functions, there is no argument about the case; but non-correspondence is not an argument against it. There might be functional change.

Wickler: Permit me to insist. You are changing the unit you are comparing: first you have only the structure, the second unit you may choose is structure-plus-function. Or does a similar function actually give you some more reason to say that structures are homologous?

van Hooff: I would say that if you look at the structural comparability, a different function is not an argument against it; that is your incidence of the rain droplets in the air.

Wickler: But can a similar function be an argument *for* it? I feel that this meant a change of the unit.

van Hooff: You will not usually expect structures to change function. It is more likely that in related species similar structures will have similar functions.

Wickler: Where does this expectation come from?

van Hooff: Does it not come from empirical data? I am talking of related species.

Wickler: That is true.

von Cranach: The difficulty of this example is that visual brain structure can be used for many purposes. It would be difficult to demonstrate that it has developed in the service of a certain signal system. Still, I would believe so if it was always present or absent along with that system. Otherwise I would assume that such a visual discriminatory capacity is a generally useful tool; it is certainly risky to assume a relation between a very specialized function and a very general ability.

Masters: Isn't there a shift in the level of analysis involved in moving from structure to function, a shift from description to hypothesis of explanation?

What one can say about parallel structures which seem to have similar functions is that it is a conceivable hypothesis for a homology. That is all one can say.

Frey: I wonder whether this process of establishing homologies could be channelled in some way, or whether it is generally open-ended and derived from the level of your research, or whether that level is always the new hypothesis of what is similar.

Senn: The difficulty I see now is this: as a criterion for homology, I prefer structure. On the other hand, a behavioural pattern is not only a function, but also a character of a group of animals. The head-bobbing of iguanids is a homologous behaviour in the different representatives of the iguanid family. In this case, my argument would be that the primary iguanids, the ancestral stock of the family, had a bobbing behaviour involved in their courtship behaviour. These primary iguanids gave origin to a radiation of more than 300 species which differ in terms of anatomy and size, in terms of colour and colour changes, and also in their bobbing behaviour. So we can describe hypothetical ancestral conditions from which the various forms differentiated.

Crook: Though this is a question of the relation of structure and function I should certainly say the criterion for homology must be structure, or at least description. If there happens to be similarity or identity of function, this aids the argument for homology, but divergence or shift of function may simply be an example of cladistic homology. The display of the various kinds of weaver birds for example shows many common components, but these components are used in different ways according to the social organization of the species. In certain colony-breeding species you will find a certain posture used to advertise a nest primary to courtship; in other species with a non-colonial structure you will find the same display in courtship away from the nest, after which the female is led to the nest. We do not, of course, have an ancestral weaver bird to look at, but we can infer from that community of components of this whole radiation that there is some common source. But it is impossible to know the direction of this radiation, that is, the direction of specialization without knowing the social organization of the ancestor.

Cain: As to the iguanid–agamid relationship, I have seen something of those groups and it is certainly true that lizard experts, who on the whole have to do without any fossil records worth considering, put them fairly close together. But as you said, one is New World and the other Old World, and *that* to me is a danger signal because it is perfectly true that the agamids occupy the general ecological niche in the Old World that the iguanids do in the New World. Two animals doing the same job obviously are bound to be remarkably similar, and this does not imply that their similarity is due to common ancestors. There are quite a few examples of Old and New World groups of birds—nectar sucking birds, for example—which are remarkably similar in all kinds of characteristics, not because of common ancestry but because of absolutely parallel evolution. I do wonder whether the example you have given of the agamid–iguanid relationship is not one the same type, and actually a convergence.

Senn: The respective brain patterns of these nectar-sucking birds were checked and found to be quite different; this demonstrates that the Old and New World forms are not related, whereas the brains of iguanids and agamids are very much alike. The two families are closely related. (In terms of time scale, this fits in with Wegener's continental drift theory, since these animals existed before the Atlantic divided the continents.)

NOTE

I am grateful to Mrs. M. Durand-Wenger for her excellent technical assistance.

REFERENCES

von Baer, C.E. 1828. *Über Entwicklungsgeschichte der Thiere,* vol. 1. Königsberg.

Haeckel, E. 1902. *Natürliche Schöpfungs-Geschichte* 10 ed. Berlin, Reimer.

Hunsaker, D. 1962. "Ethological Isolating Mechanism in the *Sceloporus torquatus* Group of Lizards", *Evolution* 16:62-74.

Northcutt, R.G. 1967. "Architectonic Studies of the Telencephalon of *Iguana iguana*", *Journal of Comparative Neurology* 130:109-148.

— 1969. "Discussion of H.J. Karten's Paper: 'The Avian Telencephalon'" *Annals of the New York Academy of Sciences* 167:180-185.

— 1970. "The Telencephalon of the Western Painted Turtle (*Chrysemys picta* belli)", *Illinois Biological Monographs* 43:1-113.

Portmann, A. 1959. *Einführung in die vergleichende Morphologie der Wirbeltiere.* Basel/ Stuttgart, Benno Schwabe.

Remane, A. 1952. *Die Grundlagen des natürlichen Systems, der vergleichenden Anatomie und der Phylogenetik.* Leipzig, Geest & Portig.

Senn, D.G. 1969. "Über die Bedeutung der Hirnmorphologie für die Systematik", *Verhandlungen der Naturforschenden Gesellschaft in Basel* 80:49-55.

— 1970. "The Stratification in the Reptilian Central Nervous System", *Acta Anatomica* 75:521-552.

— 1971. "Beständigkeit und Abwandlungen bei Reptilien", *Verhandlungen der Naturforschenden Gesellschaft in Basel* 81:207-222.

Simpson, G.G. 1950. *The Meaning of Evolution.* London, Oxford University Press.

Smith, H.M. 1967. "Biological Similarities and Homologies", *Systematic Zoology* 16:101-102.

Wickler, W. 1970. *Stammesgeschichte und Ritualisierung.* München, Piper.

WOLFGANG WICKLER

The Application of Homology-Reasoning to Behaviour

I have been asked to give my paper in English; this may create difficulties for those of us who are not native speakers. But I think the difficulties would have been even greater, had we been asked to give our talks in Bantu or Japanese. This is only one illustration of the well known fact that the existing languages can be arranged into groups of increasing dissimilarity (or similarity). Why is this so? The resemblances obviously are not just pure chance. In 1786, Sir William Jones stated that certain resemblances between languages can only be explained on the hypothesis of genetic relationship; in particular, he was convinced that Sanskrit, Latin and Greek "have sprung from a common source which, perhaps, no longer exists". A so-called Indo-European language family was established decades before scientific workers started rigorous comparison. The methods used for comparison have been discussed by Greenberg (1955): "The most convenient method is a table of pronouns, grammatical elements and vocabulary for the languages involved" (p. 108-109). Relevant to the process of establishing genetic relationships are the use of mass comparison (as opposed to isolated comparisons between pairs of languages), and the emphasis on specificity of resemblance (p. 1). Many traditional classifications made extensive use of criteria based on sound only or meaning only and thus confused typological non-historical classification with historical-genetic classification (p. 106). "The importance of resemblances recurrent in a large number of languages as plausible outcomes of some single ancestral form as elicited by mass comparison is of very great evidential power in excluding either chance or borrowing as explanations" (p. 107). Lexical resemblances (like morphemes which resemble each other both in form and meaning) are *sine qua non* for the establishment of all relations. "Under no circumstances can we reject results attained from obvious lexical resemblances in fundamental vocabulary in favor of those based on vague structural traits." Uncertainty, then, will increase the more one has to rely on such structural traits because of overall lexical similarity between closely related languages. "The problems of grouping within a linguistic family are often more difficult than the mere determination of over-all linguistic relationship. For example, there is not the slightest doubt among students of Indo-European as to which languages are to be included, but such questions as the position of Hittite and the existence of a separate Italo-Celtic branch are still the subject of lively controversy" (p. 115).

All this sounds very familiar to the zoologist, especially the taxonomist, who aims at a phylogenetic classification of animal species. He, too, starts with a table of morphological and behavioural characters, he will stress the importance of mass comparisons and will find it much more difficult to correctly arrange the species within one family than to determine which species belong to that family. He tries to avoid similarities due to chance as well as purely functional resemblances, in order to base his classification of species on their genetic relationship. To achieve this, he uses a set of methods, especially the homology criteria. Among biologists, the term "homology" has been accepted in the evolutionary definition given by Simpson: "Homology is resemblance due to inheritance from a common ancestry." In order to identify such resemblances as due to inheritance from a common ancestry, biologists have found, tested and now generally agree upon homology criteria, which have been extensively discussed by Remane (1952), who mentions that these very same criteria have long been used in comparing languages as well as cultural traits of different peoples in order to find common roots for different languages or cultural characteristics. Languages, as well as animal (and plant) species, have evolved through history. Today's differences can be interpreted as due to changes during divergent evolution, and homologous similarities as due to common historical origin.

Confusion does arise, however, with respect to the term *genetic*. Taxonomists usually deal with morphological characters that are to a large degree inherited; zoologists call these characters genetically determined. By genetic relationship, therefore, they mean resemblance with respect to the genome. Linguists, on the other hand, treat languages as if they were living species, although they recognize that languages in fact are part of the behaviour of one animal species and not genetically determined in the zoological sense; by genetic relationship they mean historical relationship.

The student of animal behaviour, the ethologist, has to deal with behaviour patterns, some of which seem to be inherited, like organs, whereas others are learnt like languages. Both show continuity over time and evolutionary changes. Every continuity can be understood as due to transmission of information. In biology, information can be genetically or traditionally transmitted; genetic transmission includes handing over of some material, tradition does not. The result of genetically or traditionally transmitted information will be similarity (or resemblance) between those characters produced by means of that information.

There is, however, one alternative, and that is the so-called independently evolved similarity as exemplified by the eyes of vertebrates, cephalopods and insects or by the wings of insects, birds and bats. The terms "eyes" or "wings" connote resemblances that are not due to common ancestry but to function and convergent adaptation. Adaptation means storage of information, via trial and error or via mutation and selection. The production of an insect's wing or a bird's wing is based on stored information; but the similarity between these types of wings cannot be traced back to the same store of information. Therefore, such similarities are called analogies, in contrast to homologies.

The methods used in linguistics to find historical relationships between languages are the same as those used in biology to find historical relationships between species: both linguists and biologists look for homologous resemblances and try to rule out analogous ones. The homology criteria extensively discussed by Remane (1952) are useful in deciding whether a given resemblance between characters is due to transmission of information from one information store to another (Wickler 1965), regardless of whether genome or memory acts as store and whether transmission occurs through geneflow or through tradition. (If someone wanted to restrict the term homology to genetically determined resemblances, he would have to add further homology criteria in order to rule out tradition.)

No character is in itself homologous or analogous. These terms indicate relations between characters; more precisely, they indicate two different causes for resemblance. Finally, it is necessary to identify precisely the characters that are to be compared: wings of birds and bats are vertebrate *forelimbs* and as such are homologous structures; nevertheless, they have independently evolved into *wings* and insofar are analogous structures (Wickler, 1967).

These considerations have proved to be important for all kinds of comparative studies, especially within the behavioural sciences. Taxonomists have shown that simple hehaviour patterns are indicative of phylogenetic relationship between species; the resemblance between these homologous behaviour patterns is due to genetic resemblance. However, many students of behaviour are not primarily interested in taxonomic problems but in generalizations or predictions from one species or type of behaviour to another. The taxonomists, by experience, have learned to concentrate upon tiny behaviour elements (wing or bill postures during display, gaits, etc.). The generalizers, by their interest, concentrate upon more complex syndromes, the names of which point to their biological effect (function) or to the physiological mechanism ensuring that effect (locomotion, nest building, aggression, etc.). They tend to overlook the fact that homologies are one source of generalizations and predictions, analogies are another. Even some ethologists, as well as some of their critics, seem to neglect analogous resemblances. I have shown (1961) that the similitarity of a simple drinking behaviour pattern in different groups of birds—in contrast to what famous ethologists have written about it—is non-homologous and that some taxonomic conclusions based on the assumed homology are wrong. More recent studies have shown that the same result, be it brood-care, monogamy or aggression, may in different animals be the outcome of quite different physiological mechanisms: terms such as drive, instinct, motivation, and so on, may refer to different mechanisms with different characteristics even within the same animal, if they are applied to different functional units, such as aggression or feeding. Studying feeding or aggression in one species allows only limited predictions about the behavioural systems underlying feeding or aggression in another type of animal; studying the behaviour system for feeding in one species allows only limited predictions about the behavioural system underlying aggression in the same species (Wickler, 1970, 1973).

Some critics suggested studying primates rather than geese in order to solve socio-ecological problems in human behaviour. Comparability, however, does not necessarily increase with phylogenetic relationship. On the contrary, the lower the probability that similarities in the mechanisms that serve the same function are due to common ancestry, the higher the probability that the characters found to be similar are truly correlated with and indispensable to this function. This application of homology-, as well as analogy-reasoning, seems to be of increasing importance if one deals with morphological structures, animal behaviour and, finally, human behaviour.

DISCUSSION

Information Store

Masters: I should like to go back to Dr. Wickler's first point: the relationship between homologies in language and homologies in biology is still, as far as I can see, an analogy. Maybe on another occasion we should discuss that. But let me restate a more immediate question: you described homology in its broadest sense as a similarity of traits which can be derived from a common information source.

Wickler: Store, not source. Source implies an external influence.

Masters: What is curious is that human language seems to have the same, or similar, structural, functional and evolutionary characteristics as an information store as the genetic material. If you compare human language and animal communication, there are some interesting differences between human languages and animal communication systems. On those dimensions, human languages are much more similar to the genetic material. That gives rise to the following problem: at what point can you really separate out analogy and homology, and do the principles of homology that you present somehow go back to a fundamental analogy, that the information store has a certain function?

Wickler: No, the important thing is the special information in that store, and not the presence of the store itself. Let me give a simple example, though it may not cover the human capacity for language. There is a special bird call which you can show is not genetically determined. You may ask why these different birds have that call in common, exactly as you could ask why they have a special bone in common. You can rule out close genetic relationship as a reason. The call is taken over; one bird has to learn it from another. If you cut off this communication, the transfer does not occur any more, and birds develop without this call. The similarity of these calls is not chance, nor is it a case of independent evolution; it is due to common origin or, in American terminology, *patristic homology*. The method you apply here is exactly the same one you would use to homologize a bone structure or something else, except that in case of the sound you already know that the similarity is not genetically fixed.

von Cranach: I could give another example, one of social structure: if you compare the legal systems of Turkey, Japan and Switzerland, you will find striking similarities which can all be traced to the fact that these three countries imported German law. This is a case of homology, and to study these structures you can apply the same kind of reasoning. Does that cover the case?

Wickler: Exactly. If you suspect that two similar phenomena must have some common origin, if you doubt that they evolved independently and that their similarity is due to pure chance, you have to look where these phenomena are derived from, and the methods to find that out are exactly the same in linguistics, in culture studies and in biology and zoology.

Masters: Let's think of the cultural example for a moment. Where the problem arises is in the sharp distinction between homology and analogy.

The problem of the legal codes reminds one of the long debates between diffusionists and evolutionists in anthropology, and of multilinear evolution, which is an analogy to what we call parallel evolution in biology. At certain points in the cultural analysis the distinction between diffusion, which would be like homology, and multilinear or parallel evolution, or analogy, becomes impossible to make. It would appear that both processes could apply at the same time.

Wickler: Of course there are problems. There are characters in biology for which the question whether they are homologous or not cannot be answered; however, that a useful method has its limits is not unusual. I do not think that by applying this method to behaviour patterns, languages or cultural traits you cross the border between homology and analogy. As I said before, the important thing is the strict definition of the unit you want to study. Let me remind you of the long and heavy debate in ethology about whether a given behaviour, for example, the nest building of the rat, is inborn or not. It finally turned out that this question cannot be answered, because the elements of nest building are inborn, but their sequence has to be learned. But the answers are clear if you choose the correct units, and that, I think, is the answer to most problems of this kind, be they linguistic, cultural or even ethological.

Concept of Homology

Foppa: I know that some comparative linguists treat language families like species, groups or units, but I know also that many linguists doubt the validity of this method. Take the developments of styles, for example, in music. Would you speak of homologies with reference to musical works as you speak of homologies between members of language families?

Wickler: Of course, and you do, too. If you have variations of one theme, it is not very difficult to find out the sameness even if most of the original characters have been changed; you can still recognize a variation on a particular theme that you have homologized,—you only have to find out by what characters.

Foppa: And what is the difference between the concept of homology and a judgment of sameness?

Wickler: Sameness has been applied in two quite different fashions: one is sameness because of origin, and the other is sameness because of function, as in a certain type of music which makes soldiers march faster; these pieces of music need not be derived from the same source, but are suited to this function.

von Cranach: One type of sameness is sameness due to origin, the other is just sameness?

Wickler: Sameness of function.

von Cranach: That is what I do not understand. What function?

Crook: Adapted significance.

Wickler: That is what makes me apply the term "wing" both to butterflies and to birds.

Leyhausen: What you are saying amounts to calling phenocopies homologous with genetically fixed characters. If you speak of a bird's song being taken over by one species from another to which it is, in a way, endogenous, then you maintain that the bird which takes it over is taking over the same kind of information, but this is not at all true. The imitator is not taking over the identical information, but the finished product, which is something very different. It is like the experiments of Bittermann and other people in which a rat, a pigeon and a human baby put in the same kind of maze all come up with the same result of learning, but they have reached this result in very different ways. The result is not homologous, although phenomenologically identical. You are starting from the premise that the criteria of homology, as we have them so far, are sufficient to express or to determine what their authors were after, but we know by now that these criteria are not sufficient and that we need one more, namely, that the phenomenologically identical is not necessarily homologous. The way in which the said identical result was achieved is the all-important problem. All this reminds one of the famous *Kulturkreis–Lehre* which fell prey to just that kind of error.

von Cranach: Do you mean that the way in which a character evolves, the actogenesis, should be made part of the definition of homology?

Leyhausen: I think by implication it has always been, and I think the other phenomenon has been neglected, or possibly just overlooked, in biology.

Information Underlying Behaviour Genesis

von Cranach: So you would include in your definition of actogenesis what Wickler has called underlying information, while in Wickler's definition this is not included. Would you be prepared to accept two structures as the same only if you knew how they had originated?

Leyhausen: No, what I really want to know as far as possible is the whole path of information including all three types of genesis.

von Cranach: But has this ever been achieved?

Wickler: This again goes back to what I said earlier: you may, of course, instead of describing the normal character of an organ such as the thymus, in addition seek those similar organs of that kind that have been developed by a special germinal layer. Then you get a different kind of thymus, which you can take as a new unit to study. You can go even further and try to find out what are the underlying genetics, and you may find, as you said, that different genes are responsible, and then you may choose one of these gene structures with its special germinal layer and its organ as a unit of study.

Concept of Homology

Leyhausen: And so this is phenocopy or tradition, and this is, in a loose way of speaking, an individual acquisition that does not come under our concept. I, at least, definitely and explicitly exclude this from my concept of homology. You may disagree, but if we have a concept of homology, I think we should take it in the meaning that has at least implicitly been accepted all the time by biologists, and I come closer to that meaning than you do. "Sameness", by the way, can be taken as a certain criterion of homology only when and if it means sameness of evolutionary and/or developmental history: it often does not apply to the finished products, witness the gill-openings of a shark and the parathyroid glands in mammals.

Wickler: But then you would restrict, as Tinbergen once tried, either the methods of finding homologies or the term homology to characters that have the same origin, or are inherited, or whatever. Then you have to add proof for this similarity of the same type of origin or inheritance as a criterion for homology, which means that you cannot speak of homologous characters unless this has been proved.

Leyhausen: Right.

Wickler: Right.

Ploog: What I wanted to say has been said. Why is it necessary to use the word homology in the context of tradition? Why can we not stick to the term homology as it has been used, that is to say, the criterion of inheritance is necessary in order to apply this word. For all the examples you have given, one has to coin a new word, be it tradition or information transfer or something like that. I think one confuses the issue, at least in biology, by introducing these examples, which, however, may be logical. I think it is just a new use of the word without new insights into what we mean by homology.

Wickler: I tried to show what may result from the homology criteria and the methods we have, and to show that these methods can give more than has usually been suggested. If you want a special result that you do not get by these methods, then you have to change the methods or add more criteria.

Cain: I think that this question of exact origin was already debated by no less than Richard Owen himself, who was aware that some bones that under any criteria were acceptable as homologous can not always arise from the same center of ossification. So there also a certain latitude was accepted quite explicitly by the father of the word homology. Later, Professor Groupovic, a great anatomist, showed very beautifully, following up some expert German work in this area, that the limbs of tetrapods which have so far been accepted as homologous are not necessarily developed from the same number of segments. I am inclined to apply to Professor Leyhausen—if it is not too unkind of me—his own very wise words that we may not be able to define very accurately in a biological example, that we may have to accept a certain amount of leeway. Yet, at the same time, I am in complete sympathy with him and I congratulate Dr. Wickler on producing his horrible example of the bird's call. My feeling is we must exclude that from homology, but I don't see how.

Crook: It seems to me there is no objection in principle to using the word homology and analogy about complex systems whose information origin is different. I think that if we talk about language or genetics, it is perfectly admissible to use the terms in either case. The issue we are concerned with is the historicity of the phenomenon, and this has to do with the determination of the phylogenesis of the structure we are interested in. I think it is an interesting and valid point that the terms homology and analogy do not have to be pinned to any particular source by genetics but can be used in a broader sense. It seems to me that some of the debate between Leyhausen and Wickler has to do with the questions of what kind of examination of historicity is to be involved, whether what is simply an examination and a comparison of phenotypes is sufficient, whether one can argue homology and analogy from similarity only, or whether one has to take into account the historical process as well. I should like to say something, too, about the introjection of strange characteristics: I would have thought that the homology or the analogy here has to do with the capacity of an organism to acquire something from other animals—this is like learning—, and the interesting thing is the capacity to do it, not the actual acquirement.

Wickler: But one normally compares characters without analysing the animal's special capacities for attaining them. The problem is what to call an entity. Imagine a rat running through grass, and grass seeds adhering to its fur; the rat may even eat the seed. Or think of Tretzel's lark hopping around and picking up a shepherd's whistled melody. Or think of a baby rat that has picked up an aberrant genetic character from its mother. The seeds may be regarded as accidental acquisition, not belonging to the rat itself. Song melodies do belong to larks, though their particular origin may be more or less accidental. Of interest here is, however, that in all cases it is possible to trace back the origin of the particular character, regardless of whether it is accidental or not.

Masters: I think the problems of terminology have also to be solved by principles of utility. One of our problems seems to be the broadening of the term homology to the acquisition of information stores which do not come genetically to the species. Because your beginning example of linguistics is an analogy, it may be useful when discussing logical inference from animal to human behaviour, to speak of homologous relations in linguistics. But since we have this problem of similarity, it may be advisable to use a different word for that to avoid confusion. I think your point was very real and useful; the problem is the way we look at homology.

von Cranach: What we have not yet done at all is to follow up Dr. Wickler's question of whether using homology in the animal–man comparison can be at all useful. In the second part of his presentation he demonstrated that it might be much more useful to start from differences, not from functional similarities. Consequently, we should perhaps be interested in functional analogies to learn about system–environment relationships. All this remains to be discussed.

Same / Similar

Foppa: We are all speaking English, and therefore demonstrating some degree of sameness in our language behaviour. Judging my language behaviour as the same as Dr. Cain's or Dr. Crook's, for example, means application of a very wide definition of sameness, because there are a lot of differences; certainly one may call these kinds of sameness homologies. But I think two things have to be demonstrated: the first is the advantage of broadening the concept of homology, the second is that this general concept of homology implies a different methodology from traditional definitional procedures. Since I am free to define our language behaviour as the same, I must pay attention only to the features of these two examples of our language behaviour. If you include only a few features, the probability that two phenomena will be judged the same will increase, but it will decrease the more features you consider. I believe that this is the real difficulty in expanding the concept of homology in the manner you did.

von Cranach: Did you not say some time ago that to state that two things are the same you have to refer to a third one?

Foppa: No, no! As similar!

von Cranach: This involves the question of whether same and similar are the same.

Wickler: What I thought about and was interested in is the reason for the similarity of characters, whatever characters you choose. What I tried to point out is that reasons for similarity which lie in history can be brought to light by the methods that have been called the homology criteria. That's it.

If I were only interested whether you have, by pure chance, invented a vocabulary and grammar that mimics English perfectly or whether you learned English somewhere from an Englishman, there is a method I could use to tell these possibilities apart. If you are interested in the specific way your learning was brought about, and how much it has been facilitated by genes, then you can elaborate other methods by which this can be achieved. My point is that the homology criteria as they stand do not allow differentiation between these different ways of handing over information, but do allow identification of the similarities due to historical reasons, not to pure chance or independent evolution.

REFERENCES

Greenberg, J.H. 1955. *Studies in African Linguistic Classification.* Reprinted from the *Southwestern Journal of Anthropology.* Branford/New Haven, Compass Pub. Co.

Remane, A. 1952. *Die Grundlagen des natürlichen Systems, der vergleichenden Anatomie und der Phylogenetik.* Leipzig, Geest & Portig.

Wickler, W. 1961. "Über die Stammesgeschichte und den taxonomischen Wert einiger Verhaltensweisen der Vögel", *Zeitschrift für Tierpsychologie* 18:320-342.

— 1965. "Über den taxonomischen Wert homologer Verhaltensmerkmale", *Naturwissenschaften* 52:441-444.

— 1967. "Vergleichende Verhaltensforschung und Phylogenetik", pp. 420-508 in: G. Heberer (ed.) *Die Evolution der Organismen,* vol. 1. Stuttgart, G. Fischer.

— 1970. "Soziales Verhalten als ökologische Anpassung", pp. 291-304 in: W. Rathmayer (ed.) *Verhandlungen der Deutschen zoologischen Gesellschaft.* Stuttgart, G. Fischer.

— 1973. "Ethological Analysis of Convergent Adaptation", *Annals of the New York Academy of Sciences* 223:65-69.

ROGER D. MASTERS

Functional Approaches to Analogical Comparison Between Species

The repeated controversies concerning the possibility and scientific validity of inferences from animal to human behaviour ultimately revolve around the issue of "analogies" between animal species. It is generally admitted that where true "homologies" exist, inference can be made from the causal relationships underlying similar behaviours or structures in a related species. But for many, such homologies are virtually the only legitimate generalizations which can be made concerning common features of animal and human life (see Montagu, 1968).

In order to clarify this hotly debated issue, it is useful to begin by placing behaviour—and bodily structure—in the perspective of general biology: all phenotypical traits must lie within the genetically programmed "reaction range", with the specific phenotype representing the product of the genotypical possibilities interacting with the environment throughout the organism's development (Dobzhansky, 1955, p. 74; Simpson, 1958, p. 14). Distinction between variables as either purely genetic or purely environmental should be avoided as a biological absurdity. What we call "instinct" is usually a quite narrow or stereotyped reaction range—a pre-programmed coordination or response that is transmitted genetically and released with little modification in a variety of environmental settings. What we call "learning" is really a broad reaction range, in which the individual acquires certain crucial components of a behavioural sequence—be it the identification of appropriate releasers, a motor coordination, a sequence of coordinations, or a pairing of coordinations with physiological states (or "appetites") —during ontogeny. When speaking of broad reaction ranges (including learning) we mean that variation in the environment is correlated with variation in phenotypical traits or behaviours. But whether the reaction range is narrow or broad, all biological phenomena presuppose an interaction between genotype and environment throughout ontogeny.

From this perspective, the debate over "instinct" versus "learning" is as trivial if not irrelevant as the older, discredited argument of heredity versus environment in genetics. It may well be that the *functional* significance (using this term in a biological sense defined below) of a behaviour pattern is independent of the degree to which it is narrowly programmed in the genotype. By looking at biological data in terms of evolutionary theory, we are led to emphasize the different ways that species "satisfy" the functions of survival and reproduction. Only in so doing can we avoid the anthropomorphic assumption that "learned" responses are somehow superior to instinctive ones—or, for that matter, the pessimistic over-reaction that "instinct" is better than learning.

HOMOLOGY AND ANALOGICAL COMPARISONS BETWEEN SPECIES

Analogical Comparison and the Concept of Function

The present paper will focus on the validity of analogous comparisons between different species. In so doing the term *analogy* will be used throughout in its precise biological meaning. While the word is also popularly used to describe any loose or superficial similarity, here we are concerned with the scientific concept which was introduced in comparative morphology by the anatomist Owen in 1843, prior to the publication of Darwin's *Origin of Species* (Dobzhansky, 1955, p. 231).

Analogous organs or structures may occur as *parallelism* or as *convergence*. Such analogies arise "where similar demands by the environment have evoked similar phenotypic responses in unrelated or at least not closely related organisms [...] If there is only one efficient solution for a certain functional demand, very different gene complexes will come up with the same solution, no matter how different the pathway by which it is achieved" (Mayr, 1963, p. 609). Or, as Simpson puts it: "The development of the same sort of evolutionary opportunity by different groups of organisms produces the evolutionary phenomena known as parallelism and convergence [...] when the element of an ultimate ancestral structure in common is lacking [...] the structures developed for the function in question are not homologous but only analogous" (Simpson, 1967, p. 181).

As these statements make clear, the concept of analogy presupposes the notion of *function*. A similar ecological niche or selection pressure, confronting different species, creates a selective advantage for similar structures or behaviours, without requiring a common ancestor as the precondition for succesful adaptation. Although the necessity of functional analysis in biology will be taken for granted here, as a logical corollary of the generally accepted theory of natural selection (see Pittendrigh, 1958), a few words are in order on the concept of "function".

Insofar as natural selection represents the differential fertility of genotypes within a species' gene pool, we may speak of "survival of the fit"—but not necessarily of the "fittest" (Dobzhansky, 1955; Simpson, 1963); selective "advantage" merely indicates that organisms vary in their capacity to satisfy one basic requisite or *function* of all living systems, namely reproduction of viable offspring. Hence functional analysis no longer implies neo-vitalism, nor does it presuppose the generally discredited form of teleology which imputes conscious or divine purpose to nature. Adaptation and selection do not mean that individuals or species have achieved the optimal mode of existence, but merely that their genotypes have a comparative advantage in terms of the function of reproduction.

The notion of natural selection thus necessarily presupposes functional comparisons between different individuals of the same—or of different—species. Whenever we speak of an evolutionary sequence in which one form supplants another, we assume that natural selection or adaptation can "explain" the observed phenomena. To do so, however, is to compare the capacity of different genotypes to satisfy the reproductive function without which a particular living form cannot persist. The study of adaptation and natural selection is thus based on the comparison between individuals or species to determine the extent to which "similar demands by the environment have evoked similar phenotypic responses" (see Mayr, 1963, p. 609). Although intraspecific selection studies compare "closely related organisms", whereas (by definition) the analysis of inter-specific selection does not, in either case evolutionary biology compares the performance of different genotypes with respect to given functional requisites.

Analogies, in the conventional sense of the *parallelism* or *convergence* of widely separated species with respect to a given structural or behavioural trait, obviously appear to deal with far less general functions than reproduction. But just as the primordial function of genetic replication can be satisfied by a wide range of reproductive systems which are, in the precise sense, analogies, so adaptations like flight or sight, which open specific ecological niches, can be satisfied in parallel or convergent ways. Since the function or adaptive significance of a trait is identified by the observer, it is necessarily a hypothetical explanation of the factors underlying natural selection. Hence virtually all comparisons of individual organisms or species from a Darwinian perspective rest—at least implicitly—on analogical reasoning.

Hierarchies of Functions: Analogy and Homology

Lest the foregoing seem excessively general, let us consider in more detail the notion of an adaptive function as a condition which maintains or increases the probability that a species will persist through time. Such biological functions constitute, in a sense, the defining characteristics of the ecological niches which the various species and genera have exploited. Each condition or process which contributes to survival in a given ecological niche becomes thereby a function whose *satisfactory* fulfillment we call adaptation.

Since some processes are extremely general whereas others are quite specialized it is possible to conceptualize the varied functions as forming a hierarchy. The most general function—the production or replication of the genetic information which perpetuates living matter—is satisfied in widely different (but technically analogous) ways by different classes and genera, though the specific chemical code for genetic information—that is, the system of DNA and RNA "codons"—seems to be homologous among species. A less general but still very extensive function, like locomotion, characterizes a broad class of organisms, conventionally defined as animals rather than plants. Still narrower functions are satisfied only by a smaller range of species; for example, mobility through the atmosphere is a functional adaptation exploited by many species among the invertebrates as well as vertebrates, though of course many animals cannot fly. Ultimately, of course, one can identify increasingly specialized functional adaptations until those exploited only by a single species, race, or even individual have been defined.

It is curious that the concept of *analogical comparison* is usually applied only to structures and behaviours at a relatively narrow level of specialization. This is not difficult to explain historically, since the discovery of closely convergent structures and behaviours in different classes or orders (*e.g.* the analogies between humming moths and humming birds—Simpson, 1967, p. 183) was an important demonstration of the principles of natural selection. It must be emphasized, however, that the logic of analogical comparison—that is, determination of a specified range of functional adaptations which are satisfied by genetically different organisms—is a necessary and common procedure in evolutionary biology.

This point has been stressed heavily here because it clarifies the relationship between homology and analogy. Homology is usually defined as a similarity of structure or behaviour which can be traced to a common ancestor (see Simpson, 1967, pp. 180-182). As such, homologies are especially revealing because they represent a similarity in the physiological causes of observed structures or behaviours. At the genetic level, homology represents a similar structure or sequence of DNA in the genomes of different species, based on descent from a common ancestral form, whereas we cannot assume that merely analogical features rest on related genotypical information. As a result, homology can best be understood as a genetic similarity of the causal mechanisms underlying comparable behaviour or structures in two or more species.

While homology and analogy are thus distinct in principle, in practice it may be difficult if not impossible to determine whether a given similarity is homologous or analogous. On the one hand, conclusive proof that two living species are descended from a common ancestor is often unobtainable due to gaps in the fossile record. On the other, a similarity between two species may appear to be homologous at one level of analysis, but analogous at a more precise level of study (Simpson, 1967, p. 182). Hence the extent to which structural or behavioural similarities between species are true homologies is necessarily a question requiring detailed research.

The difference between analogy and homology can therefore be restated in terms of the extent to which given functions are satisfied by phylogenetically related causal mechanisms or processes. In homology, common evolutionary and developmental origins are presumed to account for a close similarity or identity in gentetic and physiological characteristics. In contrast, behaviours that are analogous but not homologous result from genetically different structures or processes. This distinction between common biological functions (as identified by the observer) and common causal mechanisms or processes (as coded in the genotype) is essential to any coherent theoretical approach to the problem of inference from animal to human behaviour (Tinbergen, 1972). As will be shown in detail below, confusion between functions and causal processes—or the related confusion between homology and analogy—underlies much of the controversy arising from attempts to compare human and animal behaviour. [1]

Criticism of Analogical Comparison

The main attack on analogical comparison between species concerns the attempt to relate animal and human behaviour, particularly in the popularized works of Konrad Lorenz and Robert Ardrey. As the American economist Kenneth Boulding has written, in an assessment of "how dangerous the method of analogy may be," "the naive analogizing [...] of Lorenz and especially of Ardrey will not stand up to a moment's serious criticism" (Boulding, 1968, pp. 87-88). If we consider the grounds of this criticism, the nature of the issue will be clarified.

As Boulding (1968, p. 86) puts it, in comparisons between humans and animals "the crux of the matter, of course, is the old argument about nature and nurture, heredity and environment, instinct and learning".

> There is little doubt that the oriole's genes build into the bird itself both the knowledge and the desire to build an oriole nest and that this knowledge comes from information in the gene rather than information coming into the bird through its senses. It is the great peculiarity of man, however, differentiating him from all the other animals, that what his genes endow him with is an enormous nervous system of some 10 billion components, the informational content of which is derived almost wholly from the environment, that is, from inputs into the organism from outside. The genetic contribution to man's nervous system is virtually complete at birth (Boulding, 1968, pp. 86-87).

From a biological viewpoint, Boulding is thus arguing that different causal processes or determinants of behaviour operate in lower organisms and in humans; in this view, the behaviour of orioles is caused by "information in the gene" whereas the behaviour of humans depends on information "which is derived almost wholly from the environment".

It should be immediately apparent that this argument rests on an analogy between humans and animals. Boulding assumes that behaviour is a consequence of information available to an organism: all organisms respond to their environment on the basis of "information" which presumably establishes a "program" or pattern of possible behaviour as well as cues for the appropriateness of given responses in the behavioural repertoire (see Lorenz, 1965). Hence, although Boulding—and others critical of Lorenz or Ardrey—frequently attack the notion of analogy, this criticism itself usually depends on the assertion that human learning and animal instinct are different (analogous) causal processes satisfying the common biological function of adapting and responding to the environment in a manner that will permit survival.

A major difficulty with Boulding's criticism arises from the implication that the only valid comparisons between humans and animals would lie at the level of "instincts" defined as "information in the gene". This way of formulating the problem assumes that the function of selecting and maintaining adaptive behaviour patterns is largely fulfilled through the emergence of narrowly determined genotypical reaction ranges in lower animals, whereas for humans "almost everything that happens" after birth "is learned" (Boulding, 1968, p. 87). Since ethological research shows the situation to be far more complex (*e.g.* Masters, 1975), Boulding himself seems to be guilty of an oversimplified analogy between the genetic information in other species and a human culture as "a body of coded information which is passed on from generation to generation" (Boulding, 1968, p. 88). While analogies are possible between genetic material and human languages (Masters, 1970), it does not follow that animal instinct and human learning are as causally distinct as would appear from Boulding's argument (see Harlow, 1958).

The criticism developed by Boulding and others thus really reflects the erroneous assumption that the only possible analogies between humans and other animals concern behaviours that are the product of similar if not identical "instincts". There is a great irony in this criticism, since it is usually put forth by scientists in the so-called "behaviorist" school, which is characterized by the attempt to develop universal "laws" of behaviour based on scientific experiments. As even cursory reading of the main theoretical works in this tradition, reveals, the "laws" developed by American behaviorists rest largely on experiments with a few highly arbitrary species (especially mice, rats, and pigeons), on the assumption that all organisms share a relatively fixed number of "drives" and respond to the environment in a relatively uniform way.

Consider, for example, the following passage grom B.F. Skinner's *Science and Human Behavior* (1965, p. 69):

> When reinforcement is no longer forthcoming, a response becomes less and less frequent in what is called "operant extinction". If food is withheld, the pigeon will eventually stop lifting its head. In general when we [*i.e.* humans] engage in behavior which no longer "pays off", we find ourselves less inclined to behave in that way again.

From this behaviorist theorist's point of view, phenomena like "operant extinction" can be defined in terms that apply equally to both pigeons and human beings, since the relationships between causes and effects in different species are analogous provided that the relevant variables have been defined with sufficient precision. As Skinner (1953, p. 89) emphasizes, "the same interpretation applies to human behavior" and to the analysis of the pecking response of pigeons.

The behaviourist tradition, of which Skinner is a leading representative, recognizes of course that there are differences between species. Skinner (1953, p. 75). affirms, however, that "the fact that organisms evidently inherit the capacity to be reinforced by certain kinds of events" is not related to "predicting the reinforcing effect of an untried stimulus"; in other words, Skinner's research on operant conditioning (the effect of the "untried stimulus") is *not* based on the assumption of phylogenetic relationship in the inherited "capacity to be reinforced by events". The laws of operant conditioning and extinction are thus not presented as strict homologies, but as "convergent" or "parallel" phenomena characteristic of classes of organisms.

Paradoxically, the behaviourist tradition, from which some of the strongest criticisms of "analogies" between animals and humans is derived, itself rests on the assumption of broad analogies between species.

As the above example indicates, the criticism of analogies between animal and human behaviour does not seem to concern the possibility of comparing different species, but rather the definition of those phenomena which are deemed truly comparable. From the perspective of traditional American behaviorist psychology, the "laws" of behaviour concern causal relationships between environmental events or "stimuli" and the resulting behaviour of organisms, without reference to the genotypical peculiarities of individual species. In contrast, the comparisons between animals and humans presented by Lorenz (1966) or Ardrey (1961, 1966, 1970) concern species-specific patterns deemed to be analogous in two or more species, often representing different genera or classes, but *not* characterizing all animal behaviour as such.

In reconsidering Boulding's critique of "naive analogies" between animals and humans, therefore, it is apparent that the issue concerns the kind of comparison between species that is considered scientifically valid rather than the possibility of comparisons as a matter of principle. The "critical question", from a perspective like that of Boulding, "is the extent to which the genetic composition of the human body limits and determines what it can learn from its environment"—that is, whether one can "find in 'instinct' any explanation whatever of human behavior beyond the most elementary and primitive acts of the newly born" (Boulding, 1968, pp. 87-88).

When scientific criteria are defined in this fashion, to make valid inferences from animals to humans, it is necessary to demonstrate homologies. This follows because phrases like "the genetic composition of the human body" or "instinct" refer to the physiological or causal processes determining behaviour, rather than to functional adaptations which could be satisfied in analogous ways by different species. Hence Boulding's "critical question" would reduce the comparisons between humans and other animals to the search for characteristics that can be traced to a common ancestor. But since it is difficult to trace pure homologies between the behaviour of *Homo sapiens* and his nearest relatives (see Kurtén, 1972), and even more so when comparing humans and any other species (Boulding, 1968, p. 85), this perspective easily leads to the presumption that humans are virtually unique.

Boulding's emphasis on the difficulty of demonstrating homologies is correct. But his conclusion—like that of most critics of Lorenz and Ardrey (Montagu, 1968)—rests on the tacit but faulty assumption that analogical comparison concerns identical causal processes or "instincts", whereas analogy technically refers to similar but phylogenetically (and hence genetically) *different* means of satisfying a given function in distinct species.

The confusion between analogy and instinct—that is, the assumption that analogical comparisons seek to demonstrate the existence of identical or homologous causal processes—is understandable enough given the rather loose terminology in the popular works of Lorenz and Ardrey. Nonetheless, those who (like Boulding) have generally rejected analogies between animals and humans have failed to grasp fully the central role of analogical comparison in all biological research, including that of the behaviorist school.[2]

The Weakness of Popularized Analogies Between Animals and Humans

The above examples have attempted to show the inevitable presence of analogical comparison in the thought of those who have criticized as "naive analogies" the inter-specific comparisons presented by Konrad Lorenz and Robert Ardrey (among others). This does not mean that the analogical comparisons developed in such popularized works are always satisfactory. As I have argued in detail elsewhere (Masters, 1972, Section 4), popularizations of ethology have done a disservice insofar as they have obscured the theoretical framework of Lorenz's scientific work and led to overly hasty comparisons which perpetuate the obsolete nature-nurture dichotomy. Lorenz's own popularization—most notably, his *On Aggression*—has unfortunately contributed to this difficulty by speaking frequently of functional analogies as "instincts". Consider the following passage:

> It is superfluous to point out the analogies between the social behavior patterns of many animals, particularly wild geese, and those of man [...] the similarities in the social behavior patterns of the Greylag Goose and of man are not derived from a common ancestor but have arisen by so-called convergent adaptation [...] If in the Greylag Goose and in man, highly complex norms of behavior, such as falling in love, strife for ranking order, jealousy, grieving, etc., are not only similar but down to the most absurd details the same, we can be sure that every one of these instincts has a very special survival value, in each case almost or quite the same in the Greylag and in man (Lorenz, 1966, pp. 217-218).

Unlike Boulding, who confuses analogies and homologies, Lorenz states the logic of analogy accurately. But his use of the term "instinct" leads to a confusion between analogous *functions* and analogous *causal mechanisms*. Lorenz speaks of "falling in love, strife for ranking order, jealousy, grieving, etc." as analogous "instincts" in animals and humans because, like "analogously constructed and similarly functioning organs" in different species, we have a "right [...] to group them in the same functional conception and call them by the same name" (Lorenz, 1966, p. 218).

The difficulty arises because, earlier, Lorenz had quite properly distinguished between "a concrete function whose survival value is obvious" and "the neurological organization which we call instinct" (Lorenz, 1966, pp. 86-87). Whereas Lorenz's precise formulation (1966, p. 87) uses the concept of instinct to refer to physiological mechanisms that satisfy biological functions "in a blindly mechanical way", thereby distinguishing between functions and causal mechanisms or processes, later in the work he blurs this distinction and treats instincts as a "functional conception". Technically speaking, Lorenz treats intra-specific aggression as a critical *function* in many species, including *Homo sapiens,* even though he speaks of the various processes that satisfy this function as a common "instinct".

This is not merely a verbal quibble, but rather a fundamental point if confusion is to be avoided. As was remarked above (p. 77), theoretical clarity requires a careful distinction between adaptive functions (which are identified by the observer) and causal or physiological processes in the structure and behaviour of animals themselves. Since, in principle, a biological analogy between two species is a phenotypical trait which satisfies similar functions by means of different genetically-based causal processes, there is no reason to reject out of hand analogies between partly learned human behaviours and animal behaviour patterns that are narrowly controlled by the genes (or, in common terminology, "instincts"). But because Lorenz, in his popularized writings, occasionally confuses "instincts" and "functions", it is not surprising that his critics have made the same mistake.

Since, as Lorenz (1966, p. 86) appropriately insists, "a definite and self-contained function of an organism, such as feeding, reproduction, or self-preservation, is never the result of a single cause or a single drive", it is not necessary to assume that aggression is a functional adaptation which corresponds to a single instinct or physiological drive. Indeed, Moyer has argued that, among rats, the functional category of aggressive behaviour represents at least eight different causal mechanisms or processes (Moyer, 1968). Because the very concept of a "drive" or "instinct" blurs the distinction between function and physiological causation, it has been increasingly criticized by ethologists who thereby simply elaborate Lorenz's own conceptualization as stated more precisely in his scientific papers (*e.g.* Hinde and Stevenson-Hinde, 1971).

While the ambiguity described in *On Aggression* need not have been serious, if only because careful readers could have easily clarified Lorenz's position by consulting his experimental and theoretical writings (*e.g.* Lorenz, 1965, 1971, 1972), other popularized statements of ethology have reinforced the tendency to confuse common behavioural functions with common "instincts" or causal processes. Ardrey's (1966) notion of "the territorial imperative," for example, has confused many readers because he does not always clearly distinguish between the adaptive functions of territory and its diverse physiological causes.

One instance will illustrate the problem. When discussing the homing behaviour of salmon, Ardrey moves from the question "How do they find it (*i.e.* the salmon's home territory)?" to the question "Why? [...] What is the selective benefit to the species of maintaining territories under such difficult conditions for periods so long?" (Ardrey, 1966, pp. 136-137), without indicating that he is shifting from the level of causal processes to that of functionnal adaptiveness. As a result, it is easy to conclude that Ardrey attributes homing behaviour in salmon to a single impulse or "instinct", whereas it has been shown that a mosaic of delicately interrelated physiological processes produce the observed behavioural pattern (Fontaine, 1956). Hence territorial behaviour generally can only be considered as a *functional* category, not as a single physiological process or instinct.

This does not mean, of course, that widely observed functions like intra-specific aggression or territory can never be traced to analogous causal processes in different species. The physiological mechanisms which produce territorial defense, homing, or other spatial behaviour may well be analogous—or even homologous—in different species, but such analogies must always be demonstrated carefully. This point deserves great emphasis because the selective advantage of territorial defense need *not* be identical in all species; as a subordinate function in the hierarchy of functions, territoriality may be a means of satisfying the higher order function of population spacing in some species whereas in others it is primarily related to stabilizing social structure, regulating mating, or insuring reproductive isolation (Carpenter, 1958).

Very broad functional categories like intra-specific aggression or territory are thus difficult to use in analogical comparison without further specification of the precise biological functions involved in the species under study. Since what Lorenz (1966, Chap. 6) calls "the Great Parliament of Instincts" has a different *constitution,* so to speak, in each species, one cannot assume without detailed research that the selective advantage or function of intra-specific aggression is identical in greylag geese and humans.

Popularizations of ethology can thus be questioned for reasons that are exactly opposite to those usually stated by those critical of analogical comparisons between animal and human behaviour. Whereas behaviorists have denied the relevance of analogies which do not reflect common phylogenetic ancestry and common physiological causation, we have seen that this criticism itself tacitly presupposes analogical comparisons between species and confuses analogy with homology. The problem with popularized analogies like those of Ardrey is that they offer potentially valid hypotheses without fully specifying the precise biological functions of possibly analogous behaviour patterns.

Once this weakness of popularized analogies is made explicit, the debates concerning the validity of inferences from animal to human behaviour can be clarified if not resolved. Whereas behaviorists tend to ake tacit analogies between very generalized processes, such as conditioning, across many classes and orders of organisms, popularizations of ethology tend to describe explicit analogies between very specific behaviours in a few phylogenetically diverse species (*e.g.* Lorenz's remarks on "jealousy" or "grief" in the greylag goose and humans).

In both cases, the analogies are appropriate working hypotheses, and in both cases the hypotheses may be confirmed by empirical data. But without care, opposed schools of thought can lead to erroneous conclusions—not because the logic of analogical comparison is faulty, but rather because the precise functional and causal correlates of the analogy have not been accurately determined.

The Logic of Functional Analysis

We are now perhaps in a better position to state more precisely the logic of analogical comparison and the rules for its use in inferences from animal to human behaviour. The perennial debates between behavioural psychologists and Lorenzian ethologists are clearly obsolete, insofar as *both* schools are partially correct. Once it is recognized that all comparisons reflect a judgment concerning the adaptive role of observed structures, behaviours, or evolutionary sequences, the relationship between functional categories and analogous physiological processes or causal mechanisms in different species should become a subject for research rather than for polemics.

All biological comparisons have a fundamental logic which can conveniently be described in terms of set theory. Scientific hypotheses assume that if certain conditions are met, predictable and regular consequences will follow. Whenever two individuals, populations, or species are analysed in terms of a single hypothesis, one assumes that each of the two represents a subset of a larger set or category. Poor comparisons and faulty analogies consist in sets of phenomena which are *not* overlapping with reference to the conditions or consequences stated in the hypothesis under consideration; valid analogies presuppose that different species—be they pigeons and humans (as for Skinner) or greylag geese and humans (as for Lorenz)—are subsets of a larger set *with reference to a precise scientific hypothesis.*

The difficulty of inferences from animals to humans lies in the extent and nature of the obvious differences between the set of phenomena called the species *Homo sapiens* and that of one or more other species. Since it is evident that humans are neither pigeons nor greylag geese, it is never easy to demonstrate those elements of human behaviour that are analogous to a given aspect of animal behaviour. At most, analogical comparisons are scientific hypotheses which, if they cannot be disconfirmed by empirical data, are nonetheless only provisionally accepted within the on-going structure and process of biological science.

The precise formulation of an analogical comparison must therefore, at least in principle, specify the level of analysis on which animals and humans are being contrasted. More concretely, a comparison must not only carefully distinguish between adaptive functions and genetic or physiological causes, but also specify the units being studied. In other words, any demonstration of analogy requires specification of the entire hierarchy of functions involved in both species being compared, and the determination that the functional comparison is not vitiated by divergences at higher or lower levels of analysis.

LEVELS OF BIOLOGICAL ANALYSIS

It has been argued that the logic of analogies between species requires an explicit statement of the precise level on which comparative research is conducted. To facilitate analogical comparison between animals and humans, therefore, it is useful to classify the levels on which ethological analysis is conducted. For this purpose, biological research can be characterized in terms of two basic dimensions: the *temporal depth* of the phenomena considered, and the *spatial extent* of the organisms included. On each dimension, convenient categories can be distinguished for the most typical kinds of research—though it should be emphasized from the outset that the phenomena themselves are not as discrete as these categories.

Temporal Levels of Analysis

Following Gerard (1960), it is useful to distinguish three levels in the analysis of any system: its *structure* (or "being"), its *functioning* (or "behaving"), and its *evolution* (or "becoming"). As Gerard defines these levels, the structure or "being" of a system consists of those aspects that "recur essentially unchanged in different time-moments and constitute the enduring architecture [...] of the entity". The functioning or "behaving" of a system—also often called its "processes"—are those responses that are "adaptive (homeostatic) and transient (reversible)".[3] And whenever non-reversible events have persistent consequences for the system, "this secular or cumulative residue of change in the longitudinal time section constitutes the development or history or evolution of the system—its 'becoming' " (Gerard, 1960, p. 255).

The difference between analysis of structure, behaviour or process, and evolution has often been noted by biologists. What is not always emphasized, however, is that these three types of research rest on a different definition of the depth of *time* considered relevant to the phenomena studied. Structural analysis is essentially synchronic—that is, the object of study is the state of a system at a given, rather narrowly defined moment, on the assumption that the phenomenon in question is relatively constant. Behavioural analysis treats a deeper time frame, though it assumes the essential similarity of the processes discussed in different organisms or settings. Evolutionary or historical analysis deepens the time frame still further, by examining the similarities and differences of the state of the system through time.

These three modes of analysis are, of course, mutually interdependent: evolutionary study of phylogenetic sequences generally rests on careful comparison of structural types, bodily structure cannot be explained without reference to behaviour, and physiological processes or behaviours cannot be understood apart from their function or selective advantage (*i.e.* the principle of natural selection on which our understanding of evolution rests). While analysis at all three levels must combine the study of both physiological causation and functional adaptation, this duality is especially pronounced at the level of behaviour: on the one hand, behavioural research examines the causal processes operating in organic systems (*i.e.* the physiological or neurological causes of motor coordinations); on the other, such research emphasizes the selective advantage of some phenotypical traits or behaviours (*i.e.* the functional adaptiveness of the causal processes observed).

The distinction between causal and functional analysis thus has a particularly important role in all behavioural research. Although structures can be described without reference to their functional or selective advantage (traditional morphology), and evolutionary sequences explained without knowledge of the precise physiological causation of adaptive traits, behaviour necessarily combines both elements; as has been shown above, failure to distinguish between causal processes and adaptive functions has led to much of the uncertainty and controversy concerning interspecific behavioural comparisons. Hence the distinction between structural, behavioural, and evolutionary foci of research is particularly useful because it reminds us of the complexity of behavioural research, arising from the need to combine causal and functional criteria to a degree that is often unnecessary in structural or evolutionary studies (see Roe/Simpson, 1958).

Spatial Levels of Analysis

There is a general tendency to associate the study of deeper time-levels with the broadening of the spatial frame of reference in biological research. For example, Mayr treats an "evolutionary" approach as synonymous with the study of *populations* of organisms through time, as contrasted with a more static, physiological analysis of individual organisms (Mayr, 1958, p. 341). While this tendency can doubtless be documented in biology, it is clearly possible to utilize structural, functional, or historical approaches when studying an individual organism or an entire population. For the sake of clarity, therefore, it is useful to distinguish explicitly between the spatial range of the biological system being studied and the temporal depth of the analysis.

From the perspective of ethology, it is convenient to characterize three main cutting points along the spatial dimension: the individual organism, the social group, and the species.[4] For many purposes, of course, these categories are highly artificial; in some species, it is difficult to distinguish clearly between individual organisms and groups, particularly where colonies of organisms develop such narrow functional specializations that different organisms seem comparable to the cells or organs of a single animal. At the other end of the scale, a species may be difficult to analyse insofar as its limits are imprecise (*e.g.* the phenomenon of "*Rassenkreise*", see Rensch, 1966) and direct observation of the species as a whole is impossible (Mayr, 1963). Despite these difficulties, is should be evident that the scope of scientific focus can vary widely—particularly when studying the vertebrates—and that the individual organism, the social group, and the species provide convenient points at which this spatial dimension most clearly changes the character of research and theory.

Nine Types of Ethological Analysis

Insofar as the two dimensions described are indeed distinct, it is useful to classify levels of analysis in terms of the nine categories set forth in the table. At the risk of oversimplification, representative disciplines have been indicated in each cell; although scientific fields as a whole cannot really be categorized in this way, the table should be a useful heuristic device for clarifying the often perplexing problem of "levels of analysis". While the extreme cells in the table are most clearly discrete, it should be no surprise that the central categories tend to be most difficult to isolate from each other. In fact, this classification may be useful precisely because it indicates why the approaches represented in the four corners of the table tended to develop before the more complex, intermediary approaches: Morphology,

Ontogeny, Population Distribution (Zoogeography), and Phylogeny can be more readily studied as merely *descriptive* sciences at a preliminary stage in the development of biology. A comparable development of scientific work dealing with the more complex, intermediate levels of analysis has been more difficult, since it presupposes both an understanding of the causal mechanisms underlying organic life and the functional adaptation served by species-specific traits.

Spatial depth of analysis	Temporal depth of analysis		
	Structure ("Being")	Process ("Behaving")	Evolution ("Becoming")
Individual	Morphology	Physiology Individual psychology	Ontogeny
Group	Social structure	Social behaviour	Group dynamics Demography
Species	Population, Distribution and Population genetics	Population Ecology	Phylogeny

Whatever the utility of the above schema, it should be evident that all approaches are intrinsically inseparable. The evolution of species can only be confirmed by comparative study of individual morphology, both at the gross structural and the genotypical levels; the physiological processes or "causes" of individual behaviour are inseparable from social structure and social behaviour; group dynamics (the complex pattern of irreversible changes in group size and distribution) is clearly a central factor determining not only population distribution in the short run, but demographic patterns as a whole; population distribution not only is closely related to demographic patterns, but often plays a crucial role in morphology, physiology, social structure, and the like. What is more, of course, the genetic and cytological levels of analysis, while not specified in the table for the sake of simplicity, are of crucial causal and functional importance in the study of all higher levels.

The ninefold typology presented in the table is therefore intended solely as a heuristic device, in order to clarify the typical foci of much biological and ethological research. Since no level of research can claim autonomy, one must always consider the interaction between phenomena at different levels.

At the same time, however, it is equally important to avoid the assumption that higher level phenomena can be simply "reduced" to or totally explained by lower or less extensive levels of analysis. This point is particularly important given the current developments in molecular biology, which promise to explain much that was heretofore incomprehensible in the causal mechanisms operating at all levels of life: whatever the promise of contemporary genetics and biochemistry, it should be clear that the evolution of populations and species represents a higher order complexity which has its own "laws" or constraints. Indeed, from the evolutionary point of view, one must explain the selective pressures that produced not only sexual reproduction, but the emergence of the DNA "code" itself, speciation, and other apparently universal properties of living systems.

ANALOGICAL COMPARISONS BETWEEN HUMANS AND OTHER ANIMALS

Since accurate functional comparison between any two species requires an explicit delimitation of the level of analysis, the above classification should be useful in discussing inferences from animal to human behaviour. Ideally, a valid analogy should concern *one* of the above nine foci of research, on the assumption that variations on the other eight levels are not relevant to the traits being compared. Failure to specify the level of analysis produces confusion if not error; when the limits of presumed analogies or inferences are blurred, non-comparable features at another level can be incorrectly subsumed in or deduced from potentially correct functional analogies.

For example, a behavioural pattern or process characteristic of *individual* organisms—such as Skinner's "operant conditioning"—cannot provide an exhaustive analytical comparison of *social* behaviours—such as the "ritualizations" described by Lorenz. Not only are the phenomena on different spatial levels of analysis (individual behaviour versus social interaction), but in each case the behaviour concerned is related to different temporal levels of research. Whereas Skinner defines stimuli as environmental events in the ontogeny of an organism, Lorenz focusses on motor coordinations of one animal that have become "key stimuli" or releasers for a conspecific in the phylogeny of a species. As a result, Skinner assumes that phylogenetic differences are irrelevant or constant when comparing animal behaviours, while for Lorenz phylogenetic relationships are an explicit and necessary subject of research.

Levels of analysis must be distinguished, moreover, without the assumption that higher levels can simply be explained as the sum of lower level interactions. Causal analysis of phenomena characteristic of populations or species as a whole cannot be limited to the physiological or behavioural processes observed in the individual. To be sure, group behaviour must be mediated through the central nervous system of each individual organism. But social or species-wide regularities may represent functional adaptations that cannot be reduced to the trait's survival value for an individual organism.

For example, mating behaviour must be analysed not only in terms of the physiological and neurological mechanisms operating in the individual, but also with reference to the social and species-specific processes involved. Since there is a selective pressure in favour of behaviours which insure *both* genetic variability through recombination within the species (Bischof, 1972) *and* reproductive isolation from related species (Spieth, 1958), a complete study of mating requires consideration of all nine categories in the table. Structure, behaviour, and evolution must be analysed at the level of individuals, groups, and species as a whole before the precise causation and function of observed traits can be known.

In practice, of course, such a complete understanding of specific components in an organism's behavioural repertoire is rarely if ever available. For most purposes, analogical comparison is adequate when functional analysis at one of the nine levels distinguished in the table is combined with a precise specification of the relevant variables on the remaining eight levels of analysis. But when these criteria are satisfied, there is no reason to reject analogies merely because one of two species being compared is *Homo sapiens*. Whatever the complexity of human behaviour, whether studied at the level of individuals, societies, or the species as a whole, biological functions observed in other animals can also be observed among humans.

It would be absurd, for example, to deny the possibility and relevance of analogies between demographic patterns in other animals and humans. Population genetics is likewise an area in which analogical inferences between animals and humans have often been appropriately formulated. In neither case, does the differential role of individual learning, social custom, or evolutionary rates of change constitute an a priori ground for challenging functional comparisons between humans and other animals.

It may be well at this point to mention in some detail an instance of such a comparison which seems to provide useful inferences from animal to human behaviour. Given evidence that there is a correlation between abnormally high population density and various forms of "pathological" (or mal-adaptive) behaviour in other species, Galle, Gove, and McPherson (1972) examined the hypothesis that "density has, or at least may have, serious consequences for man." Focussing on measures of social behaviour (receipt of public assistance, juvenile delinquency, admission to mental hospitals) as well as individual behaviour (mortality, fertility), they sought evidence of a possible correlation between high density and pathology. In terms of the levels of analysis in the table, Galle, Gove, and McPherson were attempting to study the functional importance of one element of social structure, namely physical spacing, which has been found to be significantly related to individual and group behaviour in other species.

As a statistical analysis, their survey of data from the city of Chicago assumed that individual variations in morphology, physiology, and ontogeny were not crucial independent variables. Social structure was subjected to careful statistical control for both social status and ethnicity, so that density could be analysed apart from other social variables frequently associated with human behaviour. But four different measures of density were available: persons per room, rooms per housing unit, housing units per structure, and structures per acre—and clearly each of these measures has different social and behavioural implications.

When persons per acre was used as a measure of population density, social status (SES) seemed to account for virtually all of the variance in individual or social pathology. But persons per acre is an exceptionally gross index of the functional relationship between population and space: high rise luxury apartments and slums may have similar populations per acre, even though the effective space per person varies greatly. Hence Gall, Gove and McPherson reanalysed their data using persons per room as a measure of effective density—and in so doing found that this measure accounted for virtually all of the statistical variance in some of the pathological indices (admissions to mental hospitals, mortality) and at least half of the variance on the remainder. As the authors concluded: "Our study suggests that overcrowding may have a serious impact on human behaviour and that social scientists should consider overcrowding when attempting to explain a wide range of pathological behaviours" (Galle/Gove/McPherson, 1972, p. 29).

While this study is clearly a provisional exploration of the relevance of ethological findings in human populations, it illustrates the need for careful functional analysis in research comparing human and animal behaviour. Not only is it essential to control for obviously related variables, such as social status, in order to isolate a single structural variable like local population density, but the measure of density itself has important analytical consequences. Without a consideration of the functional relevance of the measure chosen, findings can be extremely misleading if not false.

When speaking of functional analogies between population density or physical spacing in animals and humans, moreover, it is important to specify the ranges of individual and social behaviour concerned. In some social contexts, increased population density results in increased rigidity of the dominance hierarchy rather than the pathologies studied by Gove, Galle, and McPherson; for example, army barracks represent different behavioural adaptations to high density spacing from urban ghettos—or prisons. Hence the definition of "pathological behaviour" among humans as well as the varied functional adaptations to high densities need further study (for a review of the literature, see Vine (1973).

However limited the findings in question, they reflect a promising kind of analogy between animal and human populations. If many mammalian species exhibit differing social behaviours under varied conditions of density, there is no reason to assume that *Homo sapiens* is immune to such effects—and, indeed, common experience suggests quite the contrary. Cultural variation and individual learning—often cited as grounds for excluding humans from analogical comparisons—can thus be viewed as crucial elements in the adaptation to such factors as differential population density. And population density may in turn be viewed as an adaptive response to ecological settings (Barash, 1974).

To be sure, specification of the precise ranges of density which have particular effects on human behaviour requires careful research. But such a study is hardly to be prejudged on the assumption that functional analogies either are or are not possible between animals and humans. In each case, such an analogy can serve as a useful working hypothesis for detailed investigation and analysis, rather than as the basis for unquestioned inferences from animal to human behaviour—or equally unquestioned claims that *Homo sapiens* is unique and cannot be compared with other species.

CONCLUSION

The foregoing discussion of analogical comparison as the basis for inferences from animals to humans can be summarized as follows:

a) Insofar as functional analysis is necessary in evolutionary biology, analogical comparison is implicit in all studies of natural selection and adaptation.

b) Structures and behaviours which serve a common function in unrelated species are called analogies, whereas homologies presuppose the origin of similar traits in a common ancestor. While in principle analogy and homology are thus distinct, in practice it is sometimes difficult if not impossible to determine whether a comparison is homologous or analogous.

c) The difference between analogy and homology thus rests on the distinction between common biological functions (as identified by the observer) and common causal mechanisms or processes (as coded in the genotype); confusion between function and cause—or between homology and analogy—must therefore be carefully avoided.

d) Most criticism of analogical comparisons between animals and humans reflect the erroneous assumption that the only possible analogies between humans and other animals concern similar if not identical instincts; this criticism is paradoxical, because many behaviorists who make it represent a theroretical tradition which rests on tacit analogies and inferences between animals and humans.
e) Popularized analogies between animals and humans often blur the distinction between function and cause, treating "instinct" as a functional category rather than a physiological process.
f) Functional analogies require exact specification of the level of analysis on which data are being compared; for this purpose, it is useful to classify nine levels or foci of research, depending on whether structural, behavioural, or developmental phenomena are studied in individual organisms, social groups, or entire species.
g) If adequate care is used in specifying the level of analysis—and if functions are not confused with causal processes—there is no a priori reason to reject analogies or inferences between animals and humans. On the contrary, this approach promises fruitful insights into human behaviour and evolution.

DISCUSSION

Usefulness of Analogy

von Cranach: You stated that analogy has a somewhat doubtful reputation. This may partly stem from the fact that there are no rules for it. In preparing this conference I found that not much has been added to what Aristotle wrote about it, to exaggerate a little. Nowadays, philosophers of science usually say that analogy was an unresolved problem. A difference between homology and analogy that has not yet been discussed seems to be that the methods used to assess homologies are restricted to the organism, and therefore cannot be applied to relations between organism and environment. This will be a problem in the treatment of motivational systems that cannot be assessed without reference to this relationship, and therefore can only be compared via analogy. On the other hand, this is one of the advantages of analogy. Functional analogy signifies comparing the situation of an organism within a certain environment, and this goes beyond what can be done by homology methods.

Meaning of the Term Function

Leyhausen: My questions are partly in the same area. You tried to make the distinction between homology and analogy on the basis that homology applies to causal processes and analogy to function. I don't think that such a distinction can be made here; something has to be said first about the semantics of the word function: you, and Wickler yesterday, used the word function in the sense of some function served by behaviour, while I used it to designate a process as opposed to morphological structure. To me, behaviour *is* function. Homology is based on common ancestry, and analogy on common ecological situations, or selection pressures; both have their own causal processes and functions, but I do not think one can relate homology exclusively to causal processes and analogy to function.

Masters: I think that is a question of terminology.

Leyhausen: The distinction that functional analogy is actually due to common ecological pressures is far more significant, and more useful, especially in view of comparisons between man and animals. An argument supporting this view was put forward by Georg Schaller, who offered an explanation for the fact that the numerous recent field studies on the social behaviour and organization of primates, and in particular anthropoid apes, revealed little or nothing about human social systems and social behaviour, aggressiveness and the like. The ecological pressure to shift from a purely fruit and plant-eating life towards hunting may have led to many adaptations in early human social life, and made it more easily comparable by way of analogy, certainly not homology, to the social life of socially living predators. This same ecological pressure, namely, to capture live prey in order to satisfy the need for food, produced social systems that were in some respects similar, although primates do not have the predatory behaviour endowment of any true predatory animal, but apply other behaviours to get their prey, as shown by a study by John Eisenberg and myself. (We preferred, for this reason, to call them hunters instead of predators.) In a case like this, the common ecological pressure is far more useful for interspecific comparisons of such elaborate systems than is actual homology; that is the reason why I should put the emphasis on the ecological situations that produce analogical systems.

Masters: I agree. The only issue has to do with the peculiar words you use. Where you speak of common ecological pressure or selection advantage, I use the word function, whereas, where you use the word function I speak of behaviour. My word usage is American, but I think that one of the benefits of a meeting like this should be to help us see that we need not stumble over pecularities in vocabulary.

von Cranach: I was surprised to hear Leyhausen say that behaviour is function; we would treat behaviour as a character, as a *Merkmal.*

Leyhausen: A character with a temporal extension which structure does not have.

Crook: I should like to add here that in English etymology there are two quite distinct and equally legitimate usages of the word function. One is to imply the function of a thing in relation to its adaptation to do something, which is an almost teleological use of the word. The other use of the word function simply means how it performs. So you can say the eye functions in seeing, but the eye has a function to detect at a distance.

Ploog: But does it not also refer to the effect of something, for example, "it functions to the effect..."?

Chance: A chewing function brings about nutrition, and the sexual function brings about reproduction. These are other uses of the word function in relation to behaviour.

von Cranach: I found, in various sources, that the concept of function has two meanings, just stated by John Crook. In most general terms, function may be understood as the behaviour of a system in relation to its environment.

Leyhausen: No, as I said before, we need the first definition to differentiate; with regard to behaviour, we are one step removed from the primary causal network in function as opposed to structure.

Masters: You say that function is one step away from structure?

Leyhausen: Function is a process in itself, a system functions in some way. Do you remember what I said about the three causal connections or networks, overlapping to some extent, but in principle mutually exclusive, of phylogenesis, ontogenesis and actogenesis? The structure finishes with ontogenesis. Once the structure is there, it remains (very generally speaking) there to be compared. Whereas function, as I use the word, though latent perhaps most of the time, and possibly dependent on the time it is actualized, appears as a process. Actogenesis is needed so that it becomes observable.

Chance: Can we not use actogenesis here?

Leyhausen: No, because actogenesis is what brings about function.

van Hooff: You could, of course, use survival value for the second meaning of function. It would cover the adaptation aspect.

Leyhausen: You could.

von Cranach: I should think that it is the second use of the word function, "the effect of a system's behaviour on the relations to its environment", which yields an additional meaning, and is, therefore, to be preferred. The relationship between the system and its environment is actually something new, while the first use implies nothing but that the thing behaves, and that does not help us. For this reason, I propose, at least for the present discussion, that we use the term function in the second way, designating the relation between system and environment.

Homology and Analogy

Ploog: I would like to go back to what Leyhausen said before we embarked on this discussion of the usages of the word function; I found Dr. Master's distinction between homology and analogy a very helpful one, based on causal processes. If we assume that similarities between man and animals, or animals and other animals, are due to causal relationships in the sense discussed yesterday by Dr. Senn, we can rely on a straight line from physiology to structure to behaviour, which permits various kinds of conclusions from that kind of causal relationship. Keeping in mind Wickler's very interesting views on information storage, and remembering Norbert Wiener's statement that information is information, not matter nor energy, then we can, whenever we cannot establish causal relationships of the kind inherent in the argumentation of comparative physiology and anatomy, turn to the whole philosophy of information transfer and make comparisons in terms of analogies, taking into account the rules of information theory. Then, if we do call something homologous, we infer that we know one or the other piece of causal relations, and if we talk about information transfer or various pieces of evidence that are analogous in terms of function, we can apply information theory. That may work both ways without violating established rules or giving away evidence for questions about various pieces of behaviour—for instance, the question of what pair bonding behaviours have in common—without having to struggle with the terms homology and analogy. I wonder, Dr. Masters, what you think of this solution?

Masters: That is very much the direction I was suggesting, and it explains why I did not want to talk about the specific relationship between the concepts of analogy and homology. I think it is much better to focus on the possible benefits of either homological comparisons in terms of causal processes, or analogical comparisons.

Functions of Analogy: Effects and Consequences

Ambrose: I should like to remind you of a useful distinction which may help the discussion of the problem of analogy and the lack of clarity about the rules of its analysis, a distinction originally made by Robert Hinde (1959) between effects and consequences. Because psychology has been largely preoccupied with the study of causation and causal processes, it has attached relatively little significance to this direction. In considering the relations between homology and analogy, however, the logical differences between these concepts become important. Let me give you an example from my own work. Years ago, in the study of infant smiling (Ambrose, 1960), my empirical results led me to consider whether there might be comparable analogous responses at other phyletic levels. One of these appeared to be the following response of the duckling. The immediate effect of this behaviour is that the duckling moves towards the mother, or if the mother moves away, the duckling will follow. This is possible because the duckling has the capacity for locomotion. The human infant, however, does not have this capacity. Instead he smiles at his mother and, when he does this, the mother smiles back, or, if she is at a distance, comes towards the infant, picks him up, and so on. The immediate effect of the infant's smile is to attract the mother and to make her spend more time with her infant than she would have done if the smile had not occurred. Thus, the duckling's following response and the human infant's smiling response have quite different immediate effects, but identical consequences: they both bring about close proximity of mother and offspring. In general, it may be true to say that, whereas homologous behaviours tend to have similar effects and similar consequences, analogous behaviours may have different effects but similar consequences. If this statement is true, it has implications for the rules of analysis of analogy. Such analysis essentially involves two steps in empirical investigation: the first is concerned with the immediate effects of the behaviour, the second step is concerned with the consequences.

von Cranach: Would you say that what happens in the environment via consequences refers to the system–environment relationship? Because these are two different things. If you regard the answering response of the mother in your example as an effect, you would say that this happens in the environment, but the consequences certainly concern the relationship between baby and mother. Would you say that the consequences concern both system and environment?

Ambrose: Yes. There is also a further distinction to be made, namely, between two kinds of consequences. Consequences can be immediate, for example the mother and baby being in close proximity, or they can be long-term, for example the subsequent facilitation of the normal development of the infant which, from a functional viewpoint, contributes to the survival of the species.

Leyhausen: So you would say in relation to Masters that the effect would be related rather to homology—and in your example the two behaviours are clearly not homologous—and the immediate consequences would have a relation to analogy, or be analogous with respect to what follows?

Ambrose: No, I am not saying that.

Leyhausen: Or do you mean that if you make this distinction and analyse behaviour along this line you can more or less do without the concepts of homology and analogy?

Ambrose: No. I am saying that in the example I have given, homology does not enter into it at all. Smiling is not homologous to the following response because it has a different effect. It is, however, analogous because it has a similar consequence, with a similar type of biological function. I am saying that, in the analysis of biological function, there are two steps: the first is analysis of effects, the second is analysis of consequences. Both are essential for the study of the biological function of analogous responses.

Leyhausen: But you could easily think of an example where the immediate effect may be identical, but not the consequences, and this would have a bearing effect on Dr. Masters' conception. I would not say it was exactly congruent with the distinction he made between homology and analogy, but it may overlap to some degree.

Ambrose: That is right.

Levels of Analysis

Foppa: I should like to ask Dr. Masters about his table: is it intended to represent more than a loose proposition of levels of analysis? Because if you take it seriously, I suppose you must say that the biogenetic law is wrong, because it is based on a comparison of two levels that are uncomparable. If you compare ontogeny and phylogeny you will, according to your example, get false results.

Cain: Biogenetic law, as normally stated, *is* wrong.

Masters: First of all, let me explain what I think is important and what is not so important: what I think important is the dimension; the identification of items within the cells is less important. Not everything in ontogeny is necessarily the study of the evolutionary time of an individual. A terribly important aspect of ontogeny is the social interaction between individuals; there is a real problem in terms of putting labels on boxes. Of course, each cell in that table presupposes all of the other cells; what I was really trying to get at in the diagram is that everything is connected with everything else. In order to get a hold on things we have to reach some clear categories of measurement, so that we may be able to measure reasonably homogenous phenomena. The only way we can begin to sort these things out is to take one point at a time, instead of combining too many problems. Let me give you an example that I touched in my paper and that might be an example of the kind of problem we would like to consider: the problem of territory. To treat territory as an instinct, as a homology or, let us say, as a homologous causal process, will not work. Instinct is not a common causal process even within a single species. It is a series of causal processes which has a very complicated sequential system. So we must say, first of all, that when we talk about territory, we talk about consequences, or, in my terminology, function. Assuming that certain forms of behaviour we are calling territorial behaviour have for certain species—not all—advantages, we then very soon come to heterogenous aspects. There are a number of different things going on. Sometimes what we call territorial behaviour is combined with what might be called reproductive isolating behaviour. Sometimes territory has to do with spacing in relationship to food supplies. These are very different selective advantages; in terms of what Michael Chance would say, two different abstractions. If we try to use one measure of territory for two different abstractions, we are bound to get into trouble. That is really the problem I had in mind, and the table does not help very much. What is so frustrating about it is that the cells do not help very much in sorting it out. The only reason for presenting this table was to remind us that we have to do some sorting; if we do that sorting, we have to avoid talking about selective advantages or consequences of behaviour. We must try to make sure that we do not measure—or confuse, as in the case of territory—the possible different selective advantages for the species. It may, of course, happen that within a given population, say of birds, something like territorial behaviour satisfies two functions, reproduction and food supply, at the same time. But it may be that there is only one, and unless we can sort this out, we are apt to make big mistakes in the analogies. We cannot say that humans have territories and birds have territories, and conclude that these are identical.

von Cranach: Would you say that this matrix should prevent us from mixing different things, but that you would not object to its use for the comparison of different cells?

Masters: Of course. That is, the ontogeny–phylogeny relations, as mentioned yesterday, can be done in terms of sequences. This will improve the analysis not only of ontegeny and phylogeny, but also of structure.

NOTES

This article first appeared in *Social Science Information,* 12 (4) 7-28.

1. To be sure, from the point of view of the philosophy of science, "causal mechanisms or processes coded in the genotype" are also "identified by the observer." The distinction here proposed is nonetheless useful, because the hypothetical identification of functions is, at least at present, subject to a greater uncertainty than comparison of physiological or genetic processes. In other words, it is harder to provide empirical tests which could disconfirm hypothetical "functions" as explanations of a trait's adaptiveness than to devise tests to disconfirm whether a hypothetical "cause" explains the structure or process of the trait itself.
2. Boulding's criticism also rests on a doubtful empirical assumption which is often made by those who attack analogies between animals and humans. "A change in human affairs comes hardly at all from changes in the gene pool and its distribution in mankind [...] it would be surprising if these factors accounted for more than a fraction of one percent of social change and social evolution" (Boulding, 1968, p. 88). While it is probably misleading to treat genetic and environmental variables as entirely distinct, additive variables, recent assessments of genetic and environmental variation in causal models of human behaviour assign a far larger role to genetic differences than Boulding indicates. As Christopher Jencks—hardly a social darwinist—has argued in a careful restudy of the data concerning genetic and environmental factors influencing IQ and intelligence, genetic differences can be said to "account" for about 45 % of the statistical variance in IQ (Jencks, 1972). If this much variation in individual traits at one point in time should be attributed to genetic differences, it seems unlikely that Boulding is correct in his radical dismissal of genetic variation as a relevant variable in social change.

3. To avoid confusion between *functioning* or "behaving" as a level on which organisms can be studied, and *functional analysis* as a mode of interpreting phenomena at any level, I will speak of the behaviour or process of a system rather than its functioning. The term *function* will be reserved throughout for the analytical approach which seeks to explain the selective advantage or evolutionary adaptiveness of a structure, behaviour, or developmental pattern.
4. In terms of biology more generally, one must of course add at least two crucial levels of analysis within the individual organism: that of cells (cytology) and that of the genome (molecular biology). To simplify this exposition, these levels of analysis will not be included in the table, though they would have to be included in a more complete analysis.

REFERENCES

Ardrey, R. 1961. *African Genesis.* New York, Atheneum.
— 1966. *The Territorial Imperative.* New York, Atheneum.
— 1970. *The Social Contract.* New York, Atheneum.
Barash, D.P. 1974. "The Evolution of Marmot Societies", *Science* 185 (August 2) 415-20.
Bischof, N. 1972. "The Biological Foundations of the Incest Taboo", *Social Science Information* 11 (6) 7-36.
Boulding, K. 1968. "Am I a Man or a Mouse—or Both?", pp. 83-90 in: M.F.A. Montagu (ed.) *Man and Aggression.* New York, Oxford University Press.
Carpenter, C.R. 1958. "Territoriality: A Review of Concepts and Problems", pp. 224-250 in: A. Roe/G.G. Simpson (eds.) *Behavior and Evolution.* New Haven, Yale University Press.
Dobzhansky, T. 1955. *Evolution, Genetics, and Man.* New York, Wiley.
Fontaine, M. 1956. "Analyse expérimentale de l'instinct migrateur des poissons", pp. 151-167 in: M. Autuori *et al. L'instinct dans le comportement des animaux et de l'homme.* Paris, Masson.
Galle, O.R./Gove, W.R./McPherson, J.M. 1972. "Population Density and Pathology", *Science* 176 (April 7) 23-30.
Gerard, R.W. 1960. "Becoming: The Residue of Change", pp. 255-267 in: S. Tax (ed.) *The Evolution of Man.* University of Chicago Press.
Harlow, H. 1958. "The Evolution of Learning", pp. 269-290 in: A. Roe/G.G. Simpson (eds.) *Behavior and Evolution.* New Haven, Yale University Press.
Hinde, R.A./Stevenson-Hinde, J. 1971. "Les motivations animales et humaines", *La recherche* 2 (May) 443-456.
Jencks, C. 1972. *Inequality.* New York, Basic Books.
Kurtén, B. 1972. *Not From The Apes.* New York, Vintage.
Lorenz, K. 1965. *Evolution and Modification of Behavior.* University of Chicago Press.
— 1966. *On Aggression.* New York, Harcourt, Brace and World.
— 1970. *Essais sur le comportement animal et humain.* Paris, Le Seuil.
— 1971. *Studies in Animal and Human Behavior,* vol. 2. Cambridge, Harvard University Press.
Masters, R.D. 1970. "Genes, Language, and Evolution", *Semiotica* 2:295-320.
— 1972. "Political Behavior as a Biological Phenomenon", (Paper presented at 1972 meeting, American Political Science Association, Washington, DC).
— 1975. "Politics as a Biological Phenomenon", *Social Science Information* 14 (2) 7-63.
Mayr, E. 1958. "Behavior and Systematics", pp. 341-362 in: A. Roe/G.G. Simpson (eds.) *Behavior and Evolution.* New Haven, Yale University Press.
— 1963. *Animal Species and Evolution.* Cambridge, Harvard University Press.
Montagu, M.F.A. (ed.) 1968. *Man and Aggression.* New York, Oxford University Press.
Moyer, K.E. 1968. "Kinds of Agression and Their Physiological Basis", *Behavioral Biology* 2:65-87.
Pittendrigh, C.S. 1958. "Adaptation, Natural Selection, and Behavior", pp. 390-416 in: A. Roe/G.G. Simpson (eds.) *Behavior and Evolution.* New Haven, Yale University Press.
Rensch, B. 1966. *Evolution Above the Species Level.* New York, Wiley.

Roe, A./Simpson, G.G. (eds.). 1958. *Behavior and Evolution.* New Haven, Yale University Press.

Simpson, G.G. 1958. "The Study of Evolution", pp. 7-26 in: A. Roe/G.G. Simpson (eds.) *Behavior and Evolution* New Haven, Yale University Press.

— 1967. *The Meaning of Evolution.* New Haven, Yale University Press.

Skinner, B.F. 1965. *Science and Human Behavior.* New York, Free Press.

Spieth, H.T. 1958. "Behavior and Isolating Mechanisms", pp. 363-389 in: A. Roe/G.G. Simpson (eds.) *Behavior and Evolution.* New Haven, Yale University Press.

Tinbergen, N. 1972. "Functional Ethology and the Human Sciences", *Proceedings of the Royal Society of London* B 182:385-410.

Vine, I. 1973. "Social Spacing in Animals and Man", *Social Science Information* 12 (5) 7-50.

O. E. RÖSSLER

A Relational Approach to the Problem of Man–Animal Behaviour Comparison

Speaking as a non-biologist, I will not, in fact, deal with the problem of comparison between man and animals, but only with the problem of comparison between analogues of man and analogues of animals, in a kingdom of abstract artificial reaction systems that can be described by chemical systems theory.

On the other hand, this approach to biology (to speak of simple conceptual systems, rather than about biological reality) has a long tradition within biological thought. It can be termed the approach of functional reductionism (as opposed to physico-chemical reductionism); it was introduced into this science by Charles Darwin. This approach always consists of three parts, like the Greek letter π. The first part is the identification, either by intuition or by empirical research, of an important functional performance shown by the system to be explained. This corresponds to the first leg of the π, so to speak. The second part is the description of a mathematical system, which, although infinitely simpler, can show the same functional performance. This corresponds to the second leg of the π. The third part, of course, is the discussion of the relationship between the abstract and the concrete world considered. This part, corresponding to the bar of the π, also contains the motivation. From here come the iterations which make the whole procedure self-consistent. Functional reductionism has recently been re-invented under the name of *relational biology,* by Nicolas Rashevsky and his school (Rashevsky, 1961; Rosen, 1972b). This name is somewhat misleading, since the described systems are not biological ones, but belong to the domain of mathematics. The whole approach becomes a biological method only when all three parts of the π are combined: the first leg, the second leg, and the bar.

EVOLUTION IN AN ABSTRACT MODEL SYSTEM

Assumptions

In the following, we will be dealing largely with the second leg of the π, the relational part of functional reductionism. We consider a set of abstract chemical reaction systems that is analogous to the set of biological species in just one simple respect: the constituents of both are autonomously growing (and autonomously self-maintaining) chemical systems. Because any autonomous process of chemical growth depends on certain conditions (*i.e.* both chemical and physical factors) such as substrate, temperature, etc., the notion of *environment* also applies to the abstract situation. It is necessary to distinguish between positive (*i.e.* growth-promoting) and negative (*i.e.* growth-impeding) environmental factors.

Just two more assumptions must be introduced: the environment is endowed with the feature of changing slowly over time; our autonomously growing chemical systems are endowed with that of undirected systematic mutability * of (in the first approximation non-interacting) subsystems.

General Result

The result is a process of *environmental selection* of those subsystems which prove "adapted" to the new environment as it is well-known (Darwin, 1859). Darwin used almost the same assumptions. Much of the controversy raised by his approach could have been avoided if the basically different methodological origin of this procedure—confining oneself to a set of simple axioms—had been understood. "Adaptation" to a new environment means nothing more and nothing less than that the original system has been modified somehow with the effect that the modified system shows the property of autonomous growth and self-maintenance—which is the only one considered in this abstract approach—also (and perhaps even preferentially, but this is inessential) in the new environment. Directions of environmental change hence automatically determine (as far as they are followed) *directions of adaptation*.

The Simplest Case

In the simplest situation, starting with a single, autonomously growing chemical system possessing the presupposed modifiability and exposing portions of it to different lines of successive environmental change, a tree of descendence-relations, determined by the applied tree of environmental changes, will be obtained.

If we had forgotten the genealogical order among the different successor-systems, we could, by virtue of the fact that each system is derived from another by a more or less small modification, reconstruct the entire tree, applying a comparative phenomenological approach to the set of systems. (The majority of the phenomenological traits shown by the systems allow for the same partial ordering. The resulting graph is the original tree.) This idea, easy to obtain in our abstract relational system, can be transferred to biological reality. This was exactly Darwin's contribution to *empirical* biology: uniting morphological, physiological and ethological methods in one approach of comparison, so that the presumed tree of partial ordering had to show up, if it existed.

Supposing further that we had also forgotten the tree of successively changed environments, we could try to reconstruct this, too, by checking a sample of each growing system against each environment and determining whether or not multiplication occurs. The result would not be unique, because less restrictive environments (lacking negative factors which could be tolerated or containing unnecessary positive factors, respectively) would also give a positive result. Even more ambiguity would be obtained if we had also forgotten the components of the set of original environments, so that they too, had to be reconstructed out of those factors that were especially growth-promoting or especially little growth-inhibiting, respectively, for a particular system. Almost complete uniqueness would be obtained again, however, if we observed the constraint that the environments of closely related systems (with respect to the first tree) must themselves stand in a close relation of similarity.

The result would be a one-to-one mapping between the individual autonomously growing chemical systems, on the one hand, and their "reduced" (*i.e.* reconstructed) environments, on the other hand. The latter would form a partial ordering isomorphic to that of the former. That is, we would have reconstructed the tree of environments. Due to the achieved parallel knowledge of both trees, any direction of change of effective environment underlying the somatic adaptation process would become evident again. Hence the functional side of our abstract evolution (concerning the property of autonomous growth and maintenance of chemical mass in different environments) would be revealed with the same accuracy by the second tree as the historical side had been by the first.

The outlined second procedure of phenomenological reconstruction is no less appropriately applied to biological reality than the first, even though Darwin did not mention this second strategy.

The More General Case

The preceding results are all implications of the single relation outlined at the outset—that a direction of environmental change determines a direction of adaptation. This theorem does not cease to be applicable if the tree of imposed directions of environmental change is *not* a simple tree (as assumed so far), but contains looped and bundled branches, also. The proposed reconstruction procedure is, therefore, still valid.

Certain irregularities are bound to occur in this situation, namely the existence of certain non-connected subgraphs of partial ordering that do not fit into the tree of the majority of ordering relations, but are "false", crossing the main trend of partial ordering. These can be explained and predicted by the second tree, by virtue of their mapping onto the *bundles* formed in the latter. The same should be observed in biological reality.

AN EVENTUAL BIOLOGICAL APPLICATION

In biology, there is a wealth of false homologies (similarities not fitting into the main tree of homological relations, referring in the simplest case to genetic connection), that are termed analogies. Also, a close relation between these phenomena of so-called convergence and the presence of certain common ecological factors is well documented in many cases (*e.g.* the somewhat controversial ecological concept of "life forms," different species showing convergent design features and living under similar environmental conditions; Kühnelt, 1970). It is a straightforward guess, therefore, that the postulated second tree of biological phenomenology (to be determined by an experimental and comparative ecological approach, as outlined) will be as illuminating for the understanding of biological convergence as the first tree is for genealogical order. The most important result to be expected is the specification of the *main common directions of evolutionary adaptation.*

Indeed, if biological evolution follows the same basic laws as does our simple relational model system, which is identical with the conceptual system used by Darwin, then any biological species must be interpretable in two different contexts of relationship, namely, one of homological relatedness (describing the historical connection with other species), and one of analogical relatedness (describing the connection with respect to the main function of survival, *i.e.* autonomous growth and self-maintenance of population biomass in a particular environment). Both relations will be rendered quantitative, although this has been effectively demonstrated, so far, only for the first one.

The demonstration of the ties linking man to the rest of nature in the *second* context will, accordingly, take time. The structure of the second tree in the neighbourhood of man's place is even less well known than the corresponding region of the first tree was a century ago.

IMPLICATIONS OF THE RELATIONAL MODEL

The pessimistic view just expressed applies only to biological reality itself. It does not extend into the domain of relational biology. Of course, any result obtained in this much more transparent field will become applicable to biology only after additional research has been done.

When playing with an abstract set of autonomously growing, modifiable chemical systems, which are thought to be placed into a set of defined environments which themselves are modifiable in arbitrary directions (an appropriate name for this relational game would be "Darwin's Chess"), it can be seen that very "simple" directions of increasing environmental restrictivity (after starting with an unrestrictive environment) determine very non-trivial design constraints, some of which are known from biological observation (Rössler, 1972). It may suffice here to point to the existence of just one such direction of straightforward environmental changes. This is characterized by decreasing accessibility in space of positive environmental factors (including a less and less uniform and regular distribution in space and time), with the opposite situation holding true for negative factors (increasing access, etc.). If this series of environmental constraints is applied to autonomously growing chemical systems consisting of spatially defined subsystems which are non-interacting in the first approximation (a structure that is determined by other, very general properties of the environment, as outlined in the cited paper), a series of design constraints are obtained that can be described as *necessary strategies of whole-body locomotion*. The strategies form a progression series, the simplest one being sufficient under the simplest degree of specific environmental restrictivity, and so on. The series starts with undirected, random locomotion, followed by arrestable locomotion and so on. The last three steps are: to trace the best sources in space, to trace the best clusters of sources in space, and to trace the best travelling routes among clusters of sources in space (Rössler, 1974).

Any one of these strategies appears to be "convergent" to a behavioural strategy observed in one or many instances in the animal world, so that the conjecture arises that those behavioural strategies may serve a similar function. Our confinement to the relational domain provides the great advantage, however, that we can derive the necessary mathematical organization of a control apparatus yielding precisely these functions without resorting to biological phenomenology. For example, we observe with astonishment that learning abilities, while playing a dominant role in any directly biology-oriented theory, merely form an accessory attribute on the abstract scene; for the degree of their being required depends almost exclusively on the assumed degree of spatio-temporal redundancy of the environment, the latter forming an independent parameter of its own.

MAN'S PLACE IN THE MODEL SYSTEM

If a single main direction of environmental change determines a series of abstract organisms that appear as abstract analogues to a series of "typical" animals, we could try to reverse the process and ask whether an abstract analogue to man can also be conceived of in the abstract stage.

Returning for a moment to the biological leg of the letter π, we may accept as a biological working hypothesis that the animal world, as far as its common design features are concerned, embodies the single main adaptation of solving increasingly intricate problems of spatio-temporal tracing, as outlined. The place of man along the same scale is not difficult to define. We know that man's mastery of the domain is rather exceptional. The degree of his skill is still progredient, due to an evolutionary process of its own which is not bound to any accompanying parallel change of the biological substrate (Huxley, 1942).

Switching back to the abstract model, we can look for the functional place of an autonomously growing substantial system whose spatially defined subsystems possess an analogous property. The answer is obvious. Its position marks both the top and the end of the underlying arrow of adaptive change. The system appears as the last representative along this line of evolutionary progress because it achieves (at least potentially, due to the second evolutionary process named) full *emancipation* from the kind of environmental restrictiveness at stake.

AN APPARENT INCONGRUITY

Ecological emancipation is a well-know notion in biology (see Kühnelt, 1970). It also plays an important role within the simpler relational framework considered. Both in biology and in the simpler analogue, the effective emancipation from a particular kind of environmental restrictivity (take the invention of homeothermy as a biological example) is followed by an "adaptive radiation": there is never a single emancipated species (or system, respectively), but always a whole variety of them. The reason (heightened affordable mutability) need not concern us here.

If our view is correct as a biological hypothesis, the absence of an adaptive radiation accompanying recent man must be surprising. A second unusual feature of this hypothetically emancipated species is, of course, the unfinishedness and nonbiological progredience of the respective organization. A third inconsistency could be seen in the number of ethological analogies linking man to other socially living animals which seems to be greater than to be expected from a species which has conquered a new space.

However, the most striking incongruity may be seen in the fact that this particular type of emancipation is *not* advantageous in evolutionary terms. Full mastery of space is being no more selected for than, for example, full metabolic exploitation of the host in a specialized parasite. Neither of these two directions of evolutionary adaptation can be expected to lead to perfection. And in both cases, any too efficient emergent solution should already be selected against at an early stage.

In this situation, it would be gratifying if an explanation for the observed discrepancies could be given in the abstract domain at least.

A HYPOTHESIS ON THE GENESIS OF ABSTRACT MAN IN ABSTRACT BIOLOGY

Indeed an abstract explanation can be offered in the form of a hypothesis. It will be sketched as a concluding point for two reasons: 1) It is based on a general abstract principle that is interesting in its own right, and 2) it suggests a specific mechanism for biological anthropogenesis which is open to empirical verification. The latter will, in the case of its approval, lead to a deeper understanding of the risks and merits of extrapolating from man to animal, and vice versa.

In general, any greater modification of design, as it accompanies a greater adaptational breakthrough, calls for the existence of a principle of its own explaining the first occurrence of the hook (or nucleation bud) from which the whole process of adaptive rearrangement of systemic structure must have taken its origin. Aside from the afore-mentioned principle of undirected mutational modification (which can hardly explain a very complex hook, as it is almost impossible to avoid postulating in non-trivial cases), the so-called "principle of function change" is the next general candidate. It refers to the fact that any real system designed to perform a particular function will, under quite unpredictable circumstances, switch to showing a highly different, non-expected type of performance; indeed, it is never possible to specify all sorts of performances that a given physico-chemical realization of a prescribed mathematical behaviour can show under all possible conditions, that is, to which it is also adapted. This principle was formulated by the relational biologist Robert Rosen (1972a) for the purpose outlined above of explaining the beginnings of major structural changes in evolution. In a less precise form, it was also invoked by Waddington (1969) and by Lorenz (1969), who used the term "fulguration". A specific example of unexpected function change in a reaction-kinetic system originally devised as a model for one particular function (spontaneous origination of spatial differentiation; Rashevsky, 1940; Turing, 1952) was recently detected: the same system shows in a trivial situation (short, reversible slackening of a single parameter) the unexpected phenomenon of an almost contrary performance (reversal of differentiation into the opposite pattern; Rössler/Seelig, 1972).

The principle of function change has two types of applications within the relational biology of evolution. In the first type, the overall effect is evolutionary (even though, as outlined, the emergence of the nucleation bud itself is in a sense revolutionary); in the second type, the effect is revolutionary throughout. The affected system undergoes (after an event of *hard bifurcation,* to use the correct system-theoretic term) a self-amplifying, autonomous change of behaviour that is not bound to any additional modification of the substrate in order to become effective. Such a *revolutionary function change* may of course have secondary and tertiary consequences of its own, including evolutionary processes, even those not implying a change of substrate.

So far, only a few instances of revolutionary function change are known in relational biology. The phenomenon is especially likely to occur if the behaviour of a larger system results from the concerted interaction of several, more or less identical subsystems. Accordingly, cancer forms the most straightforward hypothetical paradigm. It is conceivable, although not yet verified, that a type of cancer exists where the disease has no material origin, but is rather based on a novel type of interaction between the initially completely unchanged subsystems (cells). This mechanism was postulated as one possible cause of cancer by Pitot and Heidelberger (1963) and was given the name "epigenetic cancer" by these authors. (The fact that this mechanism at the same time illustrates a general principle could not be observed at that time.) The second domain of biology where the probability for detecting another example must be especially great is, of course, the social field of interacting individuals, because the basic preconditions for functioning in another sense than that so far selected for are here also fulfilled in an almost ideal manner. A specific instance (without many secondary and tertiary consequences, however) is Lorenz's observation that a bonding pair of two male greylag geese suddenly switched to individual hate, following a casually begun fight (Lorenz, 1966). It is clear that a whole variety of analogous processes can be conceived of *in abstracto,* both "tragical" and "non-tragical" ones, and both with and without determination of major secondary effects and processes. The announced abstract hypothesis, as far as the potential genesis of man's effective ecological emancipation is concerned, consists in the postulation of a positive mechanism of this type.

The preceding relational approach thus encourages looking for an interactional function change mechanism for "anthropogenesis" in biology. The latter would explain man's capability to participate in the process of overcritical tradition accumulation (Huxley, 1942; Lorenz, 1969) as an *epigenetic* acquisition (Rössler, in press).

To conclude, the singularity of man's functional position among his relatives may require a singular (function change) paradigm. Some of the apparent incongruities discussed above (which also gave rise to this conference) may then readily fall into place.

DISCUSSION

Danger of Oversimplification

Leyhausen: I cannot help feeling that the increasing sophistication and rationality of models like the one you discussed incur the danger of over-simplification of the actual phenomena. In my own experience, attempts to achieve mathematical or cybernetic models of my ideas tend to interfere with the floating equilibrium of these ideas and phenomena and facts interacting with each other. This is probably due to my limitations; anyway, I deliberately gave up. Do you think it is at all possible to avoid this danger of over-simplification in the construction of models like yours?

Rössler: I share your feelings of uncertainty. I was attempting here to present a possible relational approach. I am sure a better model would be far more difficult to achieve. Nevertheless, I do believe a sophisticated model of the parliament of propensities could be achieved; I have always wondered why nobody has ever tried to do it.

Leyhausen: In my opinion, there is another reason for this: models of the type you have in mind are quantitative models. But in order to understand the working principle of a piece of machinery, quantitative data are quite unnecessary. I can comprehend the working of a motor car perfectly without knowing anything at all about gear ratios, fuel expenditure to horsepower ratio, etc. For most of the propensity system of any one species we are, at present, absolutely unable to obtain any reliable quantitative data. But the often-heard argument that we cannot, therefore, know *how* it works is patently incorrect. It is the great advantage of Lorenz's "psychohydraulic" model that it makes us understand some of the working principles without needing any quantifications, as a cybernetic model invariably would.

Rössler: Somewhere around 1880, the mathematician Henri Poincaré created the science of *qualitative dynamics,* which enabled a rigorous treatment of "types" of dynamic behaviour. For example, the psychohydraulic model represents, as *pars pro toto,* just one such particular class (the class of monoflops, or synonymously, excitable Bonhoeffer-van-der-Pol systems). In the same way it may be hoped that a dynamic prototype can be found also for the propensity system.

Chance: If you have a chemical system behaving in the way you described, is the growth process a necessary condition for survival? And for the other changes of function?

Rössler: The capacity for autonomous growth (population regeneration) is a precondition for survival. But function changes are not bound to autonomously growing systems.

Evolution and Revolution

Cain: There is, as far as I know, no evidence of a phenomenon in evolution that requires generation by a non-adaptive mechanism.

Chance: Does this mean by pre-adaptation?

Cain: By pure selection.

Chance: Take for example the tusk that is used for fighting in seals and then used for keeping holes open in the ice during the winter; that is pre-adaptation. It involves a different exploitation of the environment. Yet the new one is dependent on the previous pattern; the change is in its use.

Cain: But not dependent on non-adaptive methods.

von Cranach: If Professor Ploog were present he would perhaps say that he was troubled by the fact that he does not know how to explain, or whether one needs to explain, certain very rapid changes, in language development, for example. Such events may be related to such revolutionary mechanisms; we are in need of models that can explain these rapid developments. Does an orderly evolutionary process require a certain amount of time, let us say, measured in generations? This would constitute the difference between evolution and revolution, would it not?

Cain: I am speaking entirely about biological evolution and not about language development and the like, because I think that is something quite different. In biological evolution, change, without a doubt, requires time. What is equally true is that in large populations, even changes made over long periods take place so rapidly by straight selection that on a geological record they would seem to happen in a flash.

Rössler: I did not speak about biological evolution itself, but only about an abstract metaphor. On the other hand, the phenomenon of evolutionary function change (in the sense of a nucleation bud generated by environmental, not mutational, chance) is well known in abstract biology, so it may be that revolutionary function change plays a role in biological evolution, too. This would be a question for Robert Rosen to answer.

Leyhausen: I should like to point out that in what you call revolutionary function change, the revolutionary part is never the change itself, but its consequences. The change itself is always in some way pre-arranged or preformed. Take, for example, the switch from a primitive swimming bladder, which is an organ already adapted to the osmotic exchange of gases, to a lung, which exchanges gases in the air. I think that one has to be very careful to explain to which processes the term revolutionary is meant to apply.

Rössler: I have confined myself to calling such events *evolutionary* function change, since they are based on evolutionary adaptive changes, even though their consequences are in a sense revolutionary. It is a matter of terminology; I agree that the invention of using a swimming bladder for a lung is not so very surprising, but there are cases that are, such as the first development of a light-sensitive cell in the skin. Once this first cell is there, the evolution of the eye is almost straightforward, if I may say so. The first nucleation bud from which evolution can proceed does in some, not all cases pose a problem.

von Cranach: Would the term revolutionary apply to the adaptation of a system to its own internal state, to its sub-systems?

Rössler: An interactional function change between subsystems can be either adaptive or non-adaptive. I would like to reserve the term *revolutionary function change* to system theoretic events of hard bifurcation which take place at the epigenetic level and are primarily non-adaptive in the selectionist sense (although they may secondarily prove adaptive in a new direction).

Leyhausen: Could you give a concrete example of hard bifurcation?

Rössler: Earth-quakes, a classic example in geology. There are shearing forces acting beneath the surface of the Earth that may have two possible effects: one is stable sliding, the other slipstick. In the latter case, the internal forces may grow until a threshold is reached where, all of a sudden, the whole energy is used up in an event of earth-quake.

Masters: Were you speaking therefore of revolutionary changes as step functions?

Rössler: In a sense, yes. Of course it is only a quasi-discontinuous transition.

Revolutionary Function Changes in Biology

Leyhausen: I still think you are after a real phenomenon in biology with your idea of revolutionary change, but I do not think that so far we have hit upon a really fitting example.

Rössler: There are two incidents, two kinds of hard function change in biology. The first is illustrated by the nerve membranes that suddenly switch to conduction. This function change is used by nature in a distinct context. Additional examples are provided by ontogenesis, in which many cases of sudden function change seem to occur. They have been termed catastrophes by René Thom.

Leyhausen: Think of the relative coordination of automatic central nervous systems as analysed by von Holst, and by "automatic", we mean that part of the nervous system which is not under direct voluntary control, but which does not belong to the autonomic nervous system proper. He has quite reasonably explained why a system keeps stable at one level for some time and then suddenly switches to another level. Although this may come as a surprise to the observer it is not surprising in the system itself. It is a quite predictable consequence of all functional and other relationships within the system.

Rössler: Yes. The same holds true in all instances. So the distinction I wanted to make is between those examples that are "known" to nature and used by nature, and those occurring for the first time, so that selection could not yet have operated on them.

Possibility of Generalization

Crook: One aspect of the new paradigm is the role of axiomatic modelling we have been hearing about in the whole process of heuristics. It seems to me that a connection exists between Wickler's paper, what Masters said this morning and Dr. Rössler's talk. All this is beginning to focus our attention not so much on evolution but on the common features of any historical process. If this type of modelling can, as it where, provide us with a functioning picture, a dynamic picture of a historical process, the question we might address ourselves to is: what are the analogies to selection pressures in different types of historical processes, for instance, in the chemical changes (from which I gather Dr. Rössler got his ideas)? In learning theory, reinforcement processes might be comparable to selection pressure, and economic forces in social change. If these different terms can be reduced to analogues within the overall model, it would be fascinating to see if they were manipulable in the same way, and whether we can strike a level of generalization. That would be very helpful in disentangling the fundamental issue of this conference.

Prediction of Revolutionary Function Change

Ambrose: Are you saying that there are several types of conditions under which revolutionary changes take place? Or is there only one condition under which they can occur when the system is one of high stability? In your geological example, for instance, it takes enormous pressures to make granite split, but only slight pressures will be sufficient to make a sedimentary block shift.

Rössler: I should say that in several cases prediction is easily possible. It is a different matter, however, with complicated engineering systems (both artificial and natural ones, if one really puts biological systems into the same class, as proposed by Rashevsky). If you devise a complete system for a prescribed function, you are lucky if it works at all. Considering the parameter space of a concrete system (the parameter space is the space of all possible variations of all parameters, that is, of all constants in the equations) and the inherent risk that by a small change of some parameter the whole system may flip into another "region" in parameter space where it behaves in a completely different way from before, it seems that there is no general strategy available by which you could predict when such an event will occur. So any design through evolutionary optimization should be bound to this liability.

von Cranach: From the point of view of social science I should say that if we had a ratio between the time scale during which internal exchange (within the system) and external exchange (between system and environment) occur, we could arrive at predictions as to how the process will go on in differentiation between evolutionary and revolutionary processes.

Rössler: Yes. It was in this context that the label "revolutionary" was confined to function changes taking place either within a single organism, or in the interaction between several organisms. So far, the whole concept is but a possibility for a solution, nothing more. Perhaps some examples will come up in the following discussions which could be matched to this concept. In an abstract ("relational") approach to biological evolution, directions of evolutionary adaptation can be defined. One such major direction refers to the pursuit of more and more sophisticated trajectories in space-time. In contrast to the other animals, man occupies a singular ("emancipated") position on this branch. Since, in the model, this position cannot be approached in an evolutionary way, the possibility of a (epigenetic) "revolutionary change" has to be taken into consideration.

NOTE

* The specific realization of this constraint of systemic mutability is a matter of complete irrelevance in this abstract context. It can, without restriction of generality be considered to stem from artificial modification.

REFERENCES

Darwin, C. 1859. *The Origins of Species by Means of Natural Selection, or the Preservation of Favoured Races in the Struggle for Life.* London, Murray.

Huxley, J.S. 1942. *Evolution, the Modern Synthesis.* London, G. Allen & Unwin.

Kühnelt, W. 1970. *Grundriss der Oekologie,* 2nd ed. Stuttgart, Fischer.

Lorenz, K.Z. 1966. *On Aggression.* New York, Harcourt.

— 1969. "Innate Bases of Learning", pp. 11-93 in: K.H. Pribram (ed.) *On the Biology of Learning.* New York, Harcourt.

Pitot, H.D./Heidelberger, C. 1963. "Metabolic Regulatory Circuits and Carcinogenesis", *Cancer Research* 23:1693-1700.

Rashevsky, N. 1940. "Further Contributions to the Theory of Cell Polarity and Self-Regulation", *Bulletin of Mathematical Biophysics* 2:65-67.

— 1961. *Mathematical Principles in Biology and Their Application.* Springfield, Thomas.

Rosen, R. 1972a. "Relations Between Structural and Functional Descriptions of Biological Systems", *International Journal of Neuroscience* 3:107-127.

— 1972b. *Relationale Biologie.* Fortschritte der experimentellen und theoretischen Biophysik vol. 15. Leipzig, Thieme.

Rössler, O.E. 1972. "Design for Autonomous Chemical Growth under Different Environmental Constraints", pp. 167-211 in: R. Rosen/F.M. Snell (eds.) *Progress in Theoretical Biology,* vol. 2. New York, Academic Press.

— 1974. "Adequate Locomotion Strategies for an Abstract Organism in an Abstract Environment: A Relational Approach to Brain Function", pp. 342-369 in: M. Conrad/W. Güttinger/M. Dal Cin (eds.) *Lecture Notes in Biomathematics,* vol. 4. Berlin, Springer.

— in press. "Interactional Anthropogenesis", *General Systems.*

Rössler, O.E./Seelig, F.F. 1972. "A Rashevsky-Turing System as a Two-Cellular Flip-Flop", *Zeitschrift für Naturforschung* 27b:1444-1448.

Turing, A.M. 1952. "The Chemical Basis of Morphogenesis", *Philosphical Transactions of the Royal Society of London* B 237:37-72.

Waddington, C.H. 1969. "The Theory of Evolution Today", pp. 357-374 in: A. Koestler/J.R. Smythies (eds.) *Beyond Reductionism.* London, Hutchinson.

HENRY HÉCAEN

Functional Hemispheric Asymmetry and Behaviour

Despite the apparent morphological similarity of the cerebral hemispheres in man and in non-human primates, a crucial difference can be clearly observed in man's functional organization. However, no experimental proof has been given of functional hemispheric asymmetry in animals,* whereas anatomoclinical and psychological experiments provide evidence of it in man.

Together with this functional hemispheric asymmetry, manual prevalence represents an equally specific character of the human species since among sub-human primates, there exists at best, depending on the individual case, a mere preference for using one particular hand; certain studies have even shown that manual preference may change according to the sensorial modality in which the task is presented. In man, functional hemispheric asymmetry and manual prevalence (right-handed prevalence in a proportion of more than 90% of individuals irrespective of the cultures to which they belong) seem, whatever the nature of the relation between them, connected to such an extent that it may be acknowledged that left-handed subjects have an inter-hemispheric functional organization different from that of right-handed subjects (Hécaen/Sauguet, 1971).

The notion of hemispheric dominance came into being with the discovery by Broca in 1865 of the constant presence of lesions in the left hemisphere among subjects suffering from language disorders (aphasia). For a very long time all the higher functions were attributed to the dominance of this hemisphere, the other so-called minor hemisphere being considered mute.

At the present time, the anatomoclinical data, to which have been added data derived from experimental psychology and even a few anatomical indices, justify the statement that each hemisphere is endowed with particular capacities so that either each hemisphere possesses a particular mode of dealing with information, or deals specifically with information in terms of its verbal or non-verbal character.

FUNCTIONAL HEMISPHERIC SPECIALIZATION

The Left Hemisphere

Left-handed dominance for the function of language, once positively stated by P. Broca (1865), was accepted almost without discussion on the basis of pathological facts.

Since 1961, this dominance has been demonstrated in normal subjects by D. Kimura (1961) by means of the dichotic listening method: verbal sounds presented to the right ear (that is to say, to the left hemisphere) are remembered more clearly than those presented to the left ear (that is to say, to the right hemisphere). It was, in fact, experimentally proved that the crossed auditory pathways prevailed over the ipsilateral pathways, which made it possible to assess the efficiency of each hemisphere in the perception of various materials.

The superiority of the left hemisphere was demonstrated with regard to the treatment of verbal material, meaningful or not, at the phonemic and even subphonemic level (Shankweiler/Studdert-Kennedy, 1967) and at the syntactic level (Zurif/Sait, 1970).

It was also possible to establish this superiority of the left hemisphere, not only in the auditory perception of verbal material but (though less clearly) in the visual modality by means of tachistoscopic methods.

Finally, the method of evoked potentials is even beginning, despite all technical difficulties encountered, to yield valid data concerning the lateralization of the physical events that occur in the course of the treatment of verbal sounds in the normal subject (McAdam/Whitaker, 1971; Morrell/Salamy, 1971; Wood/Goff/Day, 1971).

Particular types of gestural disorder also are due to lesions in the left hemisphere: *ideatory apraxia, ideomotor apraxia* and *constructive apraxia*. Ideatory apraxia concerns disorders in the manipulation of objects, whereas in ideomotor apraxia, gestures appertaining to conventional symbolics, expressive or descriptive gestures, are disturbed. The relation of these types of apraxia with language disorders is still not clearly defined. It seems, however, that they may occur, although fairly rarely, independently of phasic disorders, expressing alterations in a particular code of communication or in a function relating to the programming of activities.

Constructive apraxia which was initially described with reference to lesions in the left hemisphere, is also produced by lesions in the right hemisphere. The subject is incapable of drawing, of constructing, of copying simple, geometrical or figurative models. This disorder may be totally independent of language disorder. Although no clear-cut position can be taken in the matter, constructive apraxia due to lesions in the left hemisphere may have characteristics different from these caused by lesions in the right hemisphere; for us, it expresses a disorder of programming activities which manifests itself in the field of construction, just like that which is revealed in the gestural sequences called for by the manipulation of objects (ideatory apraxia).

Disorders of visual perception in connection with lesions in this hemisphere concern essentially colour and object recognition, while basic elementary data are perceived. Some authors, however, see the role of language disturbances as being of central importance here. According to them, the inability to recognize the significance of the object, although this inability is limited to optical presentation, arises from either the naming disorder (optical aphasia—Freund, 1889; Geschwind, 1965) or from the impossibility of associating the visual qualities of the object (colours, forms) in order to be able to identify it (De Renzi *et al.,* 1969).

The Right Hemisphere

Although this hemisphere was traditionally considered to be mute, we now know that lesions in it give rise to a specific set of symptons: impaired awareness of the body, visuospatial disorders and disorders in face recognition (prosopagnosia).

The functional specificity of the right hemisphere was revealed in terms of disorders relating to awareness of the body. When there are lesions in this hemisphere, particular forms of behaviour are to be seen in patients: the subject denies the motor impairment that has affected his left side (anosognosia), behaves as though he was unaware of the existence of this side or may even see it as belonging to someone else. Although such symptoms generally occur only in subjects with impaired consciousness and with sensory disorders of the same side of the hemi-body, the fact that they are only exceptionally observed in cases of lesions in the left hemisphere gives rise to the conclusion that a region of the right hemisphere, the *parieto-temporo-occipital-junction,* plays a role in the integration of sensory data deriving from one half of the body. Owing to the destruction of this zone, the patient is not aware of the hemi-body or else considers it to be foreign to himself.

Apart from this neglect of the corporal hemi-space, lesions in the right hemisphere frequently give rise to hemi-inattentions of the controlateral part of the space outside the body. This disorder, which is known as unilateral spatial agnosia, manifests itself in various activities: reading, writing, drawing, image recognition, orientation, etc., and may even not be restricted to the visual modality. Moreover, experimental studies have confirmed that it is clearly a case of supramodal impairment (De Renzi *et al.,* 1970). — Spatial orientation may also be globally disturbed so that either the subject can no longer get his bearings on a geographical map, on a town map or again in a maze test (loss of topographic notions), or he may even lose his way in areas with which he was previously familiar (loss of topographic memory).

— Spatial disorders caused by lesions in the right hemisphere also exist at an elementary level, possibly in a far more specific manner. This holds true for disorders of depth localization and, for some authors, impairments in the perception of direction both in the visual and the haptic modalities.

— Paroxysmal manifestations such as visual illusions consisting in distortions of objects and space are also produced with significant frequency by lesions in the right hemisphere.

— Writing, reading and calculation can be disturbed by lesions in the right hemisphere with regard to their spatial conditions when the manipulation of codes itself is not disturbed.

— Lastly, the predominance of the right hemisphere in elementary spatial tasks has been demonstrated in the normal subject by the tachistoscopic method by Durnford and Kimura (1971).

These spatial disorders are frequently accompanied by praxic disorders: *visuo-constructive disorders* in which, in contrast to what occurs when they are caused by lesions in the left hemisphere, the presence of the model and of reference points does not improve the patient's performance; *dressing apraxia* in which the impairment concerns the gestures relating to the way in which clothes are arranged on the body.

It has also been noted that lesions in the right hemisphere give rise to impairments in the discrimination of forms, particularly of those to which it is difficult to attach a verbal label, according to some, or which represent a complex task at the level of perception, according to others.

In split-brain subjects, Levy, Trevarthen and Sperry (1972) have also demonstrated that the right hemisphere has a greater perceptual capacity than the left hemisphere for all images, be they meaningful or not.

Finally, a particular type of visual agnosia, the non-recognition of the identity of human faces (Bodamer's *prosopagnosia*) seems to be connected to lesions in the right hemisphere. Investigations with tests involving the recognition of photographs of faces to which a large number of patients with unilateral lesions were subjected have confirmed these anatomoclinical findings by showing that the deficits were significantly higher when the lesion was in the right hemisphere than when it was in the other hemisphere.

Berlucchi *et al.* (1971) have also noted in the normal subject that the discriminative period of time reaction to the presentation of photos of faces or of letters in a hemi-field was lower when the presentation took place in the left hemi-field for faces, and in the right hemi-field for letters.

Although for some (De Renzi *et al.*, 1969) this deficit of recognition may simply be a sign of a more global perceptual disorder in the discrimination of all material of a certain perceptual complexity, the findings of Tzavaras *et al.* (1970) and of Yin (1970) make it possible to consider its specific nature. In this way on the right hemisphere a function which is specific to the human species, the identification of the faces of members of the species, would be represented. Language, which is another specific function of the human species, is represented in the left hemisphere.

In the auditory field, research on subjects with unilateral hemispheric lesions have shown that the deficits are significantly more marked in the perception of timbre, loudness and melodic memory in the case of lesions in the right hemisphere than in the case of lesions in the left hemisphere (Milner, 1962; Shankweiler, 1966). Vignolo (1969) has observed that, in the case of lesions in the right hemisphere, the perceptual discrimination of auditory patterns is disturbed whereas, in the case of lesions in the left hemisphere, the deficit concerns the identification of meaningful sounds.

In normal subjects, the dichotic listening method has made it possible to observe a superiority of the left ear, that is to say of the right hemisphere, in the assessment of number of clicks and in the recognition of melodies, sonar signals and familiar noises.

Although the anatomoclinical and experimental proofs of the functional asymmetry of the hemispheres were thereby provided at the anatomical level, their implications remained very uncertain.

However, in 1968 Geschwind and Levitsky demonstrated on the basis of 100 adult brains that the surface of the *planum temporale* was greater on the left side than on the right side in 65 % of cases, equal in 24 % and smaller in 11 %. This structural asymmetry was confirmed in the series of adult brains studied by Wada and Teszner. Furthermore, these two authors discovered that it already existed in the fœtus and in the newborn baby.

Finally, functional asymmetry should perhaps be seen as residing in a difference of cortical organization of the functional representations of each hemisphere as certain results suggest not only with regard to the higher functions (Hécaen/Angelergues, 1963; Semmes, 1968), but also with regard to the elementary sensory or visual functions (Semmes *et al.*, 1960; Dorff *et al.*, 1965). According to this, the representations are organized in the right hemisphere in a more diffuse manner, achieving a polysensorial integration going beyond the modalities and concerning only the controlateral side of the body, whereas in the left hemisphere they are more localized, more specific to a particular modality, concern both sides of the body or of space and are equipped to perform the most delicate sensorimotor coordinations necessary for the highest form of behaviour, such as language and manual skills (Semmes, 1968).

One might be tempted to see the left hemisphere as having an exclusive role in terms of the language function and the right hemisphere as having an exclusive role in terms of spatial functions. In fact, hemispheric dominance cannot be seen in such absolute terms. Over the past few years there have been signs of a tendency to reconsider the participation of the right hemisphere in language activities which H. Jackson always maintained. However, this participation is only a minor one, or appears only when there is a broad lesion in the left hemisphere, ensuring a certain functional compensation, or again restricting certain conditions of written language. Indeed, the study of subjects with lesions in the right hemisphere has revealed a few slight deficits in the manipulation of the linguistic code, but, above all, disorders in the spatial aspects of writing or reading. After left hemispherectomy, a fairly considerable recovery of verbal expression and comprehension has been observed. Lastly, and above all, in the experiments carried out by Sperry and Gazzaniga (1967) on split-brain patients, the right hemisphere has been able to show, in spite of its inability of expression, a certain degree of comprehension of the verbal material presented to it.

From another angle, while lesions in the right hemisphere are usually the cause of various spatial disorientations, in their most severe forms, such disorders may also be caused by lesions in the left hemisphere. According to De Renzi (1971) right-hemisphere dominance is exclusive only in elementary spatial tasks (direction, depth), but this may still be too categorical a view.

Some see the difference between the two hemispheres as deriving from the specialization of each hemisphere in the treatment of a certain type of material. Thus, whatever the type of task required, verbal material is dealt with by the left hemisphere and non-verbal material, or material that is difficult to put into verbal form, by the right hemisphere (Milner, 1967).

Others postulate the existence of a mode of treatment specific to each hemisphere. Thus, De Renzi *et al.* (1969) have attributed to the right hemisphere the role of discrimination and organization with regard to sensory data, while the left hemisphere is seen as having the task of naming and associating these data to make their significance comprehensible.

De Renzi (1969) and Vignolo (1972) thus opt for two levels of agnosia, irrespective of the sensory modality involved, thereby coming back to the dissociation proposed by Lissauer (1890) between apperceptive agnosia and associative agnosia. According to this view, lesions in the right hemisphere give rise to the appearance of the first type. The disorders appear whenever the perceptual task proposed is beyond a certain level of complexity. In the case of lesions in the left hemisphere, the disorder would consist in the impossibility of linking up attributes of a specific object (colours, sound, form).

With the study of split-brain subjects, it became possible to tackle again the question of the particular mode of treatment of the perceptual data specific to each hemisphere. Examination of these subjects makes it possible to judge from the performance achieved by each hemisphere (which may be considered as not having been subjected to a lesion), each of the two, moreover, acting in such a way as to control the other.

Thus Nebes (1971) shows in these subjects that the capacity of the right hemisphere to match the part with the whole proves to be superior to that of the left hemisphere, irrespective of the modality of presentation and the intermodal or intramodal nature of the task. J. Levy, C. Trevarthen and R.W. Sperry (1972) have compared each hemisphere's ability to perceive and to respond by making use of the tendency of each hemisphere to effect a perceptual completion of patterns across a median line. The stimuli presented tachistoscopically consisted in "chimeric" figures, obtained by connecting the right-hand side of one figure with the left-hand side of another, irrespective of whether what were involved were human faces, meaningless images, images of familiar objects or linked patterns of crosses and squares. Each hemisphere completing what it sees, there exists, when such chimeric figures are presented, a conflict between the hemispheres. In this method the superiority of the right hemisphere shows itself in cases of non-verbal response to the presentation of four series of chimeric images. A hemispheric superiority is thereby revealed in the basic perceptual capacity that is constituted by the apprehension of forms; in contrast, the left hemisphere does not seem to be sufficiently well organized to achieve this perceptual apprehension. On the other hand, if the response has to be made verbally, the superiority of the left hemisphere emerges; the information must be treated differently in this case, more analytically, seeking out the distinctive characteristics that can be accorded a verbal label.

For Levy, Trevarthen and Sperry (1972), what is involved is not a simple predominance in the competition between the two hemispheres, but rather a fundamental difference in their mode of treating information.

It must also be taken into consideration that this asymmetry of the hemispheric nervous mechanisms is not specific to all cognitive activities. We have therefore proposed a systematization of inter-hemispheric functioning on the basis of two categories of functions, instrumental functions and basic functions. Only the instrumental functions are affected by functional asymmetry. We define them as serving as tools for human beings to communicate between themselves and to know the external world or to act upon it (Hécaen, 1972).

Although certain definite regions in each hemisphere thus underlie specific functions, it must be noted that neighbouring regions also participate, although to a lesser extent. There are, therefore, grounds for postulating the existence of an intra-hemispheric functional gradient.

Finally, there exist a certain number of results produced by stimuli of the right and left sub-cortical structures which suggest that besides horizontal representation there exists a vertical representation, the sub-cortical structures participating in the realization of performances dependent on the nervous mechanism of the hemisphere with which these formations are in relation; one could thus posit a cortico–sub-cortical functional system specific to each half of the brain.

The basic functions are independent both of other functional areas (instrumental functions) and sensory modalities. It is not possible to conceive of an instrumental function that does not involve the participation of these basic functions. They are represented in a bilateral and symmetrical manner. Thus, for instance, bilateral hippocampic lesions disturb mnesic retention in its entirety without, however, affecting the instrumental activities. While in the case of unilateral ablation of the hippocampus, together with that of the neighbouring neo-cortical regions, disorder of retention and learning affects only the instrumental activity depending on the hemisphere concerned (Milner, 1967).

Perhaps one should temper this observation by admitting that the role of the factor specific to the functioning of certain symmetrical cerebral regions may be seen, albeit to a relatively slight degree, when one of these regions is affected even within the specific field of the other hemisphere.

The problem of functional and even structural asymmetry must also be posed in terms of the maturation process. The study of acquired aphasia in the child seems to show that hemispheric dominance can not be considered to set in before puberty. In children who have acquired language, left unilateral hemispheric lesions give rise only to transitory types of aphasia that do not have the same appearance as in the adult; language disorders even occur in a relatively high proportion in cases of lesions in the right hemisphere in right-handed children.

Nevertheless, before we can conclude that functional asymmetry is only the result of structural development, two separate facts must be taken into consideration: in children with lesions that occured at birth or during the first year, deficits similar to those observed in adults have been found according to the damaged hemisphere (McFie, 1961, Fedio/Mirsky, 1969); we have already seen that the structural asymmetry of the planum temporale already appears to exist in the fœtus and in the newborn child.

It could therefore be concluded that the left hemisphere, which from the beginning is potentially dominant because of a structural peculiarity, unquestionably asserts its predominance over the other hemisphere in language functions only when the nervous system has attained full maturity. In the adult, the right hemisphere could carry out certain performances, albeit at a limited level, if the left hemisphere were totally destroyed.

It appears, therefore, that prior to physical maturation and experience, in the human species there exists a structural arrangement corresponding to a particular function. This observation should be viewed in the same context as that obtained with respect to the behaviour of the infant who at the age of one month is not only capable of making subtle discriminations between verbal sounds but also appears to perceive them in terms of categories approximating those of adults (Eimas/Siqueland *et al.,* 1971).

DIFFERENCES AND SIMILARITIES IN FUNCTIONAL REPRESENTATIONS IN MAN AND IN SUBHUMAN PRIMATES.

Asymmetrical cerebral organization is a specific characteristic of the human race, but it also appears that functional representations in man are not as exact and as restricted in topography as they are in the monkey. Indeed, deficits in humans seem to be due to lesions, the location of which may vary within relatively broad areas. In the case of the monkey, on the contrary, an ever increasing body of results points to the existence of very limited critical zones, the destruction of which is enough to provoke a specific deficit. Let me refer to the findings relating to the location of the lesion responsible at the level of the frontal lobe for particular deficits. The research carried out by Mishkin (1957; 1969), in particular, made it possible to demonstrate that the dorsolateral region and the orbital cortex subserve different functions and that, at the level of the dorsolateral region itself, the region of *sulcus principalis* represents the critical zone for the deficit of delayed response and for tasks of delayed spatial alternation whereas ablations of the orbital cortex give rise to deficits in tasks of object alternation. Finally, it was demonstrated that lesions further back from the dorsolateral convexity *(arcuate sulcus)* are responsible for deficits in spatial tasks where the time factor is of but slight importance.

Lastly, even at the level of *sulcus principalis,* lesions within the middle part are responsible for the most serious deficits in delayed spatial alternation tasks and in place reversal learning (Butters/Pandya *et al.,* 1971).

For the level of the inferotemporal zone, the work of Iwai and Mishkin (1969), of Iversen and Weiskrantz (1964, 1967) and of Cowey and Gross (1970), likewise have made it possible to distinguish two zones, one anterior with a more mnesic function, underlying the possibilities of shifting visual attention; the other behind, more perceptual, making it possible to fix the attention on the relevant characteristics of the visual stimulus.

This exact marking off of the boundaries of functional areas, it should be recalled, is on a par with their symmetrical functioning. Unilateral ablations produce few or no deficits. Still, this last assertion should be somewhat amended by the results obtained by Warren *et al.* (1969) on the deficit of delayed response after unilateral frontal dorsolateral ablation, and by those obtained by Cowey and Dewson (1972) on the auditory deficit after unilateral temporal ablation. It should be stressed again that no difference has ever been observed between deficits in terms of whether or not the ablation concerns the hemisphere that is controlateral to the hand preferably used by the animal.

While taking into account intra-hemispheric and interhemispheric organization makes it difficult to directly infer from the nervous mechanisms underlying animal behaviour those underlying human behaviour, it is also necessary to consider difficulties of another kind. The deficits described in the monkey produced by ablations of specific areas are not comparable in every respect with those encountered in lesions in the corresponding areas in man. This is true with regard to the deficit of delayed response and delayed alternation, which is not observed in the case of frontal lesions in man but which has been noted in the case of diffuse lesions. (Can this discordance be accounted for by the role played by language in human performance?)

It seems interesting in this connection to refer to a study by Miles (1971) on the capacities for delayed response in different animal species and in man: first of all, with the simple customary twofold-choice technique, by increasing the length of the delay, one finds significant differences in performances from the cat to the normal child; these differences correspond to the phylogenetic position of the species studied. If the difficulty of the task is increased, while increasing the delay to more than five seconds and the possibilities of choice by as much as five and by varying the number of trials according to the species, significant differences are found: the performance of children outclasses that of animals. Finally, in man, in an experiment that involves becoming familiar with a problem with 64 choices, age and the length of the delay have an unquestionable influence. In the case of the adult, even with a problem with 128 choices and delays of as much as 24 hours, a significant amount of information concerning the spatial location is still retained.

The difference in the ability to deal with these problems between man and the best of the monkey species (the rhesus monkey) is such that it may be considered as being of a qualitative nature although it does not appear to depend on either linguistic abilities or on mnemonic ones, since it relates essentially to spatial information.

While the bilateral destruction of the internal temporal formations gives rise in man to a particularly characteristic amnesic syndrome, anterograde amnesia or *l'oubli à mesure,* no experiment carried out on animals reproduced truly comparable disorders.

It may in fact be a question of simple analogies between the behaviours observed in each case. While ablation of the inferotemporal area giving rise to disorders of the visual functions in the monkey are judged by learning, perceptual disorders in man are detected in recognition tests, or they may be revealed by the spontaneous behaviour of the patient.

Before claiming homologies in behaviour, structural homologies and the antecedents of the structural divisions that appeared with the development of the cortical formations, should be established beyond doubt.

A recent article by Killackey, Wilson and Diamond (1972) is revealing in this respect. In spite of definite similarities between the visual deficits observed in the monkey and in *Tupaia glis,* the comparison cannot be carried too far. Indeed, the twofold dissociation between the behavioural deficits produced in the tree shrew by the ablation of the temporal zone on the one hand and by the ablation of the striate area on the other, is also to be found in the monkey, but is caused by the ablation of the two parts of the inferotemporal area, that is to say, of the two extrastriate areas. Furthermore, as the authors stress, projections to the temporal area in the monkey derive from certain subdivisions of the pulvinar, which subdivisions do not seem to exist in the *Tupaia.*

Not only, then, must behaviour patterns be defined by the same methods but the details of the structural modifications must be better known for it to be possible to appreciate the relationship between brain functioning in the different animal species and their different types of behaviour.

A study by Delacour and Borst (1972) clearly demonstrates the difficulties in comparing the functions of given structures in various animal species. These authors studied the effects of the destruction of the *centrum medianum* and of the *nucleus parafascicularis* on avoiding conditioning (shuttle box test) in three species: rat, cat and monkey. While the destruction of these structures gives rise to considerable deficits in the rat, it produces no effect in the other two species. To decide once and for all upon the absence of functional homology of these structures in these three species is difficult since both the matching of the respective size of the lesions and the experimental situations (subjective intensity of the stimulus, size of the apparatus, speed of learning), despite all efforts in this direction, present difficulties when it comes to considering them in terms of similarity. But above all, the qualitative differences of the strategies in the three species can only be assessed in terms of the animals' repertoires (interference of defense reactions and of the conditioned avoidance reactions).

Finally, there are other difficulties (which can be more or less overcome) standing in the way of the establishment of valid comparisons between deficits in man and in animals. In man, the lesion that stems from illness cannot be compared to an ablation carried out on a healthy animal; we may add that it is, in fact, impossible to obtain even a limited series of lesions located exactly in the same area. Secondly, the tasks given to animals, to make them conform to the same standards, probably, if not definitely, do not correspond to the animals' natural repertoire. Meanwhile, an effort seems to have been successfully made in various laboratories to collect observations of a more natural character on the effects of cerebral lesions.

Such an effort should not be restricted to research on animal neuropsychology. It should be extended to human pathological studies although a long period of clinical observation preceded the introduction of experimental methods and these methods were sometimes derived from clinical intuitions.

The differences between man and animal with regard to behaviour and to nervous substrata thus appear to be greater than the similarities; nevertheless, it may by hoped, in view of the fairly revolutionary observations that have been made with respect to anthropoids, that studies on these species will reduce the divergence.

It was already known from the research of Jacobsen (1935), Blum (1948), and Evarts and Nissen (1953) that in chimpanzees the deficits of delayed response and of the alternation task did not exist or were only slight after bilateral frontal lesion. More recently the results obtained by the Gardners (1969) on the one hand, and by Premak (1970) on the other, raise the question, (without resolving it) of the existence of a language in the chimpanzee. Davenport and Rogers (1970) have also demonstrated that transfer between sensory modalities exists in anthropoids, despite the fact that no signs of it have been seen in monkeys. Similarly, Gallup (1970) has provided proof that chimpanzees are capable of recognizing their own image reflected in a mirror whereas the macaque does not have this capacity for self-recognition. Studies carried out by ethologists (Van Lawick-Goodall, 1968; Menzel, 1972; Struhsaker/Hunkeler, 1971) have revealed to us that chimpanzees are able to use inanimate objects as tools and in a systematic manner.

These various studies make it perhaps possible to predict that the combined development of ethological, neuropsychological and anatomical research bearing on the chimpanzee and on man will provide more accurate grounds for judging the similarities and the differences between the two species.

DISCUSSION

Foppa: May I put my question in a rather primitive form? What kind of inference is possible, from the results of animal studies of the effect of cerebral lesions, to human behaviour?

Hécaen: We are taking into consideration results obtained in experiments with ablations in animals to study whether there exist in human subjects with focal lesions in similar zones behavioural modifications comparable to those of the animal. We have to admit that perhaps these comparisons are completely irrelevant. For instance, as I reported, the experiments of Miles show that in delayed response experiments, the difference between the capacity of the human subjects and the capacity of the most gifted animal species might be a qualitative one. That is why one might better understand why clinicians fail to find a deficit of delayed response in subjects with frontal lesions. The comparison is even possible in the other direction: clinical results with humans can suggest experiments with animals. It is not enough to simply compare the disorders in two species; rather it is of interest to compare the relationship between the properties of different focal lesions in one species with those due to lesions of comparable zones in another species. This exchange of information and hypotheses should be useful if we come to know the structural differences that separate the species.

Differences and Similarities in Performance

Ploog: Could you perhaps for another minute or so comment on the problem of hemispheric specialization in terms of similarities and differences regarding performance in apes and in man?

Hécaen: It is difficult to say something about differences, because there are so many differences due to our way of examination. We deal, of course, exclusively with accidental lesions. In most cases it is possible only to be sure of the side of the lesion. Animal experiments are designed beforehand, but the tasks are not always based on the repertoire of the animal. For example, nobody knows much about the importance of the delayed response for a monkey. As to the question of similarities, the behaviour of a monkey unable to distinguish some patterns because of a temporal lesion, for example, certainly looks like some analogy (I do not like to use the word) to what is called in the clinical world agnosia; but, as I said before, when you see an agnosic patient, you instantly see his special behaviour. The monkeys with such lesions, however, usually recognize objects, and it is only by the defect in learning that one can say that they have difficulties in recognizing the pattern. For all these reasons I cannot say whether these disturbances are the same or similar. In our work, the data we get from animal neuro-psychologists are of help in the designing of corresponding tasks for our patients. From these tasks we try to exclude language, which is very difficult indeed, and that is why I consider the work on split-brain patients so important, because in these cases you can do just that by restricting the information to one hemisphere. I cannot really answer your question, Dr. Ploog; it is too difficult.

Left- and Right-Handness

Leyhausen: I should like to relate what you have been saying about right- and left-handedness to what I was saying this morning about ecological pressures. Obviously, a division of labour between the two forelimbs of an animal—and, as a consequence of this, a disparity between the hemispheres governing their movements—cannot be expected unless the forelimbs are only occasionally involved in locomotion. Besides the chimpanzee, the only mammal I can think of which has freed its forelimbs in a similar way to man from the task of locomotion is the large species of kangaroo. It might be very interesting to find out whether the hemispheres of a large kangaroo do show at least an incipient diversity.

On the other hand, I should like to say that anatomical disparity between right and left is far more frequent than is generally suspected. During many years of work on the systematics of cats, I have been measuring large numbers of skulls, taking measures of both sides. I found a considerable right–left variation within skulls of overall equal size; this applies to the brain cavity as well. This need not result in functional disparity, of course.

Chance: Is yours a variation equally from left and right or is there any bias in it?

Leyhausen: I did not pay attention to that, but I will look at the measurements again.

Chance: Kortlandt has described how chimpanzees pick up food and hold it in their hands while they move for quite considerable distances with it. There are certain activities where chimpanzees use their hands, but I do not know whether he actually noted whether they used one hand more than the other. Did you notice, Dr. van Hooff?

van Hooff: We are watching it, but we do not see any signs of it. They alternate quite easily. There are studies about a number of primates in this respect by Gaskill; there may be some preference for one hand with some individuals, but usually these differences are very small, and difficult to demonstrate statistically. Besides, in experiments these preferences may be dependent on the occasion.

Ploog: The question of right- and left-handedness is of course a very important one; it is well known even in rats. For example, Holger Hyden, who for many years investigated problems of protein synthesis and motor performance in learning, can easily distinguish which individuals prefer the right and which the left hand. But all these things are random variations of right and left, and not specializations of one hemisphere of the kind Dr. Hécaen was talking about.

Hemispheric Specialization and Emotion

Masters: My question has to do with the broader problem of the study of lesions in the interference and interruption of the performance of behaviour sequences as a whole, with a relationship between hemispheric specialization and other portions of the brain. Is there any hemispheric specialization which is related to affect, emotion or motivation?

Hécaen: I can only answer that by relating two series of facts from clinical observations: it was Gainotti who observed from a clinical point of view that patients with a right hemisphere lesion are less emotional than they were before the defect, and that they are indifferent to their defect or minimize it, while patients with left lesions show what Goldstein called catastrophic reactions; in the test situation these patients are so anxious they cannot do anything. But you must remember that this refers only to this special investigation.

The other fact is an observation by Sperry: a split-brain patient was shown slides, a picture of, let us say, a tree to his right visual field, (that is to his left hemisphere), and one of a (I am sorry) nude woman to his left visual field (that is to his right hemisphere). When asked what he saw, the patient answered that he saw a tree, but at the same time he blushed. When he was asked, why he had blushed, he made some excuse. I do not think this has been published.

Ploog: Yes, it has been published, along with some other experiments along the same lines, which seem to indicate—at least to Sperry and Gazzaniga—that the right hemisphere is more involved in the evaluation of affective situations, or emotional situations, or social situations in an emotional context. If one asks a question about localization of emotional functions, one must remember that very much depends on whether one has a onesided or a bilateral destruction of the same structure; I think it would be an overstatement to say that the right hemisphere is responsible for the elaboration and evaluation of emotional cues, and not a yes-or-no answer, but rather one of preponderance.

Hemispheric Specialization

Masters: Speaking primarily of those parts of the brain which are most distinctly species-specific, it is very striking to think about some of the differential advantages of comparing the function of the right and the left hemisphere. For instance, if language is species-specific to *Homo sapiens* and there is a certain specialization of linguistic functions in the left hemisphere, that would mean that some of the more interesting comparisons take place in the right hemisphere; thinking about that I immediately jumped to one point you mentioned. Lesions in the right hemisphere, you said just now, may result in a deficit of the recognition of the human face. That is very interesting. There are some elements in the recognition of the human face that are in some sense evolutionary in *Homo sapiens,* particularly infant reactions to some components of the face. It may be that there we might have some capacity to compare the physiological mechanism to a very important releasing mechanism.

van Hooff: I thought this recognition was individual recognition, not recognition of the human face.

Hécaen: No, the agnosic patient would recognize his friend, his wife, sometimes even the nurse. This test, given to patients with right and left lesions, was composed of photographic pictures of six different men who were unknown to the patient and a picture of one of them that had to be matched. The complexity of the task was increased during the test: the man was photographed, for instance, with a beard or a wig or a different mimic. The result was a big defect score with right hemispheric lesions and a lesser with left lesions (but compared with no lesions, there is, of course, a defect).

Wickler: Is the patient able to distinguish a photograph of the human face?

Hécaen: Yes.

Wickler: Can he distinguish a man from a woman?

Hécaen: Yes. It is a difficult question, though, because what he recognizes is not a characteristic of the face, but the hair. For instance, a patient with prosopagnosia cannot recognize an individual, but he can recognize types, like an airforceman with a moustache.

Ploog: It is a matter of differentiation of personal features and not a matter of individuality.

Masters: That is, however, terribly important. You know that in all primates personal recognition is terribly important.

Ploog: Right. In all animals with social organization.

van Hooff: Is there any indication that patients with these lesions are able to discriminate between the moods of the people in the pictures? Are they able to say, for instance, "this man is an angry man"?

Hécaen: Yes.

van Hooff: Can they easily do that?

Hécaen: Not easily. At the beginning, we always ask the patients whether the faces in these pictures are smiling or crying. In drawings, it is more difficult.

Comparative Consequences of Hemispheric Specialization

von Cranach: Dr. Hécaen, you seemed to suggest that you doubted the possibility of comparisons. Could you enlarge a little on that point?

Ploog: One of the experiments of Dr. Hécaen is a very crucial one, and that is the experiment in which a patient who used his right hand to recognize objects was able to sort out objects he had seen before, but was not able to name them. However, he was able to name an object projected into his left hemisphere, to the right visual field, but he could not sort out that same object with his left hand. This indicates very clearly that there is a cognition process in the right hemisphere that is not concerned with speech and language. That is to say, the faculty of the right hemisphere has a heavy influence on cognitive capacity, but because there is no language to express these recognitions or these percepts, there is no way to investigate it, except by that kind of trick of sorting out, or to play one against the other and so on. I do remember that Sperry stated in one of his papers that a man can do more with his right hemisphere than a chimpanzee with both hemispheres; that is to say, the cognitive capacity of the right hemisphere, regardless of language (left hemisphere) is of a higher capacity than the chimpanzee's whole brain.

In the macaque, bilateral ablations of certain cortical areas result in basic defects, for instance, ablations in the inferior temporal lobe give rise to a defect of visual learning; ablations in the inferior parietal area give rise to a defect in tactile learning. I think, Dr. Hécaen, that in regard to these specified defects in the performance of a macaque bilaterally lesioned in these major areas of the cortex there are similarities. I think that we could get the same kinds of effects in certain respects if we applied the same tasks to human beings, except for the heavy memory deficit in the temporal lobe, which is not demonstrable in macaques.

Hécaen: And the delayed response, too?

Ploog: And the delayed response, too. What I am really trying to get at is that there are certain basic faculties of cortical evaluation of our environment which are already present in monkeys and apes, but what appears to be lacking is any sign of hemispheric specialization. There seems to be nothing of that, although the hands of a chimpanzee, as everyone knows, are very similar to man's, including thumb opposition. A chimpanzee can be exploring his environment for hours standing on two legs and, of course, can walk on two legs for quite a while. So I think it is not only the freeing of the forelimbs which is needed in order to develop the human capacity, and indirectly leads to hemispheric specialization. There are many phylogenetic steps in between, from the chimpanzee's manipulatory skill to language. For instance, a chimpanzee can open latchboxes blindly in sequences of seven, eight and even twelve much faster than any human. Have you seen the movies by Ronald Myers?

Hécaen: Yes, I have seen them, the chimpanzee too.

Ploog: It is fantastic what tactile memory and skill a chimpanzee can develop with his hands, but I do not think the relationship between that and actual language is so close.

von Cranach: I am confused again. After yesterday, I had concluded that the best method of homologizing organs was to treat the morphological part and the neurological part as one system. But according to your paper, this would not lead us very far, or would it?

Hécaen: We tried to do that, but there are many difficulties, because at the beginning there are no apparent differences between any of the steps.

von Cranach: Is not the same kind of movement of the hand of a chimpanzee and the hand of a man possibly related to completely different brain structures?

Ploog: Definitely not. I think we cannot ask this kind of question in this context. You are asking for a homology to something that has no homologon. Language does not have any homologon; it is a species-specific and unique capacity, and if you would like to look for a system of relations between the morphology of certain nervous structures and a given behaviour, you would have to ask for precursors of language and do the homology on that basis. In this context, I think we have to ask questions related to non-linguistic capacities, or behaviours if you wish. We have to go back to certain characteristics which are below the human capacity to ask your question.

Wickler: What is your aim in relating overt behaviour to some part of the brain as part of the machinery that brings about this behaviour? Do you want to identify a given part of the brain, to specify it by stating what part of behaviour it influences? Or do you want to do it the other way round, to find out whether the behaviour is the same because it is produced by the same part of the brain?

von Cranach: I started from the idea that the problems resulting from the use of morphological concepts in behaviour, which is something different from morphology, would perhaps vanish if I used the behaviour as a representative instead of a neurophysiological structure that I cannot see. Actually, I would be homologizing the neurophysiological structures, but using an element of behaviour as a representative. If this could be done, there would be no difference between your application of the methods of behaviour and the methods morphologists have been using for a long time. That is why I was interested.

Brain Instead of Behaviour

Wickler: You want to look at the brain instead of behaviour?

von Cranach: I wanted at least to consider the possibility.

Wickler: I do not think that will work.

Ploog: But certainly it works.

Hécaen: Oh yes.

Wickler: Would that mean that, given morphologically similar movements can be induced by different parts of the brain, these movements can not be homologous?

von Cranach: Certainly the conclusions I would have drawn otherwise could not be drawn anymore. You would have to find another justification for treating behavioural characters as morphological.

Wickler: Actually, I doubt it. If you look for the machinery that produces a pattern instead of for the pattern itself, wouldn't that mean that you cannot homologize a series of notes, a melody, because it is not a structure, unless you know how this melody was produced? So that whether you whistle or sing it decides whether it is a homologous pattern?

von Cranach:. This is something we did not discuss yesterday because we had no time left. If you homologize a melody or a language, you actually homologize the effects of something behind it, the outcome instead of the behaviour.

Wickler: But melodies are perfect behaviour patterns.

von Cranach: No, they are not. The melody, the sound is a wave of the air which you homologize instead of the movement of the organ that creates it. Thus, you are one step behind the behaviour. I think one may as well be one step before the behaviour and use the behaviour as representative of the actual morphological structure.

Wickler: You missed the question. If you whistle or sing, there are quite different organs involved, so the same melody or the same pattern is produced by different organs. So why cannot these melodies be homologous?

von Cranach: That again suggests neurophysiology, where the same representation of the melody should be found to induce the movements of the different organs.

Leyhausen: Let me give a practical example: in a fish you will find that object recognition is localized in the *tectum opticum* of the midbrain roof. If in a dog, you remove the visual cortex, there will be no more object recognition, but the tectum will still be capable of avoiding obstacles when the animal moves around. If you do the same to a human, he is no longer capable of moving around under optical guidance. Would you call object recognition in these three cases homologous? Or, because the structures which serve the function have been changed while the function has been retained through evolution, would you call it an analogy? I am afraid that in this particular case I have to come down heavily on the side of Wickler. You have to keep your concept of homology to the level of complexity and the particular function, to the particular kind of process you are investigating and it will not be transferable one level up or down. There, you have to do your homologizing anew. Within the structure or functional system which you are investigating, you cannot start at the bottom, register there a homology with another simlar structure, and then conclude that you are on the same track of homology all the way up the tree. It is not so. You have to do the homologizing anew at each level. This is known in evolution theory as the mosaic type of the evolution of systems, sub-systems and characters.

von Cranach: What does the word *mosaic* refer to?

Leyhausen: Well, things have a way of independently moving around in the kinship, so to say, at each level.

Cain: The mosaic law is made of many pieces of law. Mosaic evolution is made up of many characters. And like any individual character in the law they can be changed separately and individually.

NOTE

This article first appeared in *Social Science Information,* 12 (6) 7-23.

* However, very recent studies on split-brain animals, particularly those by Buther and Francis *(Neuropsychologia,* 1973) and by Hamilton *et al. (Neuropsychologia,* 1974 (12) 193-198) while not yet providing conclusive proof, no longer permit such positivism.

REFERENCES

Berlucchi, G./Heron, W./Hyman, R./Rizzolatti, G./Umilta, C. 1971. "Simple Reaction Times of Ipsilateral and Controlateral Hand to Lateralized Visual Stimuli", *Brain* 94:419-430.

Blum, R.A. 1948. "The Effect of Bilateral Removal of the Prefrontal Granular Cortex on Delayed Response Performance and Emotionality in Chimpanzee", *American Psychologist* 3:237-238.

Broca, P. 1865. "Sur le siège de la faculté du language articulé", *Bulletins et mémoires de la société anthropologique de Paris* 6:337-393.

Butters, N./Pandya, D./Sanders, K./Dye, R. 1971. "Behavioral Deficits in Monkeys after Selective Lesions Within the Middle Third of *Sulcus Principalis*", *Journal of Comparative Physiology and Psychology* 76:8-14.

Cowey, A./Dewson, J.H. 1972. "Effects of Unilateral Ablation of Superior Temporal Cortex on Auditory Sequence Discrimination in *Macaca mulatta*", *Neuropsychologia* 10:279-290.

Cowey, A./Gross, C.G. 1970. "Effects of Foveal Prestriate and Infero-Temporal Lesions on Visual Discrimination by Rhesus Monkeys", *Experimental Brain Research* 11:128-144.

Davenport, R.K./Rogers, C.M. 1970. "Intermodal Equivalence of Stimulus in Apes", *Science* 168:279-280.

Delacour, J./Borst, A. 1972. "Failure to Find Homology in Rat, Cat, and Monkey for Functions of a Subcortical Structure in Avoiding Conditioning", *Journal of Comparative Physiology* 80:458-468.

De Renzi, E./Faglioni, P./Scotti, G. 1970. "Hemispheric Contribution to Exploration of Space through the Visual and Tactile Modality", *Cortex* 6:191-203.

De Renzi, E./Faglioni, P./Scotti, G. 1971. "Judgment of Spatial Orientation in Patients with Focal Brain Damage", *Journal of Neurology, Neurosurgery and Psychiatry* 34:489-495.

De Renzi, E./Faglioni, P./Spinnler, H. 1968. "The Performance of Patients with Unilateral Brain Damage on Face Recognition Tasks", *Cortex* 4:17-34.

— 1969. "Perceptual and Associative Disorders of Visual Recognition: Relationship to the Side of the Cerebral Lesion", *Neurology* 19:634-642.

Dorff, J.E./Mirsky, A.F./Mishkin, M. 1965. "Effects of Unilateral Temporal Lobe Removals in Man on Tachistoscopic Recognition on the Left and Right Visual Fields", *Neuropsychologia* 3:39-52.

Durnford, M./Kimura, D. 1971. "Right Hemisphere Specialization for Depth Perception in Visual Field Differences", *Nature* 231:394-395.

Eimas, P.D./Siqueland, E.R./Jusczyk, P./Vigorito, J. 1971. "Speech Perception in Infants", *Science* 171:303-306.

Evarts, E./Nissen, H.W. 1953. "Test of the 'Abstract Attitude' in Chimpanzee Following Ablation of Prefrontal Cortex", *Archives of Neurology and Psychiatry* 69:323-331.

Fedio, P./Mirsky, A.F. 1969. "Selective Intellectual Deficits in Children with Temporal Lobe or Centrencephalic Epilepsy", *Neuropsychologia* 7:287-300.

Freund, C.S. 1889. "Über optische Aphasie und Seelenblindheit", *Archiv für Psychiatrie* 20:276-297, 371-416.

Gallup, G.G. 1970. "Chimpanzee: Self Recognition", *Science* 167:86-87.

Gardner, R.A./Gardner, B.T. 1969. "Teaching Sign Language to a Chimpanzee", *Science* 165:664-672.

Geschwind, N. 1965. "Disconnexion Syndromes in Animals and Man", *Brain* 88:237-294, 585-644.

Geschwind, N./Levitsky, W. 1968. "Human Brain: Left-Right Asymmetries in Temporal Speech Regions", *Science* 161:186-187.

Hécaen, H. 1972. *Introduction à la neuropsychologie*. Paris, Larousse.

Hécaen, H./Angelergues, R. 1963. *La Cécité psychique*. Paris, Masson et Cie.

Hécaen, H./Sauguet, J. 1971. "Cerebral Dominance in Left Handed Subjects", *Cortex* 7:19-48.

Iversen, S.D./Weiskrantz, L. 1964. "Temporal Lobe Lesions and Memory in the Monkey", *Nature* 201:740-742.

— 1967. "Perception of Redundant Cues by Monkeys with Inferotemporal Lesions", *Nature* 214:241-243.

Iwai, E./Mishkin, M. "Further Evidence in the Locus of the Visual Area in the Temporal Lobe of the Monkey", *Experimental Neurology* 25:585-594.

Jacobsen, C.F. 1935. "Functions of the Frontal Association Area in Primates", *Archives of Neurology and Psychiatry* 33:558-569.

Killackey, H./Wilson M./Diamond, I.T. 1972. "Further Studies of the Striate and Extrastriate Visual Cortex in the Tree Shrew", *Journal of Comparative and Physiological Psychology* 81:45-63.

Kimura, D. 1961. "Cerebral Dominance and the Perception of Verbal Stimuli", *Canadian Journal of Psychology* 15:166-171.

Levy, J./Trevarthen, C./Sperry, R.W. 1972. "Perception for Bilateral Chimeric Figures Following Hemispheric Deconnexion", *Brain* 95:61-78.

Lissauer, H. 1890. "Ein Fall von Seelenblindheit nebst einem Beitrag zur Theorie derselben", *Archiv für Psychiatrie* 21:222-270.

McAdam, D.W./Whitaker, H.A. 1971. "Language Production: Electroencephalographic Localization in the Normal Human Brain", *Science* 172:499-502.

McFie, J. 1961. "Intellectual Impairment in Children with Localized Post-Infantile Cerebral Lesions", *Journal of Neurology, Neurosurgery and Psychiatry* 24:361-365.

Menzel, E.W. 1972. "Spontaneous Invention of Ladders in a Group Young Chimpanzees", *Folia Primatologica* 17:87-106.

Miles, R.C. 1971. "Species Differences in 'Transmitting' Spatial Location Information", pp. 165-179 in: L.E. Jarrard (ed.) *Cognitive Processes of Nonhuman Primates*. London, Academic Press.

Milner, B. 1962. "Laterality Effects in Audition", pp. 177-195 in: V. Mountcastle (ed.) *Interhemispheric Relations and Cerebral Dominance*. Baltimore, Johns Hopkins Press.

— 1967. "Brain Mechanisms Suggested by Studies of Temporal Lobes", pp. 122-132 in: C.H. Millikan/F.L. Darley (eds.) *Brain Mechanisms Underlying Speech and Language*. New York, Grune and Stratton.

Mishkin, M. 1957. "Effects of Small Frontal Lesions on Delayed Alternation in Monkeys", *Journal of Neurophysiology* 20:615-622.

Mishkin, M./Vest, B./Waxler, M./Rosvold, H.E. 1969. "A Re-examination of the Effects of Frontal Lesions on Object Alternation", *Neuropsychologia* 7:357-364.

Morrell, L.K./Salamy, J.G. 1971. "Hemispheric Asymmetry of Electro-Cortical Responses to Speech Stimuli", *Science* 174:164-166.

Nebes, R.D. 1971. "Superiority of the Minor Hemisphere in Commissurotomized Man for the Perception of Part–Whole Relations", *Cortex* 7:333-349.

Premak, D. 1970. "A Functional Analysis of Language", *Journal of the Experimental Analysis of Behavior* 14:107-125.

Semmes, J. 1968. "Hemispheric Specialization: a Possible Clue to Mechanisms", *Neuropsychologia* 6:11-26.

Semmes, J./Weinstein, S./Ghent, L./Teuber, H.L. 1960. *Somatosensory Changes after Penetrating Brain Wounds in Man*. Cambridge, Mass., Harvard University Press.

Shankweiler, D. 1966. "Effects of Temporal Lobe Damage on Perception of Dichotically Presented Melodies", *Journal of Comparative Physiology and Psychology* 62:115-122.

Shankweiler, D./Studdert-Kennedy, M. 1967. "Identification of Consonants and Vowels Presented to Left and Right Ears", *Quarterly Journal of Experimental Psychology* 19:59-63.

Sperry, R.W./Gazzaniga, M.S. 1967. "Language Following Surgical Disconnection of the Hemisphere", pp. 108-118 in: C.H. Millikan/F.L. Darley (eds.) *Brain Mechanisms Underlying Speech and Language*. New York, Grune and Stratton.

Struhsaker, E.T./Hunkeler, P. 1971. "Evidence of Tool-Using by Chimpanzees in the Ivory Coast", *Folia Primatologica* 15:212-219.

Teszner, D./Tzavaras, A./Gruner, J./Hécaen, H. 1972. "L'asymétrie droite–gauche du planum temporale. A propos de l'étude anatomique de 100 cerveaux", *Revue Neurologique* 126:444-449.

Tzavaras, A./Hécaen, H./Le Bras, H. 1970. "Le problème de la spécificité du déficit de la reconnaissance du visage humain lors des lésions hémisphériques unilatérales", *Neuropsychologia* 8:403-417.

Van Lawick-Goodall, J. 1968. *The Behaviour of Free-Living Chimpanzees in the Gourbe Stream Reserve.* Animal Behaviour Monographs 1. London, Baillière, Tindall and Cassell.

Vignolo, L.A. 1969. "Auditory Agnosia. A Review and Report of Recent Evidence", pp. 172-208 in: A.L. Benton (ed.) *Contributions to Clinical Neuropsychology.* Chicago, Aldine.

— 1972. "Les deux niveaux de l'agnosie", pp. 222-240 in: *Neuropsychologie de la perception visuelle... sous la direction de H. Hécaen.* Paris, Masson et Cie.

Wada, J. Clarke, R./Hamm, A. 1975. "Cerebral Hemispheric assymmetry in Humans", *Archives of Neurology* 32: 239-246.

Warren, J.M./Cornwell, P.R./Warren, H.B. 1969. "Unilateral Frontal Lesions and Learning in Rhesus Monkey", *Journal of Comparative Physiology and Psychology* 69:498-505.

Wood, C.C./Goff, W.R./Day, R.S. 1971. "Auditory Evoked Potentials During Speech Perception", *Science* 173:1248-1251.

Yin, R.K. 1970. "Face Recognition by Brain Injured Patients: a Dissociate Ability", *Neuropsychologia* 8:395-402.

Zurif, E.B./Bryden, M.P. 1969. "Familial Handedness and Left-Right Differences in Auditory and Visual Perception", *Neuropsychologia* 7:179-188.

Zurif, E.B./Sait, P.E. 1970. "The Role of Syntax in Dichotic Listening", *Neuropsychologia* 8:239-244.

DETLEV PLOOG

Similarities and Differences of Behaviour as a Function of Cerebral Evolution and Dissolution

There are those who stress the similarities between man and the remaining primates, and those who emphasize the differences; agreement with either one of these positions would place one on the wrong side of the fence. Although it became evident during the past decade that questions related to this issue cannot be asked in a dichotomous manner, it still seems to be difficult to avoid this sort of conceptual thinking. This will be called the layer cake concept. It has a neurological as well as a behavioural aspect. It conveys the idea that the phylogenctic heritage—be it brain structures or behaviours—is on the bottom of the cake and the most recent products of evolution—human language, arithmetic and reason—on the top. Consequently, reflexes, fixed action patterns, drive states, conditional responses, learned behaviour, etc., and their neuronal correlates are layered on top of each other. And on top of them, is placed the cultural development of man. However, a more detailed analysis of the brain structures as well as of the inferred functions of these structures shows that new developments in evolution are not brought about by adding new elements to old ones but by structural changes of the whole, with the new elements intimately imbedded. Hence, the evolution of the brain and of behaviour cannot be understood as an additive process of increasing faculties. This may sound rather trivial, but the consequences of this are fourfold:

— New faculties in man have their own special biology which is characteristic of *Homo sapiens;* for instance, the biology of language and speech with its special brain mechanisms and the appropriate vocal apparatus.

— The new faculties are functioning within a structural framework of older functions whose biology has been adapted to the new situation.

— Wherever we encounter behaviours in man which share common features with the behaviours of other primates we have to take into account that the observed behaviour is still part of the total structure of human behaviour.

— The result of this is that there is no pure "animal" behaviour in man which can be dissected from human behaviour, and, vice versa, there is no pure "human" behaviour which can be separated from animal behaviour or for that matter from subhuman primate behaviour.

Considering the enormous variabilities and the astounding modification of size, weight, forms, and shapes among vertebrates, the brain is by far the most conservatively developing organ of all structures. The brains of the living vertebrates represent the frozen documents of their natural evolution. This is to say, that the elements of the organization (the *Bauplan*) have been preserved over time, whereas the behaviour has not and, as the consequence of adaptive processes, may have changed drastically. The inference from animal to human behaviour can, therefore, be supported by comparative investigations into cerebral functions and correlated behaviour.

To illustrate this viewpoint, I will first report on some well-known experiments on sensorimotor behaviour in cats and man, with emphasis on visually guided behaviour and its neuronal correlates. Secondly, I will present data on vocal behaviour in subhuman primates and man, with special reference to subcortical and cortical control over the voice. Thirdly, I will close with some remarks on the dissolution of cerebral functions which may shed light on pathological behaviour in animals and man.

PLASTICITY OF CEREBRAL FUNCTIONS AND BEHAVIOUR

Held (1965, 1968) conducted experiments on visually deprived kittens, which were placed in pairs on opposite ends of a rotating swing in a vertically striped cylinder. One kitten could move around on its own legs in a little harness while the other was dragged around by the first in a gondola so that walking on its own legs was made impossible. The visual input for both was the same. The cats which moved on their own legs could recognize their stereotyped world of stripes around them and were able to avoid obstacles in a test cage, while the cats in the gondola could not recognize obstacles and bumped into them when placed in the test cage.

In a corresponding experiment two human subjects were placed in an unstructured, homogeneous visual environment. Both wore goggles that distorted perception under normal environmental conditions in such a way that vertical straight lines appeared as curved lines. One subject sitting in a rolling chair was moved around by the other. As in the cat experiment, the man in the rolling chair could not adjust himself to the distortion of the goggles while the actively moving man became rather quickly adapted to the goggles, that is, the distortion disappeared and he could orient himself quite normally.

Both experiments show that visually guided behaviour, in both cat and man, needs motor feedback in order to adapt to new environmental conditions. We might, therefore, conclude that visual behaviour in man and cat is similar in this fundamental aspect of visuo-motor feedback control, while other aspects of their visual behaviour are different (as there are similarities and differences in the anatomical features of the visual systems in cat and man).

Recently, these experiments have been extended by neurophysiologists. Again kittens were individually raised in striped cylinders, one part with horizontal, the other with vertical lines. The kittens wore large collars so that they could not see their own bodies. After some weeks of experiencing this, single neurons of the visual cortex (area 17, the area of the retina in which light stimulation causes discharge of the corresponding cortical cell) were tested for their retinal receptive fields. It could be shown that the orientation of the receptive fields corresponded to the two types of environments in which the cats were raised: the cortical neurons of the kittens raised with vertical stripes showed predominantly vertically oriented receptive fields whereas the neurons of the kittens raised with horizontal stripes showed a predominantly horizontal orientation of their receptive fields (Blakemore/Cooper, 1970). This experiment (which was conducted in various laboratories under varying experimental conditions with the same results) shows the plasticity of neuronal functions under different environmental conditions, at least up to a certain age. (Whether or not this phenomenon is related to "imprinting" is not yet sufficiently clear.) It is likely that the visual system of cat and man functions similarly in this respect and also corresponds to similar adaptive behaviour on otherwise different levels of behavioural complexity.

From this general example of sensory information processing, I want to proceed to a more specific aspect of motor behaviour, namely, vocal behaviour and human speech. It is obvious that this behaviour has a very long evolutionary history in the vertebrates and was employed very early for communication purposes. A fine example is a living fossil, the tuatara *(Sphenodon punctatum)* of New Zealand which has remained probably rather unchanged over the last 250 million years, and produces a variety of calls.

VOCAL COMMUNICATION

It is certainly not just a whim of nature that language is transmitted by sound. Therefore, I share the opinion of Philip Lieberman (with Crelin and Klatt) that the phonetic aspect of language may contribute substantially to the question of the evolution of language.

There is no doubt that language serves social communication. There is also evidence that phylogenetic steps in the evolution of social communication systems in mammals have a bearing on mechanisms that are involved in the emergence of human language. Hence, it is legitimate to study social communication and especially vocal communication in subhuman primates in order to compare their vocal signalling with the human speech mechanism and its ontogeny.

In ethology, there are very well documented examples that show parallel evolution of the apparatus which emits a set of signals and the apparatus which receives the signal and decodes it. If this were not so, it would be difficult to understand how selective pressure could shape the behaviour of the communicating individuals. To quote from an early paper of Julian Huxley (1923): "If it is of advantage to a sender for a recipient to perceive the disposition or state of the sender by a specific behaviour, that pattern is likely to be modified until it becomes conspicuous. This transformation of a behaviour into a signal is called ritualization. Animal ritualization constitutes an adaptive canalization of expressive behaviour, produced by natural selection."

From the neurological point of view a signalling system must consist of at least four parts. These are: first, the peripheral apparatus (with its anatomical properties), which generates the species-specific signals; second, its cerebral motor control system, which patterns the signal; third, a sense organ, which receives the species-specific signal (in terms of physical energy); and fourth, the cerebral decoding apparatus which transforms the signal into a message that may or may not produce a modification of behaviour on the receivers' side. How the cerebral motor and decoding systems are tied together is an open question and is probably the keystone for the understanding of communication processes in general. These four logical parts, involved in every animal communication system, form the framework for our studies of the communication system in squirrel monkeys. I will mainly dwell on the vocal system and its ontogeny.

BRAIN STIMULATION STUDIES

In 1962, in a paper on the cerebral representation of penile erection in the squirrel monkey (MacLean/Ploog) it was noted that species-specific calls, such as chirping and cackling, were elicitable from certain brain structures, *e.g.* the septum, the preoptic area, and the periventricular structures.

In subsequent brain stimulation studies, these findings were substantiated (Jürgens *et al.*, 1967), and in 1970 (Jürgens/Ploog), an extensive and systematic study on the cerebral representation of vocalization revealed,

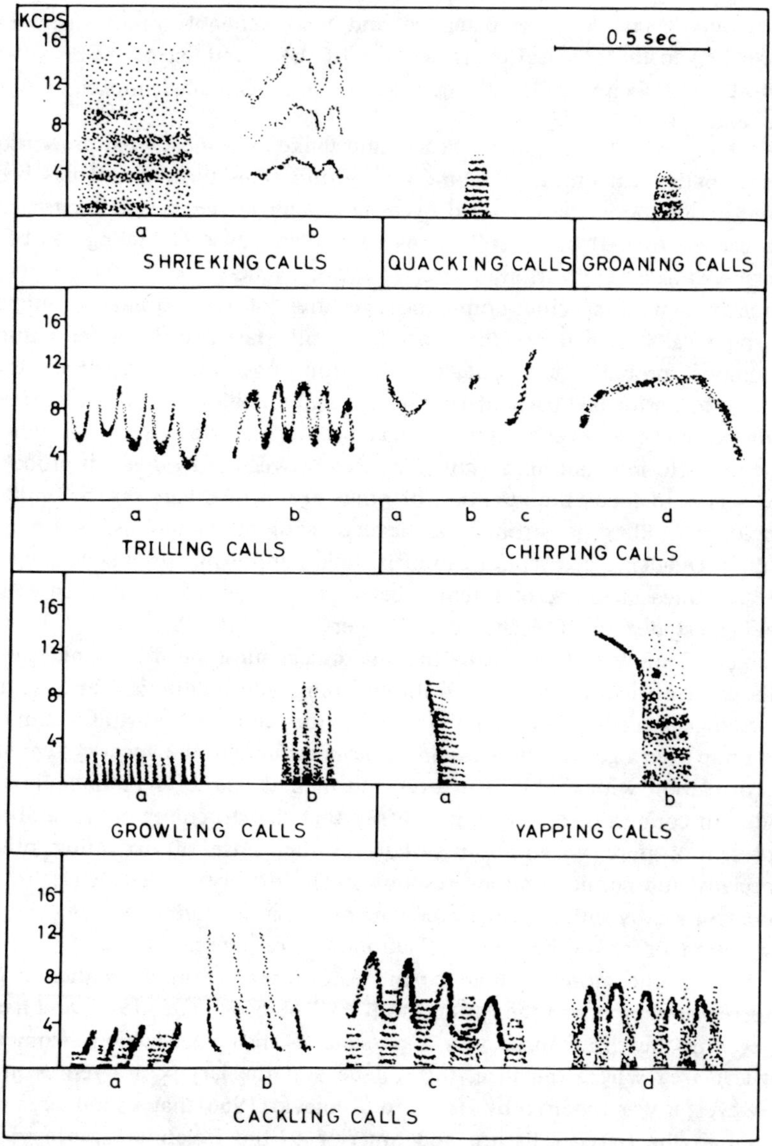

FIGURE 1. *Schematic representation of the eight fundamental vocalization types according to which all electrically elicited vocalizations were classified*

briefly stated, the following results: the whole brain (including the neocortex and excluding the cerebellum) was explored. A large variety of calls, spectrographically and otherwise indistinguishable from natural calls, were electrically elicitable from extended and well definable brain structures. According to the physical characteristics of the sound spectrograms of the reproducibly elicited calls, all vocalizations were classified into eight types (see Fig. 1).

Cackling calls are frequently heard, and make up a substantial proportion of the total vocal output of a monkey group. This type of call has been found in situations characterized by general aggressiveness and excitement; it is usually uttered by several animals simultaneously. Growling has been interpreted as an expression of directed aggressiveness, for it often occurs in connection with specific dominance gestures of the vocalizing animal. Chirping calls, which are the most frequently heard calls under natural conditions, probably aid contact and group cohesion; furthermore, they seem to be performed particularly when the attention of the animal has been attracted by new events. Trilling has been interpreted as a call whose function is to maintain minimum distances between individuals; it probably also serves to focus the attention of other group members on the emitter. Moreover, trilling is strongly associated with all situations concerning feeding. Quacking is a typical call of irritation and expresses a state ranging between uneasiness and threatening behaviour. Shrieking, finally, represents the highest degree of excitement (Winter *et al.,* 1966).

Figure 2 shows schematically the anatomical substrate from which these calls are electrically elicitable. Without going into anatomical and neurophysiological details and considerations, the cerebral substrates can be shown to give a general impression of the specificity of the vocal system and its correlation with limbic structures and their thalamic, hypothalamic and midbrain connections. This is not to say that the structures outlined above represent a pure vocalization system in the sense of excluding other, especially autonomic stimulus responses. The majority of elicited calls are accompanied by either penile erection, micturation, salivation, changes in respiration or any other type of autonomic response.

There is one more point to be made. Contrary to man, neither in the squirrel monkey nor in the rhesus monkey (Robinson *et al.,* 1967) and hardly ever in the chimpanzee was it possible to elicit vocalizations from the cortical area where the face, the tongue and the larynx are represented. However, it was reported by Hast and Milojevic (1966) that a small area just dorsal to the Sylvian fissure and anterior to the Rolandic fissure yields movements of the vocal folds. This finding has been confirmed recently in my laboratory by Jürgens (unpublished data). According to cyto-architectonic studies of Rosabal (1967), this field is homologous to Brodmann's area

44, and has also been identified in various New World and Old World monkeys and apes.

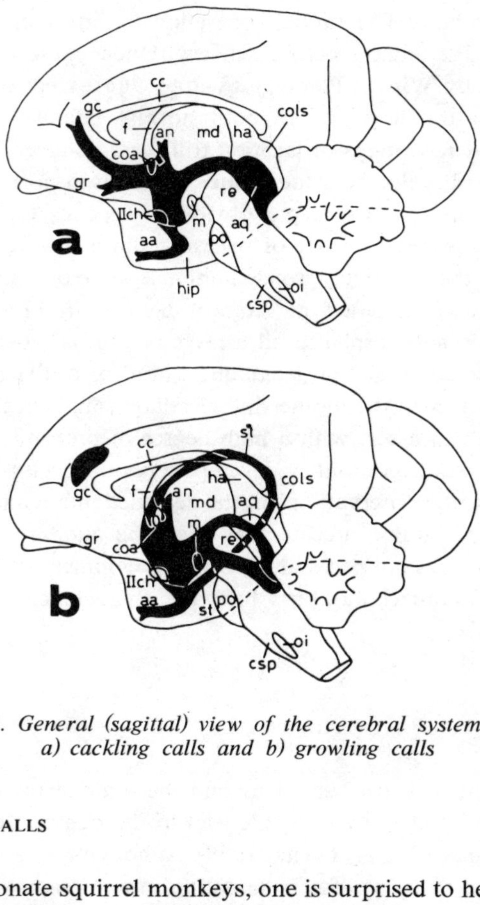

FIGURE 2. *General (sagittal) view of the cerebral system yielding a) cackling calls and b) growling calls*

ONTOGENY OF CALLS

In observing neonate squirrel monkeys, one is surprised to hear a substantial part of the vocal repertoire already during the first days of life (Ploog *et al.*, 1967). It is obvious that the vocal apparatus and its neuronal control is rather well-developed at birth. The question then arises of whether or not an auditory input of species-specific calls is needed for a full-fledged vocal signal production. If one deafens adult squirrel monkeys, no apparent difference in the quality of the vocalizations can be detected, even after several years (Talmage-Riggs *et al.*, 1972). From this, one might conclude that the squirrel monkey is not in need of his own differential acoustical feedback for maintaining the proper species-specific vocalization. However, these deafening experiments do not settle the question as to whether an early species-specific input is necessary for establishing the proper calls. Recently, the hypothesis of an innate vocal repertoire was put to a direct test by raising infants born of muted mothers in acoustic isolation (Winter *et al.,* 1973). The vocalizations were compared with neonates isolated with their intact (non-muted) mothers. The muting operation (removal of the vocal folds) was performed 1-2 months before delivery. Under these conditions all call types as defined by Winter, Ploog and Latta (1966) except twitter/trill, were recorded within the first day in both normal infants and the acoustic isolates. The first recordings of location trill were obtained at the third day. The latest call to develop was the twitter call both in normals and isolates. The first occurrence of call types obviously depends on the appropriate conditions—that is, the presence of releasing stimuli—and on maturational factors. Among the selectively isolated animals were one hand-raised infant and one that was deafened at its 5th day of life. Even under these conditions, the infants displayed all aspects of normal vocalization. By the standards of our observational techniques, mutism of the mother does not interfere seriously with the mother-infant relationship. On the basis of these experiments we can assert with a high degree of probability that the vocal repertoire of the squirrel monkey develops without species-specific auditory input. Moreover, the observations on the deafened infant and the supporting data on deafened adults indicate that even the auditory feedback of the animal's own voice is not necessary for the development of proper calls. The vocalizations of squirrel monkeys bear the characteristics of fixed action patterns.

VOLUNTARY CONTROL OVER THE VOICE

The development of vocal behaviour and the brain structures from which vocalizations are electrically elicitable lead to the question of the voluntary control of the voice (Riklan/Levita, 1969). Although monkeys (and more so, apes) are known to be skillful in imitating motor gestures, they are quite inapt at imitating vocal gestures which do not belong to their repertoire. While the vocalizations of cats, dogs and dolphins can be shaped and trained it was doubtful, until very recently, whether the voice of monkeys were conditionable at all. Leander *et al.* (1972) and Sutton *et al.* (1973) have demonstrated quite convincingly that cebus monkeys and rhesus monkeys, respectively, can control their vocal behaviour contingent upon conditional stimuli and reward. The latter authors have shown that rhesus monkeys can even control the length and the intensity of a call (the "coo"). However, so far no one has shown that a monkey has voluntary control over a variety of his natural calls nor the modification of the sound patterns itself.

Although man appears to gain almost unlimited control over his voice (along with the pronounced developmental changes of the vocal tract) during the various phases of language development, foreign accents cannot be overcome easily after puberty (Lenneberg, 1967). The foreign accent phenomenon could have contributed "to the establishment of a highly adaptive population structure involving a degree of endogamy in sub-units of the total population, in which rapid and flexible adaptive response to selective pressures would then be possible" (Hill, 1972). This hypothesis seems especially plausible if one takes into account that "dialects" in closely related but physically distinguishable monkey populations can be demonstrated (Winter, 1969).

In this section on vocal behaviour of subhuman primates and certain phonetic aspects of human speech, I wanted again to demonstrate similarities and differences of behaviour as a function of cerebral evolution. Although it was not possible to state the similarities and differences explicitly, one can sum up as follows: vocal behaviour as a means of social communication has a very old evolutionary history. The sound-producing peripheral organs (larynx, pharynx and tongue) evolve together with cerebral structures which are involved in the patterning and decoding of the vocal signals. The sound production and the patterning of calls become increasingly refined with cerebral evolution. Whereas auditory feedback of the voice is not necessary in monkeys (squirrel monkeys), this control is highly necessary for human speech development. The control of the voice is possible in monkeys as far as length and intensity of certain calls are concerned, but impossible with respect to the voluntary control over the variety of their natural calls or the modification of the sound pattern itself.

Man appears to be the only primate in whom the cerebral representation of vocalization has its cortical substrate in the motor area. It is thought that this arrangement serves in part as the neuronal basis for the voluntary control of the human voice; that is, herein lies the species-specific capability to imitate and learn phonemic combinations and sequences during ontogeny. This evolutionary course is paralleled with the evolution of the vocal tract, especially the supralaryngeal and pharyngeal area.

Besides the many similarities among all primates, there are already very important differences between subhuman and human primates on the phonetic level of sound production that can be related to structures and functions as well. Various recent experiments on newborn human babies (Eimas *et al.*, 1971, 1973) have shown that they are able to distinguish phonemic sound patterns from non-phonemic ones, which leads to the assumption of a special human decoding mechanism for certain phonemes represented universally in all human languages.

DISSOLUTION OF FUNCTION

Sometimes the cerebral organization of behaviour can be revealed by the "dissolution of function" as the great neurologist Hughling Jackson has termed certain aspects of pathological behaviour. He stated in 1884 "that there should be a loss in the order from voluntary towards automatic in what the part diseased represents". I will give but one example which is concerned with oral behaviour. Human babies in early phases of development tend to gape at objects and when brought in proximity to grasp them with their mouths. This behaviour is called oral grasping and is an extension of the nipple searching behaviour of the newborn. After a few months, when the hands come under better control, this oral behaviour disappears and cannot be elicited in conscious adult subjects. In patients with cerebral atrophy (or some other diseases), this behaviour reappears during certain phases of the disease. The patients gape at objects and, when these are brought close to their mouths, try to grasp them with pursed lips as babies do. This behaviour is stimulus bound and automatic, *i.e.,* not under voluntary control. I have observed this behaviour not only in a variety of neurological cases but also in certain stages of catatonic schizophrenia. This homologous behaviour during the ontogenic evolution and dissolution of oral functions tells us a great deal about inhibition and disinhibition of functions and about voluntary and automatic behaviour.

In my last example, I should like to show that the dissolution of function can also take place under abnormal conditions which are not related to brain pathology. We have been observing rather carefully the behaviour of newborn squirrel monkeys. The case I want to discuss here was a baby which had lost its mother after a few days of normal nursing experience. After the death of the mother, the baby was given a woollen sock which was quickly treated like its mother (comparable to the "mother surrogates" in Harlow's experiments). The baby was now trapped in an enormous conflict: when it wanted to drink it moved down to the "belly" of the sock where the nipples, the natural site for food, were supposed to be. The feeding bottle, however, was presented above the "back" of the sock and in easy reach for the baby. Nevertheless it was almost unable to drink. When it became hungry it emitted the "milk-purr" and moved towards the natural drinking place where milk was not available. It then climbed up hastily for a few sucks on the bottle and went down again. It could not learn that the source and site of food had changed. One of the consequences of this conflict was intense thumb sucking, which had never been observed in babies with nursing mothers. Here we see the dissolution of a function by a simple experimental manipulation of the stimulus conditions which are necessary for a normal functioning. In this case, the early nursing experience is probably negligible because newborn squirrel monkeys seem to find their way to the breasts right after birth, usually without any trial and error. In other cases early learning (or "mislearning") or imprinting (or "misimprinting") may play an important role in the dissolution of a function, perhaps especially in instinctive (automatic) behaviour patterns.

If we think in terms of systems and subsystems of behaviour which are functioning in an hierarchical organization, we can adopt the Jacksonian concept of the evolution and dissolution of functions on the phylogenetical as well as on the ontogenic level. In evolution, functions of subsystems tend to develop from more automatic to more voluntary control. This is sometimes reversed in the dissolution of functions. As Piatelli-Palmarini (1973) has phrased it in his synopsis prior to the conference, this concept channels into "the twofaced tendencies of local systems to preserve their autonomy and at the same time to sacrifice a fraction of their degrees of freedom in order to participate in higher levels of organization".

This leads back to my introductory remarks wherein I reasoned that there is no pure "animal" behaviour in man which can be dissected from human behaviour, and, vice versa, there is no pure "human" behaviour which can be separated from animal behaviour. This should always be taken into account when similarities and differences of behaviour in animals and man are discussed.

DISCUSSION

Dissolution of Brain Function

Crook: I wonder whether you could enlarge a bit on the generalizations which seem to be implied in the discussion of the dissolution of function. If I understood you rightly, you were saying that lower functions reemerge in impaired brains. I am thinking of what you said about the catatonic patient showing the mouth-grasping reflex. I was wondering what set of evidence for actual brain disfunction there would be in a case like that, or how far regression of this kind might be interpreted in terms of programming errors. The brain may, in a sense, be functioning correctly, but the programming of the brain may be affected, which is thought here, in the course of this regression. My last question is whether in extreme senility you get the emergence of different patterns?

Ploog: Yes, you do; even Babinski's reflex comes back. These are difficult problems, and I would have to go back to Hughling Jackson's papers and his ideas that in the case of dissolution not the whole evolution is reversely repeated. We have agreed already on the fact that the biogenetic law is not quite true, but somehow it is sometimes. What Jackson thought is that the local areas that are affected can in themselves, independently from the rest of the brain, go through that phase from more voluntary to more automatic. He said in 1884 that there should be a loss in the order from voluntary towards automatic, in terms of what the diseased part represents. This view has been challenged many times in neurology, but every once in a while it has been supported again by new evidence. It is, of course, an old-fashioned concept in the form in which it is stated, but it may be a very modern concept in the light of modern neurophysiology.

Regression

Crook: How would you relate what you have been saying to the phenomenon of regression in psychoanalytic terms?

Ploog: Well, I did not answer your question completely. You have asked whether we found any neuronal correlate in this catatonic patient, and of course we did not. But some people have found, and this again is being challenged, a widening third ventricle around the areas which I have shown to be involved in vocalization. There are various cases of speech disorders which are not due to cortical, but to thalamic damage, so, somehow, the organization of the oral motor apparatus is not dependent on the pyramidal tract, but on the striate complex. I think that under certain conditions these states of regression may well functionally imitate the same kind of impairment that would occur if these cerebral structures were disturbed. You don't need to employ psychoanalytic terminology here, the term regression and these notions are older and broader than is psychoanalysis.

Hécaen: I am not quite satisfied by your answer about the question of regression, because the suction reflex may appear very quickly with a frontal lobe tumor, for instance. We cannot say your catatonic has nothing in the brain, he may have an atrophy of the brain.

Ploog: Oh no, we did a pneumo-encephalography, and there was no atrophy.

Hécaen: Anyway, it is difficult to think that psychic relations are sufficient to cause this imbalance between two motor reactions.

Ambrose: John Crook's question about regression is very timely because it makes us face up squarely to the fact that, so far, we have been discussing homology and analogy in terms of phylogeny only. In fact the distinction may also be useful in ontogeny. In comparing regression with dissolution of behaviour, for example, it is interesting to consider whether the behaviour patterns involved are analogous or homologous.

Ploog: What would you call a regression to that stage of infancy which one can induce by hypnosis?

Ambrose: The hypnotic reappearance in adults of certain patterns of behaviour, or tendencies which normally occur in infancy but not in normal adults are due not to structural changes in the brain but to psychodynamic factors.

Ploog: Oh yes, I know that. You see, I did not want to claim that this type of oral grasping was a differential diagnostic means for diagnosing schizophrenia. I have shown this slide to demonstrate that certain early behaviours under certain conditions reappear in the sense Jackson stated, and of course Jackson never thought of regression in psychoanalytic terms, he thought in terms of straightforward neurological diseases. This should be made clear.

von Cranach: In what way do dissolution phenomena apply to learned behaviour? In a longitudinal study on the language of depressed patients, we related the depth of depression, measured by psychometrics, to the grammatical components they chose, and as far as we can tell (the evaluation of the data is not yet completed), the depth of depression seems to be directly related to the primitiveness of the language used. In deep states of depression these patients used very simple sentence structures which we would call primitive, and which can also be found in the language of children early in the process of language development. Is there any more evidence that this happens with learned behaviour?

Leyhausen: There is at least something comparable to your results, and that is the tendency of old people speaking a foreign language, however perfectly, to gradually revert to the language, especially to the pronunciation and intonation, of their youth. So I think that such behaviours learned in early youth have a similar tendency to reappear with senility.

Ambrose: A similar type of thing can also happen during extinction procedures. If an individual during the learning process comes to expect reinforcement of his behaviour under certain conditions, and you then start an extinction procedure such that reinforcement is no longer given under these conditions, you can sometimes then see the occurrence of more primitive behaviour patterns.

Ploog: I thought that my last case of the squirrel monkey and early learning was another type of dissolution of function, in that the experimental separation of naturally combined stimulus properties (source of nutriment and site of nourishment) caused a breakdown of sucking behaviour.

Homonomy-Homology of Successive Developmental States

Wickler: The methods taxonomists use in finding out whether two things which look the same actually are the same or are different, are exactly the same methods by which you can find out whether you correctly identify a reappearing character with an earlier one; that is, whether the behaviour of these senile people actually is a reappearance, or something new which just looks the same. If you do that, it is again a question of homology. If you are not interested in that, the whole question is purely academic.

von Cranach: Dr. Senn has provided us with a name for cases in which you homologize two elements of successive development.

Cain: Homonomy.

Wickler: But this term has only been used by taxonomists, and I do not think it makes much sense.

Ploog: In the case of comparisons between embryonic structures and adult structures, it was also important for comparative anatomy. I think that all secondary insessorials have the grasping reflex which is certainly in all grasping in early infancy related to grasping the nipple, and bound to the nipple-searching, head-turning movement.

de Schonen: Dr. Wickler's remark concerned the problem of interpretation of reappearance or early appearance of behaviour. If you look at the way children of four or five years manipulate articles, you will find that they use them in some situations but cannot use them in others. Part of their behaviour of saying "the" in the right place exists, but another part of it does not exist. If you want to know what behaviour appears and how early it appears, you have to specify the conditions of appearance and the meaning of the behaviour with regard to the whole behaviour of the animal; this goes for vocalization as well.

Ploog: Yes, I agree, but this is the difference between a subhuman and a human primate: the subhuman primate, or at least the squirrel monkey, does not need to learn in what situation he has to emit what vocalization. This is innate. The vocal apparatus, the whole machinery that generates the signal and, at the same time, the receiving apparatus that decodes this signal, is one and the same system. This was my point.

de Schonen: All use of the system is innate.

Ploog: In essence, yes.

von Cranach: We have other examples, like writing, in which movements have to be learned while the perception apparatus is there and can be used.

Ploog: Well, since you mention it, we know about human cases. There are cases in which reading seems to be impaired from the beginning. So this again seems to be a species-specific capacity that locally can be disturbed. These children can speak and use their language, but they cannot read.

Evolution of Language

Leyhausen: I want to add to your answer to Madame de Schonen that although the emitting of the signal and the analysis of its meaning may be innate, this does not mean that the animal cannot learn something additional about it. Animals may, for instance, use elements of their communication system to convey something, especially with intent to deceive. Bergmann recently reported the case of a gull which deliberately deceived other gulls by giving the appropriate cry announcing the approach of a skua. The skua is a big predatory gull that chases smaller gulls and forces them to drop their prey. When a skua is approaching many gulls instantly drop what they have in their bills, to escape being harassed. So this gull gave the warning cry, looked about to see which birds dropped their prey, and caught it. Tinbergen gave another similar example about a polar fox.

van Hooff: You said that in the squirrel monkey both the production and the understanding of vocalization seem to be innate, but I wonder to what extent this is the case in all primates, at least in the higher ones. As for facial expressions, there are good indications that things are more complicated in these species. Certainly the communicative behaviour shows deficits when the animals have been reared in isolation. The production of facial expressions seems indeed to be largely innate, but their proper understanding requires a certain amount of experience.

Ploog: Yes, I know about Mirsky's and his co-workers results on macaques and I am sure they are true for other primates, too. In the case of innate or unlearned recognition of certain vocalizations, I cannot be so sure as in the case of their performance. But I do know that the responding reaction to the call is appropriate even the first time. Peter Winter examined a series of deafened adult squirrel monkeys, and one baby who was deafened on its fifth day. The vocabulary of the adult animals did not deteriorate even after years, and the baby exhibited the total vocal repertoire of the species. The only change—and that is very similar to man—is a decrease in frequency of calls, and an increase in loudness.

Leyhausen: You cannot really conclude from the kind of experiment Dr. van Hooff quoted that the recognition of facial expressions is not innate. In recent years evidence has been accumulating that certain innate faculties, motor or cognitive, have maturational curves increasing to a peak when they are ready for performance. But if they are not "called upon" by that time and subsequently kept in practice, they deteriorate, and if you make the critical Caspar Hauser experiment too late, you get a negative answer. I am quite sure that in a number of such experiments the conclusions that have been drawn are not quite correct.

von Cranach: We should perhaps take into account the fact that facial expressions are normally embedded in social situations, and it is very difficult to find out which parts of this total stimulus situation have to be learned, and which parts are understood innately.

Wickler: I should like to give an example of birds which seems to come closer to what we think of as language and speech than does the uttering and understanding of innate elements. We have been studying for several years now the special behaviour of duetting in birds. We know already that the sounds that these birds can make within their limits are not innate. Of course, the limits are. They have quite a number of calls, and the use of these vocalizations depends on the situation. In the extreme case that we found in the drongo, this situation is determined by the partner. Two drongo birds give a performance of a very rapid alternating uttering of sounds. They have about twenty-eight elements they *can* use; the sequence they produce is neither fixed nor random, there is something like a grammar in what each bird says. We can prove that whichever element is uttered at a given moment *a)* depends on what element preceded it according to the bird's own grammar, and *b)* depends on the one element that the partner has given just before. There is a very close linkage between the utterances of the two. It is very difficult to find out whether there is some meaning in this. One could say that they talk, but one does not know what they are talking about. This is only to show a clear analogy, but again, the analysis of a clearly analogous system may give us some insight into what is called language in another group of animals. One may even turn to birds to better understand what we mean when we talk of language.

Chance: When you talked about the cerebral representation of vocalization in man, were you referring to the neural regions controlling the supralaryngeal region or to the actual muscular region?

Ploog: To the whole area that is modifiable in man, that is, above the larynx.

Chance: In the mouth?

Ploog: Yes, all the articulators, including the lips. It seems that the vowels *a, i* and *u* are the limiting articulations of a vowel-triangle that is language-universal.

Chance: And how does one know about the Neanderthal man?

Ploog: That is not quite easy to explain. What they did in the Haskins laboratories was to take measurements, by casts, x-rays and other means, of the so-called vowel triangle in neonate human babies, chimpanzees and of reconstructions of the supralaryngeal area and the jaws of the Neanderthal man—La Chapelle-aux-Saints it was in this case. By means of a computer program containing these measurements in terms of a formula that pictures and synthesizes the vowels you can get from these structures, comparisons were achieved which showed that the Neanderthal man, the human neonate and the chimpanzee cannot produce the vowels *a, i, u* as children and adult man can. Philip Lieberman and his co-workers are quite convinced that the Neanderthal man did not possess the neural and anatomic abilities that are necessary for human language.

Chance: You know about the work of Clifford Jolly, of course, and the importance he ascribes to the mobility of our tongue with reference to our ability to speak?

Ploog: I did not go into that for lack of time. What I think important is the genetic change over 50,000 or 100,000 or 200,000 years.

Ambrose: Is the brain area for vocalization in birds at all similar to that of man?

Ploog: No. Some birds can imitate some human words or even sentences, but, of course, they cannot decode them. A bird's brain is so completely differently organized that there is no possibility of homology with regard to the primary temporal convolution.

Ambrose: They cannot decode, but surely the capacity to imitate is a major step forward?

Ploog: Yes, and what is really amazing is that primates which are so capable of imitating gestures and many other things are hardly able to modify their voices.

Leyhausen: Are we to conclude then that the Neanderthal man did not speak? This would be most interesting in regard to cultural development; how could they have developed their bear-cult without some sort of language?

Ploog: We do not know, of course. It might well be that the Neanderthal man had developed a language which is now extinct, and which was very different from what we now consider, in terms of Chomsky's universals, the bases of modern languages. These universals, which of course refer mainly to the deep structures of language, can, in my opinion, be put much closer to the type of phylogenetic development I have been talking about, from a neurological and an ethological point of view.

Bresson: There are many behaviours during the first six weeks of a baby's life, such as the imitations of vowels and other imitations, which completely disappear by the sixth or seventh week, only to reappear integrated into other behaviours at the age of six months. What kind of significance do you ascribe to this disappearance and reappearance in another form?

Ploog: I think this is a perfect example of biological clockwork, or milestones in language development where features appear in sequential order and sequential complexity. Lenneberg reports a case where a baby in the babbling phase underwent a tracheotomy, and therefore was mute for some weeks until the tube was removed. As soon as the baby was able to voice again, it went on babbling as if it had been doing it for all these weeks. The answer to your question is that it is not uncommon, even in animal behaviour, that the performance of a behaviour may be missing for a while and then reappear in a context of behaviour that is put together by various mechanisms, or triggered by hormones.

de Schonen: Homologous?

Ploog: Yes.

Leyhausen: At what age does the baby then attain the conditions slightly different from Neanderthal man?

Ploog: That I cannot answer. But I am pretty sure that if you take the same measurements (and I believe the Haskins laboratories are working on these problems) stage by stage and compare them with the phonemic capacity of the baby you would find that the vowel triangle that is so important for the generating of our phonemes would come closer and closer in the course of descending from the larynx.

Foppa: I have been very much impressed by the demonstration of the responsiveness of different cells in the reception of sounds in the squirrel monkey. You said that this is one of the necessary conditions for the evolution of language. Another necessary condition for it is the distinctiveness of features of specific sound patterns as demonstrated by Lieberman. But I cannot understand why you don't categorize a parrot's capacity for analysing and emitting human-like speech sounds as an example of language decoding mechanisms. The parrot's behaviour requires a decoding mechanism which makes possible the differentiation of certain phonemes and sound patterns. Why do you say that this is something quite different?

Ploog: This is a good argument which I cannot fight. But I know that the cerebral organization of a parrot is so different from a primate's cortex that I am inclined to claim, although I cannot prove it, that the decoding mechanism is different. We have certain evidence on feature detection in the visual system that points in the same direction. I am inclined to believe that the meowing of a cat, the bark of a dog, the song of a bird and the call of a primate must each have in its own right a specialized detection apparatus.

Foppa: You mentioned Chomsky's assumptions of language universals and their relation to languages' so-called deep structures. McNeill takes a similar position. He assumes that the steps in children's language development are universal, at least insofar as the first language structures reflect exactly the deep structure of the respective community's language. This seems to be not absolutely wrong for English-speaking children, but it has been shown that it is not true in German. The development of syntax in the German infant's language does not start with "deep structure sentences." What I want to say is that in speaking of language universals, one has to be very cautious.

Ploog: Well, I hesitated in adopting the term universals of language in regard to deep structures, and I said that, to me, the universals of language seem to lie much more in the generation of speech and the generation of phonemes—the whole phonemic aspect which is so obviously related to morphological changes rather than to this very complex subject called deep structure. I have a faint notion about these investigations of language development in German children, but I think that at the present time we have not gotten quite far enough to really compare grammatical development in different languages. I know, for instance, that investigations in Japan confirmed Chomsky's notions on this subject. Anyway, I am not competent to talk about it, so let us forget about universals and deep structure; all I can say is that there seems to be a continuum of phylogenetical acquisitions which eventually made language development possible.

NOTE

This article first appeared in *Social Science Information,* 13 (3) 83-95.

REFERENCES

Blakemore, C./Cooper, G.F. 1970. "Development of the Brain Depends on the Visual Environment", *Nature* 228:477-478.

Eimas, P.D./Corbit, J.D. 1973. "Selective Adaptation of Linguistic Feature Detectors", *Cognitive Psychology* 4:99-109.

Eimas, P.D./Siqueland, E.R./Jusczyk, P./Vigorito, J. 1971. "Speech Perception in Infants", *Science* 171:303-306.

Hast, M.H./Milojevic, B. 1966. "The Response of the Vocal Folds to Electrical Stimulation of the Inferior Frontal Cortex of the Squirrel Monkey", *Acta Oto-laryngologica* 61:199-204.

Held, R. 1965. "Plasticity in Sensory-Motor Systems", *Scientific American* 213:84-94.

— 1968. "Dissociation of Visual Functions by Deprivation and Rearrangement", *Psychologische Forschung* 31:338-348.

Hill, J.H. 1972. "On the Evolutionary Foundations of Language", *American Anthropologist* 74:308-317.

Huxley, J.S. 1923. "Courtship Activities in the Red-Throated Diver *(Colymbus stellatus,* Pontopp) Together with a Discussion of the Evolution of Courtship in Birds", *Linnean Society of London. Journal of Zoology* 53:253-292.

Jackson, H. 1958. *Selected Writings of John Hughling Jackson,* Vol. 1-3. New York, Basic Books.

Jürgens, U. nd. *Unpublished data.*

Jürgens, U/Maurus, M./Ploog, D./Winter, P. 1967. "Vocalization in the Squirrel Monkey *(Saimiri sciureus)* Elicited by Brain Stimulation", *Experimental Brain Research* 4:114-117.

Jürgens, U./Ploog. D. 1970. "Cerebral Representation of Vocalization in the Squirrel Monkey", *Experimental Brain Research* 10:532-554.

Leander, J.D./Milan, M.A./Jasper, K.B./Heaton, K.L. 1972. "Schedule Control of the Vocal Behaviour of Cebus Monkeys", *Journal of the Experimental Analysis of Behaviour* 17:229-235.

Lenneberg, E.H. 1967. *Biological Foundations of Language.* New York, Wiley.

Lieberman, P./Crelin, E./Klatt, D.H. 1972. "Phonetic Ability and Related Anatomy of the Newborn and Adult Human, Neanderthal Man and the Chimpanzee", *American Anthropologist* 74:287-307.

MacLean, P.D./Ploog, D.P. 1962. "Cerebral Representation of Penile Erection", *Journal of Neurophysiology* 25:29-55.

Piatelli-Palmarini, M. 1973. *Manuscript.*

Ploog, D./Hopf, S./Winter, P. 1967. "Ontogenese des Verhaltens von Totenkopfaffen *(Saimiri sciureus)*", *Psychologische Forschung* 31:1-41.

Riklan, M./Levita, E. 1969. *Subcortical Correlates of Human Behavior: A Psychological Study of Thalamic and Basal Ganglia Surgery.* Baltimore, Md., Williams and Wilkins.

Robinson, B.W. 1967. "Vocalization Evoked from Forebrain in *Macaca mulatta*", *Physiology and Behaviour* 2:345-354.

Rosabal, F. 1967. "Cytoarchitecture of the Frontal Lobe of the Squirrel Monkey", *Journal of Comparative Neurology* 130:87-101.

Sutton, D./Larson, C./Taylor, E.M./Lindeman, R.C. 1973. "Vocalization in Rhesus Monkeys: Conditionability", *Brain Research* 52:225-231.

Talmage-Riggs, G./Winter, P./Ploog, D./Mayer, W. 1972. "Effect of Deafening on the Vocal Behavior of the Squirrel Monkey *(Saimiri sciureus)*", *Folia Primatologica* 17:404-420.

Winter, P. 1969. "Dialects in Squirrel Monkeys: Vocalization of the Roman Arch Type", *Folia Primatologica* 10:216-229.

Winter, P./Handley, P./Ploog, D./Schott, D. 1973. "Ontogeny of Squirrel Monkey Calls Under Normal Conditions and Under Acoustic Isolation", *Behaviour* 47:230-239.

Winter, P./Ploog, D./Latta, J. 1966. "Vocal Repertoire of the Squirrel Monkey *(Saimiri sciureus),* its Analysis and Significance", *Experimental Brain Research* 1:359-384.

JAN A.R.A.M. VAN HOOFF

The Comparison of Facial Expression in Man and Higher Primates

Man may have differentiated from a common stock that also gave rise to the African apes, perhaps 15 million years ago (Kortland, 1972). Farther back in time lie the points of divergence between this stock and the other primates. Thus man has gone his own way already a relatively long time and during this period he has adapted to a rather unprimatelike ecological niche, as member of a hunting-gathering society. This has brought him, among other things, some behavioural adaptations that make him fairly unique. Nevertheless, his phylogenetic relationship with the apes and monkeys is beyond doubt, thus making behavioural comparisons a matter of great interest.

Although primate behaviour, notably human behaviour, seems quite flexible when compared with that of other animals, especially the non-mammalian animals, it is possible, also here, to distinguish within the stream of behaviour, recurrent patterns of movement and posture that can be called the elements of behaviour. When comparisons between different species are made, these can be concerned first with the form and structure of behaviour, which are the manifestations of its organizing mechanism. One can roughly distinguish two aspects: the first is the morphology of the behaviour elements, based on mechanisms of motor coordination; the second is the reason for the occurrence of these elements and their integration in the hierarchical organization of behavioural systems (see van Hooff, 1971, 1973).

With respect to their form, many elements are very similar in many species; practically all primates *run, bite, grasp, hit,* etc., in much the same way. This is particularly true with respect to behaviour elements used in interaction with the non-social environment, which can, in a flexible manner, be integrated into functional complexes.

Greater interspecific variability in morphology may be found with respect to the elements used in social interaction, especially the communicative behaviours. It is particularly the second aspect, the causation of the behavioural elements, that is, their conditional programming (for instance, when is aggression shown, against whom and why) that may reveal interspecific (and also intraspecific) differences.

The mechanisms determining both the morphology and the conditional programming of behaviour have been specialized in order to achieve those effects that yield a relatively optimal contribution to the survival of the animal and its reproductive success. Thus the comparison can also be directed at the system of functions and the ways in which certain functions are accomplished. Similarities, based on such comparisons are, by definition, analogies.

FACIAL EXPRESSIONS AS COMMUNICATIVE BEHAVIOURS

Among the communicative behaviours, the facial-vocal displays occupy a special position. This category of displays, prominent in some mammal groups, such as the higher carnivores, the *equidae* and the higher primates, has reached its greatest differentiation in the last group. Although Darwin discovered the expressive behaviour of animals and man as a subject for comparative and evolutionary research about a century ago, little interest in these comparative aspects has been shown until about twenty years ago. Subsequently there has been a great upsurge in behavioural studies of the primates. Information on their facial-vocal displays is widespread, in more or less detailed form, in a number of both naturalistic and laboratory studies of particular species (see Altmann, 1968, for an extensive bibliography). Apart from these, there have been a few broadly comparative studies dealing especially with this subject (Andrew, 1962, 1963; Bolwig, 1964; van Hooff, 1962, 1967). Furthermore, some experimental studies have been undertaken concerning the relative efficiency of facial displays in causing affective responses in conspecifics (*e.g.* Miller, 1967). Such studies have shown that a partner can properly interpret the information conveyed by the face about some motivational states. In studies in more natural settings in which the significance of the facial-vocal displays is inferred from changes in the concurrent behaviour of both the actor and the reactor, it is difficult to prove that the changes in the reactor are indeed the result of the facial-vocal display (Cullen, 1972; van Hooff, 1972). However the elaborateness and conspicuousness of some of these displays, particularly in species with a complex social organization, and the fact that they are almost exclusively shown during social interactions has never left any doubt about their communicative function. Researchers, therefore, readily assume that the observed changes are brought about by these expressions.

Apart from a few efforts to categorize the facial-vocal displays in terms of emotions in a rather subjective and anthropomorphic manner (*e.g.* Kohts, 1937; Bolwig, 1964) investigators have tried to infer the causal and functional aspects of displays from the observable behavioural changes of the interacting animals and from the situational context.

Since behaviour is not a random occurrence of elements, any behaviour element of an animal is potentially informative. The elements appear in more or less stereotyped sequences and groupings that have been, in time, adapted to certain functions. On the basis of these contingencies in the behaviour of the sender, a receiver may predict the other's behaviour and react accordingly. A particular element has signal value if a regular relationship is observable between its occurrence in a certain context and the subsequent line of action of the animal that performs it. The probability that a certain action is going to follow can be considered the *message* of this behaviour element. If the message is actually understood, that is, if the receiver manifests a constraint in its own probabilities of behaviour in reponse to it, this means that the behaviour indeed acts as a signal; the receiver attributes *meaning* to it. If this is again advantageous to the sender in that it promotes, perhaps quite indirectly, its reproductive success, the conditions are given under which natural selection can operate to maximize the chance of understanding by *ritualizing* into a display those elements of the behaviour that have the releasing effect. The behaviour is now adapted to function as a communicative signal (see Smith, 1968). Of course, this is not to say that elements with a primary function in non-communicative spheres may not be informative and understood. In primate communication research, however, attention has been given primarily to those behaviours which by their conspicuousness and elaborateness are assumed to be displays.

THE ELEMENTARY PATTERNS AND THEIR ORIGINAL FUNCTION

The facial-vocal repertoire of the more advanced higher primates forms a system of continuously intergrading displays. This is in contrast to the situation found in some more primitive types of catarrhine and platyrrhine species. In these predominantly arboreal animals the facial displays are poorly developed, but the vocal repertoire is rich and consists of a number of discrete vocalizations (see Marler/Hamilton, 1966; Struhsaker, 1967). In the higher catarrhine primates a number of facial-vocal displays, typical combinations of relative positions and movements of the facial-vocal elements, can be distinguished. These can be designated as the modal points in the continuum of variation. The elements involved are the eyebrows, the forehead skin, the eyeballs, the eyelids, the ears, the lower jaw, the mouth-corners, the lips and the tongue. Some of these facial displays are characteristically accompanied by certain vocalizations. However, not all vocalizations (*e.g.* low intensity grunts and groans) are necessarily accompanied by elaborate facial changes other than some necessary movements in the immediate neighbourhood of the mouth (van Hooff, 1962, 1967).

In trying to understand the signal value of the movements and postures of these facial elements it is necessary to know what the original function of these elementary patterns was, before they became adapted to their signal function in the process known as *ritualization* (modifying their form and perhaps their conditional programming). These original functions most probably have to do with preparing for action, regulating sensory input and protecting vulnerable parts (Darwin, 1872; Frijda, 1956, 1965; Andrew, 1963, 1972). The classification of these functions by the authors mentioned can be summarized as follows:

— General openness and receptivity to stimuli (opening the eyes, lifting the eyebrows, dilating the pupils).
— Concentration on a given input (steadying and directing forward the face and the eyes, frowning).
— Protection and rejection (closing the eyes, frowning, flattening the ears, retracting the mouth-corners and lips, closing the glottis and coughing, eliminating nasty material from the mouth by tongue movements and head shakes).
— Aspects of the process of vocalization (there may be muscle contractions to check violent expiration or to reduce vibration of tissue and vessels of the throat, resulting in mouth-corner and lip retraction as well).
— Manifestations of the level of activation (responses evoked in connection with and anticipation of exertion and immobility).
— Intention movements of specific acts (*e.g.* opening the mouth to bite, smacking the lips to groom).

Although these expression elements may still produce their original adaptive effects, the great interspecific differences in the degree to which they may be performed in social interaction (*e.g.* extreme eyebrow raising in man, extreme eyebrow raising and ear flattening in macaques, but neither very markedly in apes; extreme frowning in man in comparison with most other primates; strong "grinning" in the chimpanzee, but not in the orang-utan or the gorilla) indicates that their programming has been extensively influenced by their incorporation into displays. Such ritualization may go so far as to obscure the original functional context. Thus laughter and crying seem to defy an explanation in such terms (Frijda, 1965, 1969). Also the strong differentiation of mouth-corners up (in smiling) and down (in sorrow) in man can only be understood as an antithetic ritualization (see Lersch, 1957).

A CATEGORIZATION OF PRIMATE FACIAL-VOCAL DISPLAYS

The following global characterization of the basic facial-vocal displays of the catarrhine primates in terms of their morphology and their causal context and message is based on van Hooff (1967). In addition, the possible function of the displays is indicated. For recent reviews see Ploog (1972) and Chevalier-Skolnikoff (1973). For a schematic representation of the expressions see Figure 1.

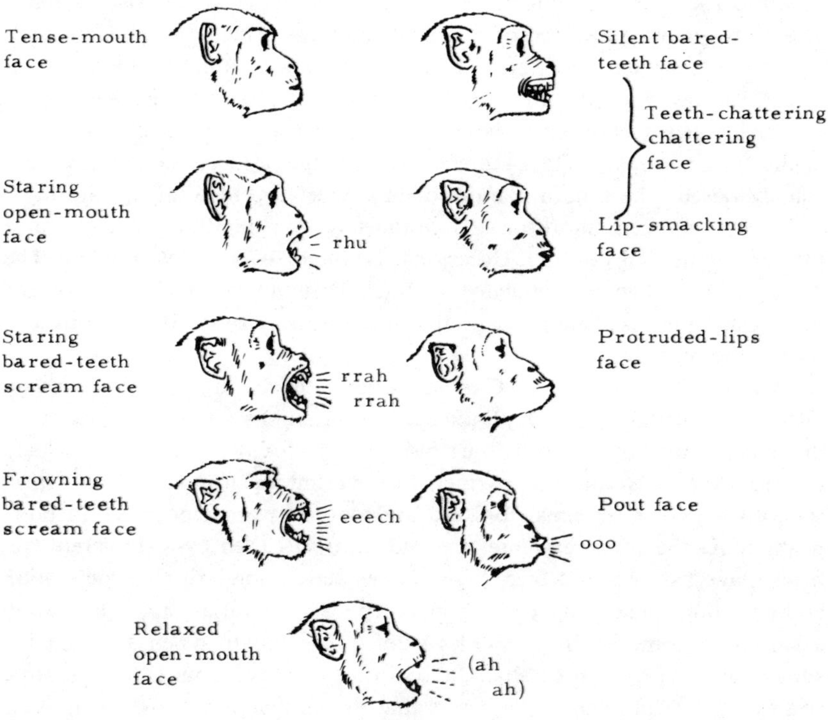

FIGURE 1. *Schematic representation of the main compound facial expression in a macaca type primate (after van Hooff, 1967).*

All facial elements can be considered to be in a neutral position in the *relaxed face,* shown when the animals are completely at ease. Gradually closing the eyes, lowering of the tonus of the muscles of face and body and sagging, slowing down of the body and eye movements *(drowsy face,* also shown when ill or sedated) leads finally to the posture of sleep. Further opening of the eyes, higher muscle tonus, brisker eye and body movements lead to the *alert face*.

These variations are clearly associated with general changes in responsiveness, occurring frequently both in and outside social contexts. The expressions that follow have more the character of displays, since these occur typically in social interactions.

Tense-mouth face. Its morphology is characterized by high overall muscle tonus; the eyelids are well parted, but perhaps also antagonistically tensed. The eyes are fixed on the partner. Determined forward tendencies in body posture and movements; lips tensed, nearly closed; in some species the mouth-corners may be pulled forward and the lower jaw thrust forward. The brows may be lowered or raised. Its message is deliberateness and self-assurance, a strong tendency to attack. It obviously functions as a strong threat, usually resulting in subjection or displacement of the partner.

Staring open-mouth face. The forward tendencies may be less strong than in the former display. The eyes are fixed on the partner; the eyebrows are usually raised. The mouth is more or less widely opened; the lips are tense and cover the teeth more or less completely; the mouth-corners are in a forward or neutral position. Occasional grunts or barks may be made. The display is indicative of a tendency to attack (biting intention mouth posture), but it can be checked more easily than the former display. It functions as a threat and warning.

Bared-teeth scream face. Characteristic is the baring of the teeth and usually the gums, and high-pitched vocalizations. The accompanying movements are comparatively variable. At one extreme we may find widely opened eyes fixed on the partner (see staring open-mouth face), shrill staccato barks or screams, forward and/or backward tendencies in body posture. At the other extreme we find (nearly) closed eyes, lowered eyebrows, averted eyes and face, high-pitched screeching, cringing, defecation and urination. In accordance with its morphological variability, it is given in a variety of contexts. In all species studied it is shown when an animal is subject to severe threat or attack, discomfort or frustration. In an agonistic context the staring version indicates that the animal is prepared to make a last stand; it functions as a defensive threat. In the other contexts the function is less clear; it may mark surrender and thus avoid attack; it may release support by the mother or allies; in the frustration context, it may make other animals give in.

Silent bared-teeth face. Via forms with low-intensity, high-pitched vocalizations like yelps and cackles, this display is morphologically related to the former. Likewise characterized by retracted lips and mouth corners, it is associated with backward tendencies in most species, the gaze varying from steady to evasive. General mobility seems to be inhibited. In all species that possess this silent bared-teeth display it is shown towards dominant animals in contexts of (possible) contest and marks a submissive attitude. In some other, more advanced species, it can, moreover, be performed by confident animals towards uncertain fellows. It obviously signals a tendency to submit and to give in, so in general its message is non-hostile intent. It may have an appeasing function; in some more advanced species, it may also have a reassuring or perhaps even an attachment reinforcing function.

Lip-smacking face. Present in a great many species, particularly in those that possess the silent bared-teeth face only as a submissive gesture. It is characterized by rapid smacking movements of the kind that also accompany grooming, however, with an exaggerated frequency and intensity. Posture and movements usually exhibit forward tendencies, but occasionally also backward tendencies, both in body and eyes. Movements are inhibited and smooth. The message is one of non-agonistic and friendliness-inviting intent; the display may accordingly have an appeasing, a reassuring or an attachment-increasing function. In some species a clear intermediate between this and the former expression exists, the *teeth-chattering face*. The *protruded lips face,* found in only a few species, might perhaps be thought of as an exaggerated version of the closed phase of the lip-smacking. It seems to convey a confident positive attitude.

The pout face. Characteristic are the funnelled lips and the plaintive sounding of drawn-out, clear calls. The corners of the mouth are pulled forward. There is low muscle tonus and a low tendency to locomote. In most species it is a typical infantile expression, given when the youngster is separated from the mother and stopping when she reestablishes contact. It conveys helplessness and is very effective in bringing the mother to "mother" it. In the chimpanzee, it is also used as a more general begging display, by adults.

Relaxed open-mouth face. In this display associated with play, particularly with social play, the mouth is widely open; the mouth corners are in a neutral or slightly retracted position and the lips mostly cover the teeth (or at least the gums) in the majority of species. The eyes (not widely open) are relaxed, but move vividly. Breathing comes often in a fast staccato manner; short inhalations and expirations alternate rapidly. In the chimpanzee it is regularly accompanied by rhythmic grunting. There is boisterous but tensionless mobility. Its mouth posture can be regarded as an intention movement of the "gnawing", which is such a characteristic element of primate play. This display seems to function as a metacommunicative signal emphasizing the non-serious character of the interaction.

CATEGORIZATION OF HUMAN FACIAL EXPRESSIONS

Let us now turn to human facial expressions and see to what extent the human repertoire is comparable with that of the other primates. The data from comparative anatomy indicate that we may expect a more diversified system of facial expressions in our own species than in our primate relatives. Although the mobility of our scalp and our ears has been reduced, the musculature of the face proper is the most differentiated amongst the primates. However, it clearly shows the same *Bauplan* (Huber, 1931).

Our everyday experience teaches us that human facial expressions play an important role in our social relations. Yet the many studies on this fascinating aspect of human behaviour show that it has not been easy to scientifically interpret its significance. Attempts to describe and explain the patterns that reflect our "passions" date back to the seventeenth and eighteenth centuries (*e.g.* La Chambre, 1648; Engel, 1785; Camper, 1792). As early as 1862 Duchenne used electrophysiological methods to study the contractions of the facial muscles and to establish their participation in emotional expression. This work greatly inspired Darwin (1872) in his descriptive studies on the subject. Later on, however, psychologists became concerned not so much with the morphology of the facial expression movements, but rather with questions relating to the judgement of emotions.

Recently the interest in the morphology of the facial expressions has caught on again. Thus Ekman *et al.* (1971) have devised a scoring procedure to measure accurately the components of facial expression. New in the field is also the ethological approach, attempted by a number of investigators (*e.g.* Grant, 1969; Blurton Jones, 1972a). These investigations testify to the great diversity of facial expression that can actually be observed in "free field" behaviour. Thus an appreciable number of subtly different mouth shapes is the result of different positions of the mouth-corners and the lips, which are acted upon by several largely independent muscles. The relative independence of the facial elements, indicated, for instance, by the existence of asymmetrical expressions, makes it certainly less easy to establish a number of basic displays. Nevertheless, the investigations indicate that, notwithstanding the greater differentiation, the form and the occurrence of the different facial elements is in accordance, *grosso modo,* with the principles outlined for non-human primates.

The extent to which the face can portray certain emotions has been the subject of the classical judgement studies. These were concerned with questions like what categories of emotions observers can distinguish, how accurately they can do this (if at all), and in terms of how many dimensions the "emotional space" can be explained. The common procedure was to have observers judge still photographs or posed behaviour in terms of more or less well defined category lists or scales.

These studies seemed to show disappointingly little agreement, to such an extent even that some investigators began to doubt whether the face was a vehicle of unambiguous emotional expression at all. In a recent critical review Ekman *et al.* (1972) have argued that many studies, especially the earlier ones, show appreciable methodological differences and short-comings. Taking these into account, they could clarify some of the contradictions and thus reveal some consistent results. Seven categories of emotion, namely 1) happiness/joy, 2) surprise, 3) fear/terror, 4) anger/rage, 5) sadness, 6) disgust/contempt and 7) interest, emerge consistently in some major studies, namely in Woodworth (1938) categories 1 to 6; Plutchik (1962); Tomkins/McCarter (1964); Osgood (1966) and Frijda (1968) categories 1 to 7. For each of these categories, patterns could be found that represent these emotions unambiguously (Tomkins/McCarter, 1964).

Comparison of these data with those on non-human primates is complicated by the fact that the latter refer to units of display and their meaning in terms of the observable changes in behaviour in actor and reactor and of the context. The human emotion categories are defined on the basis of judgements which, besides the action anticipation and the situational reference, include a subjective element, an empathy based on emotional experience on the part of the observer (see Frijda, 1969). The following comparison is therefore necessarily a somewhat partly tentative one. Recently Chevalier-Skolnikoff (1973) presented an extensive review of the literature concerning this subject and made elaborate comparisons. The reader who is in need of more detailed information, is referred to this study.

Interest. This category of emotion finds expression in a way that is comparable to the alert face described before, characterized as it is by attentive watching or listening, open, tensed eyes and a slight frown when concentrating, and surprise when the emphasis is on open receptivity (open eyes and slightly raised eyebrows).

Surprise. In its characteristic and intense form, with the wide-open eyes, the strongly raised eyebrows and the limp open mouth, it does not seem to have a ritualized socially used equivalent in the non-human primate repertoire. It is not seen even in the macaques and baboons which, as part of some other displays, show marked eyebrow raising and scalp retraction.

Anger/rage. The determined expression with frown, tense mouth and lips, protruded lower jaw and staring, slightly squeezed eyes corresponds well with the tense-mouth face. The more intense forms of rage may find their equivalents in the continuum of the staring open-mouth face and the staring bared-teeth scream face. The furious sneer with the squared lower lip is a differentiation that does not occur in the non-human primate repertoire. It is certainly not equivalent to the "horizontal figure-eight grin" of *Mandrillus,* as suggested by an illustration in Eibl-Eibesfeldt (1971); the latter is the silent bared-teeth display of this genus, a comparatively tensionless display of a confident positive nature (van Hooff, 1967). Whereas frowning seems to be the rule in human expressions in this category, we sometimes find in human "warning" a strong raising of the eyebrows, together with staring and a tense open-mouth face; this is reminiscent of the staring open-mouth face with strongly lifted eyebrows seen particularly in macaques and baboons.

Fear/terror. This emotion is expressed in ways that correspond well with the backward tending forms of the silent bared-teeth and bared-teeth scream faces.

Disgust/contempt. This is characterized by a lifting of the upper lip and by an associated wrinkling of the nose, futhermore by nearly closing the eyes, by frowning and by a tendency to turn the face up and away. It does not seem to have a ritualized, socially used equivalent in the non-homan primates, although I noticed a morphologically similar pattern in a chimpanzee peeling a citrus fruit. In origin it may, therefore, be a gesture to cut off unpleasant incoming stimuli.

Sadness/anguish. At low intensities, these emotions are expressed by the arched "sad frown" and a mouth posture in which the mouth-corners are slightly retracted and lowered and in which the lips may tend to curl outwards. The latter element of expression becomes more manifest, particularly in children, when they are whining or sulky. At higher intensities it turns into weeping and is accompanied by sobbing, crying and secretion of tears. In children its function seems clear; there it is an expression of helplessness, probably aimed at bringing to the actor comfort and support. Thus it might

A B C

FIGURE 2. *In the chimpanzee there exists an intermediate between the* bared-teeth, *display* (*portrayed is the* bared-teeeth yelp *display,* A) *and the* pout moan (C), *both of which occur in the other non-human primates as well. It is the* stretched pout whimper (B) *which bears resemblance to human crying. (After van Hooff. 1971, 1973.)*

be the functional equivalent of the pout face, reported for most non-human primates. Although the element of lip funnelling may still be seen in the outward curling of the lips in the human expression of sadness and crying, the latter differs markedly from the pout face in respect to position of the mouth-corners. These are drawn forward in the pout face, but retracted in sadness and crying. In the chimpanzee, however, we also find a modified form of the pout face, namely the *stretched pout-whimper face,* which bears a great resemblance to the human expression of weeping. Both with respect to its morphology (Fig. 2) and its place in the causal organization of behaviour (Fig. 3) it is an intermediate between the pout face and a particular form of the bared-teeth scream face of that species (van Hooff, 1971).

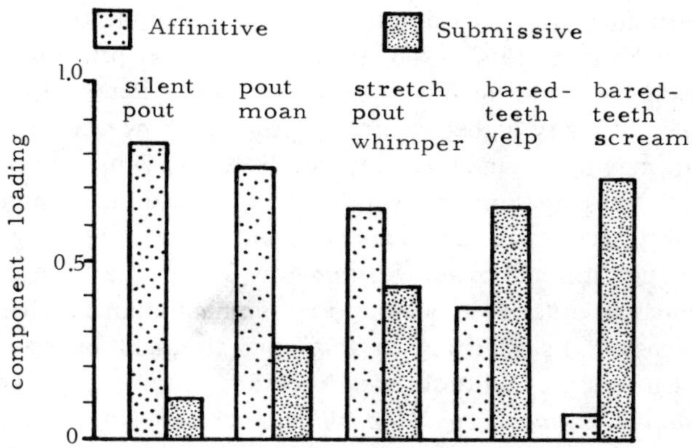

FIGURE 3. *In the chimpanzee the* stretched pout whimper *occupies an intermediate place in a range of facial-vocal displays which almost continuously vary with respect to both their morphology and their loadings on the "affinitive" and "submissive" systems, as established in a multidimensional analysis (component analysis) of the organizational structure of chimpanzee social behaviour (from van Hooff, 1971, 1973).*

Happiness/joy. The mouth is broadened, the mouth-corners going backwards and upwards in an otherwise relaxed face, and the eyes are squeezed into a "smile" as well. There may be various degrees of teeth-baring and mouth-opening. It may change into laughter with its typical staccato vocalizations and its boisterous body movements. Like crying, the expression of laughter seems to defy a simple functional explanation. Because of its association with humour, laughter has been considered by many as an exclusively human expression. Van Hooff (1972) has extensively discussed the comparative aspects of smiling and laughter and suggests that their equivalents in other primates are the silent bared-teeth face and the relaxed open-mouth face respectively. Both expressions will be discussed in more detail.

TWO INSTANCES OF MORE DETAILED COMPARISON: THE SILENT BARED-TEETH AND THE RELAXED OPEN-MOUTH DISPLAYS

The silent bared-teeth face can be considered as a ritualized low-intensity version of the vocalized bared-teeth display. The latter, which is accompanied by strong expiration or loud vocalization ("spitting", hissing or screaming), forms part of the expressive repertoire of many, even lower mammals. With respect to its original causation, it can be thought of as a response to strong aversive stimulation, serving a protective or defensive function (see Andrew, 1962, 1963, 1972). In the lower primates, the baring of the teeth is still mainly associated with comparatively strong vocalizations. In many higher catarrhine primates, however, intense baring of the teeth may appear without any vocalizations. Thus it seems that the low intensity voiceless form has become an expression in its own right. In many monkey species (*e.g.* in the guenons and in the more "original" long-tailed macaques) it is still mainly a submissive gesture, performed by subordinate animals in order to appease more dominant animals when involved in social tension. Its meaning in this context might be transcribed as "Please, let me alone, I am not going to bother you". In a few species (*e.g. Macaca maurus, Cynopithecus, Mandrillus, Theropithecus, Pan),* however, it can also be shown on the part of a confident animal towards a subordinate in a context which suggests a reassuring function ("Don't be nervous, I am not hostile") or even as a sign of positive intentions ("I am not hostile; on the contrary, I want to be friends"). Note that the same emancipation of the meaning has occurred independently in some species of New World primates (see Andrew, 1963). Together with this functional emancipation of the silent bared-teeth face, the lip-smacking face, which in the other species performs the reassuring and attachment increasing functions, seems to have lost importance. In *Pan* (the chimpanzee) it still occurs only in the direct context of grooming. In *Homo* the display is virtually absent.

Thus it appears that, at some stages in the ascending scale of the primates, there has been progressive broadening of the meaning of the element of baring the teeth. From an originally protective pattern of behaviour, in social contexts serving as a signal of submission, it has developed in some species into a reassuring and, finally, a friendly signal. At the human end of the scale, it has even replaced lip-smacking. Our human smile clearly seems to fit at the end of this range of emancipation.

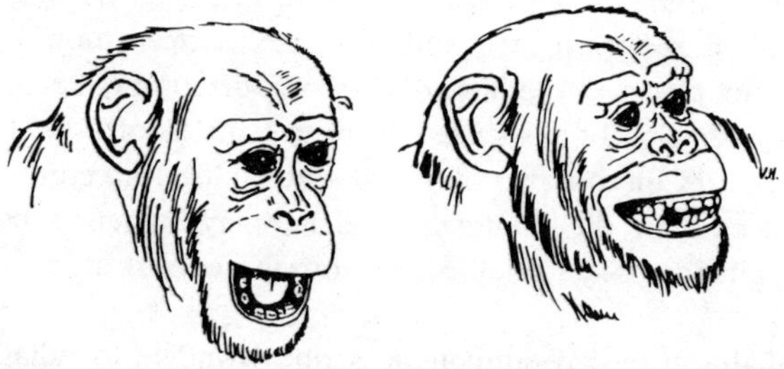

FIGURE 4. *The* relaxed open-mouth *display and the "horizontal" version of the* silent bared-teeth *display of the chimpanzee (from van Hooff, 1971, 1972).*

Besides the silent bared-teeth face, the relaxed open-mouth face is of interest. As described before, it typically accompanies the boisterous but tensionless mock-fighting that is known as social play. The fast, repetitive, shallow breathing may be the correlate of the actively sought and expected unpredictability of, on the one hand, exertion (normally associated with an emphasis on the inhalatory component) and, on the other hand, relaxation (normally associated with emphasis on the expiratory component, as in a sigh of relief).

Particularly in our nearest relative, the chimpanzee, the resemblance of the relaxed open-mouth face to our human laughter is striking. In the chimpanzee the rapid breathing is often vocalized, notably the expiratory phase of the rapidly alternating inhalations and expirations; it sounds like "ah-ah-ah". Outbursts of such rapid breathing occur often at sudden turns in the playful interactions whereas the more "silent" phases of this relaxed open-mouth display seem to be associated particularly with the anticipatory phases of the interaction. Thus there is a striking similarity in both form and causal context between this display and human laughter (see plates 1, 2, 3, and 4). A major difference in human laughter is that the inspiratory phase is almost absent. It consists largely of a series of interrupted staccato expirations, in intense laughter leaving the person gasping for breath after excessive expiration.

The comparative data suggest that our smiling is the equivalent of the non-human primate's silent bared-teeth face and our laughter, of the relaxed open-mouth face. Previously, however, many authors have regarded laughter and smiling solely as different intensities of the same motivational state (*e.g.* Lorenz, 1963). This opinion is clearly based on the fact that both can occur in close temporal association. They appear to shade into each other quite easily and to be contextually interchangeable, at least to some extent.

This strongly contrasts with what is found in the catarrhine primates. In these the silent bared-teeth face and the relaxed open-mouth face occupy relatively distinct places in the hierarchical organization of social behaviours, being hardly ever closely associated in time (van Hooff, 1971, 1973). The first display occurs in contexts of social uncertainty, greeting and the like; the second occurs in the context of social play, when a positive social relation has already been established and there is thus a relaxed social atmosphere.

In view of the proposed equivalences one wonders to what extent, if at all, it is possible to distinguish situations where specific causal and functional aspects reveal themselves, aspects that would be in agreement with such equivalences.

The only ethological data pertinent to this problem are found in a study by Blurton Jones (1972a) who analysed the temporal contingencies of behaviour items of small children and found that these can be grouped into four main motivational categories: *rough and tumble play, aggression, crying* and *social*. Laughter belonged to the play, and smiling to the social category. Another indication in this direction is found in a study by Frijda (1970). This study, in which subjects had to judge photographs on bipolar scales, led to a hierarchical model of the system of recognized emotions. The cluster of pleasant emotions contained two subclusters, one characterized by terms like "sweet", "loving", the other, by terms like "amused", "roguish". A word judgement study based on these data, indicated that smiling is judged to be more appropriate in an affinitive mood than a laugh. The latter is judged to be characteristic of a playful mood, though smiling is also judged to be rather likely in such a situation (van Hooff, 1972). These results report the supposed equivalences.

I also stated that the different forms of laughter and smiling cannot be explained in terms of a one-dimensional morphological scale of intensity. One must recognize the existence of at least two dimensions of morphological variation. On the one hand, there is a range that leads to a broad ostentatious smile with marked teeth-baring and mouth-corner retraction. This display may be used as an emphatic expression of a non-hostile, friendly attitude. On the other hand there is a range leading to the widely opened mouth "playface" with scarcely any vocalization (Blurton Jones, 1967) and the hearty wide-mouth laugh. Although this wide-mouth laugh may often alternate with broad smiles, there is also a diminutive of the laugh, in which the mouth is only slightly opened and mouth-corner retraction is slight. It is, moreover, associated with diminutives of the staccato laughter vocalizations in the form of a series of chuckles or snorts.

Thus there are some strong arguments in support of the view that there are forms of laughter and smiling that correspond with the non-human

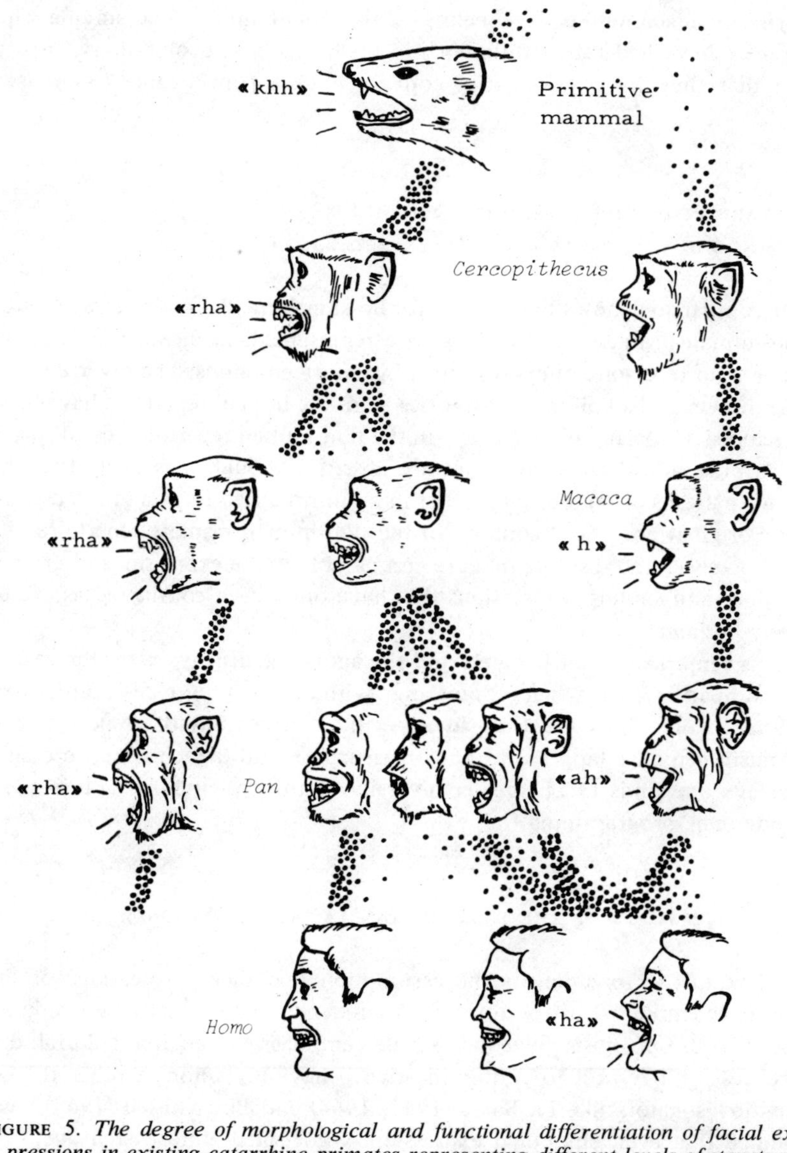

FIGURE 5. *The degree of morphological and functional differentiation of facial expressions in existing catarrhine primates representing different levels of structural and behavioural complexity suggest a model of the phylogenetic development of these displays. On the left is the speciation of the* silent bared-teeth *display and the* bared-teeth scream *display. The* sbt-*display, initially a submissive, later also a friendly response, seems to converge with the* relaxed open-mouth *display (on the right), a signal of play. N.B.: the model does not imply a linear genealogical relationship between these present-day genera. (From van Hooff, 1971, 1972.)*

primate displays under discussion. If this correspondence is the result of phylogenetic continuity (see below), this would imply that smiling and laughter have had rather different precursors in their evolutionary history and that they have secondarily converged to a considerable extent (see Fig. 5).

COMPARISON OF THE REPERTOIRES OF HUMAN AND NON-HUMAN PRIMATE FACIAL EXPRESSIONS

Our comparison shows that, except for lip-smacking, the major categories of non-human primate facial displays are represented, in their basic form, in the human repertoire of expressions of primary emotions. The reverse is not true, however. Not all main categories from the human repertoire have been developed as socially used displays in the non-human repertoire; the displays of surprise and disgust/contempt are absent, although it is likely that the motor patterns of both displays exist in an unritualized form and are used in their original non-social context. In the non-human primate repertoire, we do, however, find displays that are comparable to the expressions of crying, laughter and smiling, expressions that have often been considered as exclusively human.

It is remarkable that the alert and the agonisitic displays were the easiest to compare. Also when comparing within the group of non-human primates, one finds the least interspecific diversity. With respect to the agonism-denying (appeasing and reassuring) and attachment-increasing displays one finds much greater diversity, both in their form and in their conditional programming.

THE UNIVERSALITY OF THE BASIC HUMAN FACIAL EXPRESSIONS

Implicit in the foregoing is the assumption that these expressions of the "primary emotions" are performed and understood universally by mankind. Social anthropologists have, as a rule, emphasized the intercultural differences with respect to communicative behaviour, amongst other things; thus investigators like La Barre (1947, 1964) and Birdwhistell (1963) have claimed that even emotional expression is governed almost exclusively by cultural rules and that there are, therefore, no a priori universal symbols of emotional state. However, there are recent indications that at least the so-called primary emotions are performed and understood in practically the same way by people of otherwise very different cultures, even when they have had a minimum of previous contact outside their own culture (Eibl-

1

2

3

4

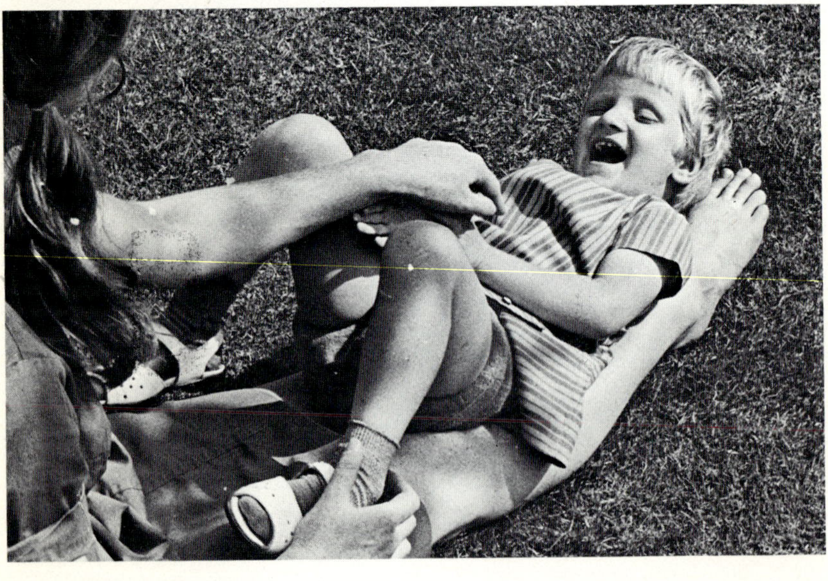

Eibesfeldt, 1972; Ekman *et al.*, 1972; Ekman, 1973). Moreover, studies of those born deaf and blind show that their basic repertoire of emotional expression develops in an approximately normal way even if the cultural influences that could be held responsible for its shaping are not in effect (Eibl-Eibesfeldt, 1973).

This constancy and uniformity indicates that the basic coordinating and programming mechanisms of these behaviours are not differentially influenced by the variable aspects of the environmental conditions under which they developed. This may be due to the development of quite specific neurophysiological "hardware" during early ontogeny (see, for instance, Ploog, 1972); it may also be due to "necessary" experiences of learning, common to all species members, later on, to the development of a universal "software" so to speak. Since similar constancy has been found in other primate species and since primates of several species have been shown to develop species-specific expressions even if reared in the social milieu of another species (*eg.* human), the differences between species with respect to their expressive repertoire must be largely genetically determined.

As more or less similar facial expressions are present in all higher primates and as differences between species tend to be larger on the whole, the less closely related these species are (see the above example of the silent bared-teeth display), it is likely, though not certain, that there is phylogenetic continuity with respect to the genetic information involved in the development of the mechanisms both of motor coordination and conditional programming of these expressive behaviours. To the extent that this is the case, one may regard the interspecific similarities in the morphology and the programming of expressive behaviours as homologous.

Andrew (1962, 1963) has argued that even in those instances where there is an appreciable differentiation with respect to the functional significance of an expressive behaviour, it is not necessary to assume great differences in the causal mechanisms. Such functional differentiation is particularly manifest in the case of the silent bared-teeth display. In many of the species where it occurs as an independent display, it functions almost exclusively as a submissive gesture of appeasement. In a number of other species it occurs also in more positive, confident versions as a gesture of reassurance and friendly greeting. The original causation of the elementary patterns involved in this display could have been a disagreeably high level of stimulus contrast (see p. 176). Its occurrence towards a dominant or in response to a threatening approach could be based on a similar causation. Andrew argued that its occurrence in some species towards a nervous fellow or towards a partner upon meeting after a temporal separation could be explained in terms of changes in the relative thresholds of the releasing stimulation of the elementary patterns constituting the display. In all cases the message is one of non-hostile intent and tolerance. Its meaning and function in the regulation of the social relation (whether it be a gesture of appeasement, reassurance, or attachment) is specified by the context. If phylogenetic continuity can be assumed to exist with respect to the information involved in the development of the expressive behaviours, it is of interest to consider the evolution of these expressions that show interspecific differentiation.

Since we will never know what the expressive behaviour of ancestral species was like, the only inferences we can make, must be based on the distribution of the similarities and the differences in such behaviours in extant species. Fortunately, the primates are unique among the mammals in that this order represents a graded series of forms which correspond approximately with the phylogenetic stages of the fossil record. Thus the order provides a model of primate phylogeny, gradually leading from original type creatures to the highest stages of structural and behavioural organization (Napier/Napier, 1967).

As has been explained above, the silent-bared teeth display offers an interesting case, because of its functional differentiation. The fact that its submissive message and appeasing function correspond most closely to the original functional context of the characteristic elementary pattern of this display (protective, avertive), and the fact that this submissive meaning is almost universally distributed among the genera of the catarrhine primates which possess the display, whereas the reassuring and friendly versions occur in a few of the higher species, has lead to the conclusion that the former is the phylogenetically older form (Andrew, 1963; van Hooff, 1967, 1972). Similar arguments have lead van Hooff (1972) to propose the hypothesis that our laugh and smile represent two expressions with a different ancestry which have converged in our species (see Fig. 5). It may be stating the obvious to stress that such phylogenetic models, based on the distribution of an expression in existing primates, should not be taken to mean that there is a linear genealogical relationship between these forms, for the existing species present the terminal stages of many diverging evolutionary lines.

DIFFERENCES IN USAGE OF FACIAL EXPRESSIONS BETWEEN THE HUMAN AND THE NON-HUMAN PRIMATES

The demonstration of species-specific displays in human facial expression with a universal basic meaning, which are comparable with those of non-human primates, does not mean that there are no cultural differences in expressive behaviour. On the contrary, there is an appreciable diversity. These differences seem to be concerned with the degree to which and the circumstances in which emotions are expressed, in other words "with the display rules that govern the management of facial behavior in particular settings" (Ekman *et al.*, 1972).

It is in this respect that the most noteworthy differences of the facial-vocal displays between non-human primates and man can be observed. Most animal displays are used particularly to regulate social relations. Their main function is to inform a fellow of the actor's social attitude towards him. Of course such an interaction may be informative also to a third animal. In the socially alert, higher primates with their complex polyadic relations, this is certainly the case. A juvenile who has got involved in an agonistic interaction may elicit help from, for instance, its mother against its opponent by showing the bared-teeth scream display towards this opponent. It is conceivable that displays even become specialized in order to release specific actions by third individuals (van Hooff/de Waal, 1975). However, it seems that primates use their facial displays only to a limited extent to communicate to fellows their attitude towards the non-social environment. In humans, however, facial expressions are used extensively to convey one's appreciation of the environment. As such they support and complement verbal behaviour. Not only do they regulate the process of verbal interaction, they also illustrate or modify the verbal information (*e.g.* Ekman/Friesen, 1969; Vine, 1970; von Cranach, 1971; Argyle, 1972). One's approval of a dish or one's disapproval of a tactless remark can be shown emphatically by a facial expression, not only at the confrontation itself, but also when reporting about it later. Maybe it is in this context that expressions like those of surprise and disgust have evolved, expressions which have not been observed, at least not in a ritualized form, in non-human primates. This emancipation of facial expressions has clearly been favoured by the same selective pressures that have also promoted the evolution of speech.

CONCLUDING REMARKS

The contributions and discussions during this symposium bear testimony to the problems involved in establishing behavioural homologies. It has been argued that similarities in behaviour between higher species, for example, human and non-human primates, can best be regarded as analogies, that is, as being caused by the independent action of similar selective pressures and thus as being due to the adaptation to similar functional requirements. This applies particularly to the non-social behaviours of the higher primates. The functional applicability of these under variable conditions brings about a rather great flexibility and thus intraspecies variability with respect to the way these behaviours are performed. The fact that the different species have to perform similar functions makes it understandable that, across species, there is a great similarity in the behaviours performed.

Interspecific differences will reveal themselves especially in the causation and integration of behaviours, as adaptations to the specific ecological conditions of the species. These are often demonstrably determined by individual experience and tradition. In such cases the idea of homology is hardly applicable, if it is not to become a trivial concept (Atz, 1970).

A species specificity that is not obscured by intraspecific variability can more easily be found in connection with social behaviours, in particular in connection with facial expressions. Here we have, so far, the only convincing example of more complex behaviours of higher primates, the species characteristic aspects of which are almost certainly genetically determined. The distribution of these behaviours over the existing species and the fact that the interspecific differences in the repertoire tend to be larger the less closely the species are related are indications that phylogenetic continuity of the genetic information involved can be assumed. Thus it may appear that facial expressions may offer the best, if not the sole example of behavioural comparisons between man and his relatives, where the concept of homology can be justifiably applied and where its application is more than a mere triviality. It is, once more, a tribute to Darwin's genial insight that he suspected this, more than a century ago.

DISCUSSION

Innervation of Facial Expression

von Cranach: May I ask Dr. Ploog and Dr. Hécaen whether they could add to this material from brain research? What could you add as to the instigation and innervation of these movements from certain brain centers?

Ploog: This is an almost complete equivalent to the modulation and innervation of the voice. In both cases, we know that the pyramidal innervation is not sufficient. If someone has a pyramidal lesion—I do not want to go into details—it results in a paralyzed face, but this paralyzed face is still able to reflect emotions. That means that the emotional expression and the neuronal pattern responsible for it are not neocortically controlled, but tied to the limbic and extrapyramidal system. The contrary is seen in cases of Parkinson's disease, where the patients can do everything voluntarily, but are no longer able to show emotional expressions; their faces look like masks. Of course the anatomical correspondent to Parkinson's disease is a complex one; I'll confine myself to the system that is called extrapyramidal to indicate that it is not under the control of the pyramidal system. Here, we have not only the very complicated innervation patterns to the many muscles that form emotional expressions, but we also know quite a bit about the anatomical structures that modulate and form this pattern.

Hécaen: I should like to add two points: patients with cortical lesions, who cannot carry out facial gestures to verbal commands or in imitation, can smile or laugh spontaneously. In aphasia, patients are generally good at imitations of facial gestures.

van Hooff: I think that, together with the material reviewed by Eibesfeldt on deaf-blind children and also from the data concerning the universality of the basic expressions in very diverse cultures, there is good evidence that we are dealing with species-specific behavioural characteristics which are largely genetically determined.

von Cranach: May I again ask whether this genetical determinedness concluded from the morphological apparatus can also be backed up by data from brain research? Could you say that similar brain processes would be responsible for facial expression changes in the phylogenetic line that has been demonstrated just now in terms of morphology?

Ploog: In psychiatry, there is an interesting phenomenon concerning certain drugs with which one can induce Parkinson's disease. One can also generate the disease in chimpanzees and macaques; we know not only where the anatomical organization for this is, but also the neurochemical background in the synapses. We know what happens there under certain drugs. So the system is rather well understood by now, and there is no doubt that it is tied in with emotional expression, at least in apes and the family of the macaques.

Physiological Correspondents of Affect

Crook: These elaborate sets of facial expressions are concerned with the communication of affect, but what exactly is affect? Its background undoubtedly lies in endocrine involvement, certainly as regards the adrenal complex, and I am wondering what is known about the endocrine background for the different categories of emotional expressions? Do we know, for example, what differences may relate to anguish, anger and so on?

Again, how about the involvement of different breathing rates in these expressions? Alternations in breathing are often used in neo-reichian therapy. We can use changes in breathing to induce changes in affectivity. For example, one slows breathing to control anxiety, and so on. Is there any systematic study of the way in which breathing patterns correlate with facial expression?

van Hooff: As far as I know, there has not been much detailed research, concerned specifically with primate breathing. Other autonomic response and physiological correlates have received more attention. We recently started investigations on the relation between heart rate and heart rate changes and social behaviour. We cannot say very much yet, but it is clear that such relations are far from simple. Thus we found both heart rate increases and decreases associated with the same expressive behaviour. It is interesting though that such changes are a very sensitive indicator of the social involvement of an animal. A collaborator of mine found momentaneous changes in animals observing interactions between fellows, while the observer did not manifest any overt behavioural change. About the role of endocrine factors in facial expressions one cannot say very much. Apart from the more general influence of such factors on behaviour, there do not seem to be any specific direct influences on particular facial expressions.

Crook: And about breathing?

van Hooff: Well, the relaxed-open-mouth display is the best example of an expression which is characterized by a specific form of breathing, namely a rather fast alternation of short inspirations and expirations. Deep inspiration has been explained as preparation to exertion ("taking a deep breath"), expiration as a correlate of relaxation ("a sigh of relief"). The rhythmic breathing in the play situation is clearly related to the ambivalence which is characteristic of play, namely the essential unexpectedness in changes between boisterous exertion and more passive expectancy.

Ambrose: I agree that the study of respiration is fundamental for the understanding of affective expression. What little research has been done on humans in this respect does give some indication that positive and negative affective expressions are associated with differences in the inhalative and exhalative components of respiration. It suggests that positive affect is associated with muscular work being done primarily on the inhalation phase, while negative affect is associated with such work on the exhalative phase. In the human infant, for example, it has been observed that the baby shows heightened inhalation when approached by the mother: this has been measured and is quite conclusive. By contrast, when crying the muscular work is done essentially on exhalation.

von Cranach: In humans, of course, the faculty of speaking is related to facial expression, emotions, and breathing. Thus it is much more difficult to analyse facial expressions, of the mouth region especially.

Leyhausen: As regards the influence of hormones and the like I should say that it is a very indirect one. The only case we have information about is the adrenalin complex and its occurrence together with excitation in getting the organism ready for quick and strong action. What kind of action this will be is not defined from the start, but depends on the situation, which may quickly change, and thus results in quick fluctuation of emotional states, which cannot be related to endocrine states point by point. If there is a consistent relationship, it must be a rather indirect one.

Ploog: I think that the adrenergic system and all the rest of the hormones or rather neurotransmitter substances are so to speak disconnected from the emotional expressions of the face in higher animals, and certainly in primates. Of course there is a certain underlying motivation but, for instance, the fact that babies can smile in their sleep and while awake and the quick changes of mood Leyhausen mentioned seem to rule out the possibility of a fixed relationship. In extreme states of emotion, let us say in the highest stages of aggression, facial expressions disappear. They have their purpose on the lower intensity levels of emotion, where they can influence the social situation. But when emotions run high, and not only in monkeys but in humans too, you don't see much. In monkeys, where one can compare the accompanying vocalization, one finds that they make an enormous noise before the actual fight starts, but when it comes to biting, they are silent and bloody.

Chance: When taking the blood sugar of mice which had been fighting for five minutes, I found that the one that had shown the maximum of submissive gestures had a normal blood sugar, while the winner had a higher one. So there is a definite distinction between the submissive aggression and assertive aggression, and this distinction presumably could be endocrinologically engendered.

Pitcairn: In Birmingham, we have been trying to correlate electro-physiological data with facial expression. Overall, it is impossible because of the complexity, variation in individuals due to various points in the situation, and so on. What we found is that in monkeys gaze direction, "looking at", is often very strongly correlated with increasing heart rate. Our second attempt was to correlate the chimpanzee's fleeing response with fleeing distance. We gave it up, though there is, without doubt, a correlation; it is too complicated and too difficult to get at.

von Cranach: I think all instances of orienting behaviour show very well that you cannot distinguish the signals of some behaviour from its other functions. Both are contained in that behaviour, especially in social situations.

Senn: There are interesting new findings in press by the physiologist Jacob Steiner in Jerusalem. He finds that facial movements can be influenced by gustatory sensations; the application of sweet, sour or bitter fluids on the tongue produce typical gusto-facial expressions; human neonates have this response, and so do anencephalic neonates; this shows that it is a reflex not dependent on cortical activities. Unfortunately, the experiments have not yet been extended to monkeys or apes, but I think this would be a good field for comparison with humans.

Leyhausen: In this discussion, we have hitherto been looking at brain structures when looking for the substrate of a behaviour. In the case of facial expressions, however, we must look at muscular differentiation as well, at the corresponding bone structure and at the allometric relationship between the elements of the face. The human upper lip, for instance, is so much shorter than the chimpanzee's that any opening of the mouth will far sooner result in a baring of the teeth. Furthermore, the differentiation of the single muscular fields in the human face is far more complex than it is in any subhuman primate, and this of course also influences the resulting expression. So what looks "on the face" very similar may be very different when considering innervation patterns; and as muscular shape cannot transmit the difference, it may well be that things which look homologous are not, and vice versa. There is something else I want to say: I think that implicitly you have made an unwarranted transfer of criteria. The relative or correlative position or occurrence of one element in regard to others must not be transferred to the functional context, as in expression and its meaning. The latter is, by definition, irrelevant to the question of homology.

van Hooff: I think you have made a good point with regard to the influences of differences of the facial skeleton. An example may be the "horizontal-figure-eight" grin of the mandrill and the "square teeth-baring" shown by humans in a fit of anger. The morphological resemblance between both species in this respect has lead to present these as an instance of homology. In this case, however, there is no evidence of phylogenetic continuity, for in none of the other primates do there seem to be morphological equivalents with respect to this conspicuous squaring of the mouth. This should make one rather reluctant to use the term homology. In this case, the resemblance may not be due to a similar muscle action pattern. The great difference in the facial skeleton between these two species may cause different action patterns to yield the same result.

von Cranach: If we consider that the decoding capacity of the receiver guides the evolution of the signal, we could assume the case that a certain signal is taken over by another muscular group; it would then be very difficult to speak of homologous signals, although the function of the thing would be the same. This leads to the conclusion that the remarkable work Dr. van Hooff has presented should be complemented by investigation of the decoding mechanism.

van Hooff: Yes, and of the actual muscle movements.

Ambrose: My question is related to Leyhausen's remarks: are there, over the primate range, any major trends in the development of facial musculature, and if so, how are they related to the progressive flattening of the face?

van Hooff: The latter part of your question I find difficult to answer. As to the first part of your question, there clearly is a developmental trend as Huber has shown in his anatomical studies many years ago. When you follow the ascending scale of primates, you find more and more differentiation of the facial musculature, especially in the mouth region. Muscle bundles split up, and it seems that they become independently innervated as well. With respect to human facial expressions we see the effect of this quite clearly. We are able to turn the corners of our mouth up or down, and this is a markedly different movement that we do not see to that extent in any other primate. As regards smiling and laughter, it would seem that these are adaptations particular to expressive behaviour, because this may be related to an antithetic development resulting in stronger differentiation of the facial expressions, indicating happy or sad moods. There is a marked progressive differentiation of facial musculature.

Laughter and Smiling: Comparison of Function

Pitcairn: My point is still related to the difficulties involved in homologizing expression. We heard a lot about the possible homologies between the primates' facial expressions and human laughter and smiling, but these would have to be based on analysis also of the functionality of expression in both groups, and the difficulty is that there is a far greater diversity in human smiling and laughter. If you watch children at play at a stage that is near rough-and-tumble play, you will also see a very wide opening of the mouth showing both sets of teeth. However, this is not accompanied by staccato laughter, but by continuous vocalization. So in a sense this goes against the strict division between playful and affinity behaviour. I wonder whether you could comment on that.

van Hooff: A distinction between behavioural systems is, of course, never a strict dichotomy. Moreover, as I explained, the expression of playfulness and friendliness is much less different in man than in most non-human primates. Nevertheless, we can distinguish different forms of laughter and smiling in man. What one would like to see, however, are more detailed analyses of the meanings of the different versions in this continuum of variation.

Pitcairn: So the functionality must still be investigated in the two groups?

van Hooff: Yes, both the *functional significance* and the *functioning* in the sense discussed before, namely the processes and the "internal programs" determining the occurrence of these behaviours. With respect to the second aspect we are back to the question of whether this has to be drawn into the comparisons on which the establishing of a possible homology is based.

Pitcairn: We could do it the other way round, and ask what having established a homology does tell us.

van Hooff: This is something really different to which we have not come round yet.

Risks in Comparisons

Frey: I should like to enlarge a bit on Dr. Leyhausen's argument. I think that some of the problems we have are caused by the fact that we don't really focus on the question which types of error we would tolerate in the decision concerning what phenomena are to be considered homologous. In making this decision for any particular behaviour, we could have two types of errors: either we could say it is homologous while actually it is not, or we could say it is not homologous while actually it is. The statistical inferences of these types of error are called type 1 error or type 2 error. I think the examples given so far about what behaviours might be homologous vary all along that range to the point where we are obliged to tolerate either type of error. As Dr. Foppa pointed out yesterday in regard to the dimensions of sameness, the situation is similar as regards the number of criteria we apply in order to come to a conclusion. If we decide that only a few criteria have to be met to permit the conclusion that there is a homologous relationship, it will be very easy to find examples of homology. But if, on the other hand, we want to meet a big number of criteria, it will become almost impossible. My question is: is it possible to focus on the problem of which type of error we are willing to risk in our statements on homology?

von Cranach: One of the aims, at least for me, in this conference is to find out what different types of risks people want to take.

Masters: If we claim a behaviour is homologous in *Homo sapiens* and any other species, and it is not, we have made what could be a very serious political error in the field of science, because we might thus discredit any inference from animal to human behaviour. So, type 1 errors are dangerous and may have high costs.

Type 2 errors, in which homologous behaviours are not called homologous, may not be serious at all, provided we can make good inferences on the analogical level. That, too, would indicate that the strategy of comparison between animal and human could very appropriately focus on analogical levels. This would have one reservation which I think is crucial: the existence of some good homologies which play some role in social signals is essential, at least to show to people like Boulding that it is important to remember that men are also animals. It seems to me that once we get some good homology, our best strategy is to avoid the danger of type 1 errors by focussing on analogical studies, particularly as it is there that we get to the problem of how systems of signalling function and the constraints that are built into the design of signals.

von Cranach: I should like to note that the morphology of our scientific conscience is obviously far more differentiated than in most people, because most people just decide whether to compare or not to compare, and most decide not to compare. That is why we are having this conference.

Differences between Species

Ploog: I want to bring up a completely different question, that of the selective value of facial expressions. If one compares New World and Old World monkeys, and one also compares New World monkeys like *Cebus* and *Saimiri,* things become rather complicated. Dr. van Hooff, could you comment on the surprising similarity between certain facial expressions of *Cebus* and macaques? They are certainly not homologous. Is it a convergent adaptation? Or think of this contrasting evolution: *Cebus* monkeys live in trees, and so do *Saimiri. Saimiri* have a very poor repertoire of facial expressions and a rich vocal one, whereas *Cebus* has a rich repertoire of facial expressions and a comparatively poor vocal repertoire.

van Hooff: Looking at the Old World primates one sees a difference with respect to the role of facial expression as in the New World primates. In the guenons, facial expressions are comparatively poorly developed, whereas there is a differentiated system of more or less discrete vocalizations. In, among others, the macaques, the baboons and the chimpanzees, vocalizations form a graded system and the repertoire of facial expressions is comparatively strongly differentiated. The greater part of the latter species live in more open habitats, where facial expressions can indeed be more effective. By contrast, a greater reliance on vocalization seems to offer an advantage in a more dense habitat. However, this is not the whole story. The two black species of Celebes, *Cynopithecus* and *Macaca maura,* the drill and the mandrill of Africa, and also the African mangabeys, which live in more dense habitats, are comparable to the other macaque and baboon species which live more in the open, with respect to the relative emphasis on visual communication. Together with your example taken from the New World primates, this indicates that explanations in terms of the general nature of the habitat are insufficient. Obviously other aspects, for instance, the complexity and differentiation of the social organization, should be taken into account as well. There is another puzzling aspect, that has to do not so much with having facial expressions or not, but rather with the functional significance of having a particular one. Thus, lip-smacking plays an important role in many species, for example, the baboons and the macaques, in establishing positive relations. One wonders why in a number of species this function is (partly) taken over by the silent bared-teeth display. In the chimpanzee, lip-smacking plays a role of only minor importance; it is only shown in the direct context of grooming. As for humans, it would be easy to say that lip-smacking would considerably interfere with speech, but this cannot be the sole explanation since it is hardly present in the chimpanzee. As I said, there are quite a number of unresolved questions about the adaptive value of having one particular expression instead of another.

With Prof. Frijda in Amsterdam I discussed on occasion the problem of whether one can regard human facial expressions as ritualized displays in the ethological sense, that is, phylogenetically modified or exaggerated in behalf of a signal function. He argued that they are not; that they are just manifestations of general states of activation, epiphenomena of intentional states, and their changes. Comparative evidence on the other hand yields a number of arguments that, indeed, our expressions are ritualized. If, for instance, frowning and lifting of the eyebrows in facial expressions were solely to be explained in terms like these, one might wonder why these movements occur in such an exaggerated form in our species and a few others, whereas these movements are only performed in an inconspicuous manner in, for instance, the chimpanzee and the gorilla. I think that the great interspecific differences in the extent to which these facial elements are used indicate that ritualization has at least begun. To go back to the platyrrhine primates, the original functions are undoubtedly there, forming easy bases for analogous development for similar expressions. There cannot be much doubt that the *Cebus'* expression and the macaque's are not homologous, but developed independently using the same preadaptations.

von Cranach: Cultural ritualization can well be studied in those cases where facial expression has been combined with certain additional movements. As an example for such an additional movement, we could use the covering of the lower part of the face by the hand, or hands. According to Ekman, this movement may have the function of hiding the actual feelings, but it has become ritualized and can be seen as a common cultural ritualized gesture, even in small children. Do primates also have the movement of covering the lower part of the face in order to hide emotions, or is this a specific human gesture?

van Hooff: I have seen primates covering their eyes, but that probably is a gesture meant to ward off unpleasant stimulus input.

von Cranach: I mean to prevent a signal being seen.

van Hooff: I have never clearly seen that; it is not an argument that the facial expression is ritualized.

von Cranach: Yes, it is a combination of two things.

Chance: I have seen three cases of raising the hand to the face; I have seen it in the macaques several times.

REFERENCES

Altmann, S.A. 1968. "Primates", pp. 466-522 in: Th. A. Sebeok (ed.) *Animal Communication, Techniques of Study and Results of Research.* Bloomington, Indiana University Press.

Andrew, R.J. 1962. "The Situations that Evoke Vocalization in Primates", *Annals of the New York Academy of Science* 102, 296-315.

— 1963. "The Origin and Evolution of the Calls and Facial Expressions of the Primates", *Behaviour* 20:1-109.

— 1972. "The Information Potentially Available in Mammal Displays", pp. 179-206 in: R.A. Hinde (ed.) *Non-verbal Communication.* Cambridge University Press.

Argyle, M. 1972. "Non-Verbal Communication in Human Interaction", pp. 243-269 in: R.A. Hinde (ed.) *Non-verbal Communication.* Cambridge University Press.

Atz, J.W. 1970. "The Application of the Idea of Homology to Behavior", pp. 53-74 in: L.R. Aronson *et al.* (eds.) *Development and Evolution of Behavior, Essays in Memory of T.C. Schneirla.* San Francisco, Freeman.

Birdwhistell, R.L. 1963. "The Kinesic Level in the Investigation of the Emotions", pp. 123-139 in: P.H. Knapp (ed.) *Expression of the Emotions.* New York, International Universities Press.

Blurton Jones, N.G. 1967. "An Ethological Study of Some Aspects of Social Behaviour of Children in Nursery Schools", pp. 347-368 in: D. Morris (ed.) *Primate Ethology.* London, Weidenfeld & Nicholson.

— 1972a. *Ethological Studies in Child Behaviour.* London, Cambridge University Press.

— 1972b. "Non-Verbal Communication in Children", pp. 271-296 in: R.A. Hinde (ed.) *Non-verbal Communication.* Cambridge University Press.

Bolwig, N. 1964. "Facial Expressions in Primates with Remarks on a Parallel Development in Certain Carnivores (A Preliminary Report on Work in Progress)", *Behaviour* 22:167-193.

Camper, P. 1792. *Redevoeringen over de Wijze om de onderscheidene Hartstogten op onze Wezens te Verbeelden.* Utrecht.

Chevalier-Skolnikoff, S. 1973. "Facial Expression of Emotion in Non-Human Primates", pp. 11-39 in: P. Ekman (ed.) *Darwin and Facial Expression.* New York, Academic Press.

Cranach, M. von. 1971. "Die nichtverbale Kommunikation im Kontext des kommunikativen Verhaltens", *Jahrbuch der Max-Planck Gesellschaft zur Förderung der Wissenschaften:* 104-148.

Cullen, J.M. 1972. "Some Principles of an Animal Communication", pp. 101-125 in: R.A. Hinde (ed.) *Non-verbal Communication.* Cambridge University Press.

Darwin, C. 1872. *The Expression of the Emotions in Man and Animals.* London, J. Murray.

Duchenne, G.B.A. 1862. *Mécanisme de la physionomie humaine.* Paris, Baillière.

Eibl-Eibesfeldt, I. 1971. *Liebe und Hass.* München, Piper.

— 1972. "Similarities and Differences Between Cultures in Expressive Movements", pp. 297-314 in: R.A. Hinde (ed.) *Non-verbal Communication.* Cambridge University Press.

— 1973. "The Expressive Behaviour of the Deaf- and Blind-Born", pp. 163-194 in: M. von Cranach/I. Vine (eds.) *Expressive Movement and Non-Verbal Communication.* London, Academic Press.

Ekman, P. 1973. "Cross Cultural Studies of Facial Expression", pp. 169-222 in: P. Ekman (ed.) *Darwin and Facial Expression.* New York, Academic Press.

Ekman, P./Friesen, W.V. 1969. "The Repertoire of Non-Verbal Behaviour: Categories, Origins, Usage and Coding", *Semiotica* 1:49-98.

Ekman, P./Friesen, W.V./Ellsworth, Ph. 1972. *Emotion in the Human Face.* New York, Pergamon Press.

Ekman, P./Friesen, W.V./Tomkins, S.S. 1971. "Facial Affect Scoring Technique (FAST), a First Validity Study", *Semiotica* 3:37-58.

Ekman, P./Sorenson, E.R./Friesen, W.V. 1969. "Pan-Cultural Elements in Facial Displays of Emotions", *Science* 164 (3875) 86-88.

Engel, J.J. 1785. *Ideen zu einer Mimik.* Berlin, auf Kosten des Verfassers. Later ed., Berlin, Mylius, 1804.

Frijda, N.H. 1956. *De Betekenis van de Gelaatsexpressie.* Amsterdam, van Oorschot.

— 1965. "Mimik und Pantomimik", pp. 352-421 in: R. Kirckhoff (ed.) *Handbuch der Psychologie* vol. 5: *Ausdruckspsychologie.* Göttingen, Hogrefe.

— 1969. "Recognition of Emotion", *Advances in Experimental Social Psychology* 4:167-223.

— 1970. "Emotion and Recognition of Emotion", pp. 241-250 in: M.B. Arnold (ed.) *Feelings and Emotions: The Loyola Symposium.* Personality and Psychopathology Series 7. New York, Academic Press.

Grant, E.C. 1969. "Human Facial Expression", *Man* 4:525-536.

Hinde, R.A. 1972. *Non-verbal Communication.* Cambridge University Press.

Hooff, J.A.R.A.M. van. 1962. "Facial Expressions in Higher Primates", *Symposia of the Zooligical Society of London* 8:97-125.

— 1967. "The Facial Displays of the Catarrhine Monkeys and Apes", pp. 7-68 in: D. Morris (ed.) *Primate Ethology.* London, Weidenfeld & Nicolson.

— 1971. *Aspecten van het Sociale Gedrag en de Communicatie bij Humane en Hogere Niet-humane Primaten.* Thesis, Utrecht.

— 1972. "A Comparative Approach to the Phylogeny of Laughter and Smiling", pp. 209-241 in: R.A. Hinde (ed.) *Non-verbal Communication.* Cambridge University Press.

— 1973. "A Structural Analysis of the Social Behaviour of a Semi-Captive Group of Chimpanzees", pp. 75-162 in: M. von Cranach/I. Vine (eds.) *Expressive Movement and Non-Verbal Communication.* London, Academic Press.

Hooff, J.A.R.A.M. van./de Waal, F.B.M. 1975. "Aspects of an Ethological Analysis of Polyadic Agonistic Interactions in a Captive Group of *Macaca fascicularis*", pp. 269-274 in: S. Kondo/M. Kawai/A. Ehara (eds.) *Contemporary Primatology,* Proceedings of the 5th International Primatology Congress, Nagoya 1974. Basel, Karger.

Huber, E. 1931. *Evolution of Facial Musculature and Facial Expression.* Oxford University Press.

Kohts, N. 1937. "La conduite du petit chimpanzé et de l'enfant de l'homme", *Journal de psychologie normale et pathologique* 34:494-531.

Kortlandt, A. 1972. *New Perspectives on Ape and Human Evolution.* Amsterdam, Stichting voor Psychobiologie.

La Barre, W. 1947. "The Cultural Basis of Emotions and Gestures", *Journal of Personality* 16:49-68.

— 1964. "Paralinguistics, Kinesics and Cultural Anthropology", pp. 191-220 in: T.A. Sebeok/A. Hayes/M.C. Bateson (eds.) *Approaches to Semiotics.* The Hague, Mouton.

La Chambre, C. de. 1648. *Les charactères des passions.* Paris, P. Rocolet.

Lersch, P. 1957. "Zur Theorie des mimischen Ausdrucks", *Zeitschrift für experimentelle und angewandte Psychologie* 4:409-419.

Lorenz, K. 1963. *Das sogenannte Böse.* Wien, Borotha-Schoeler.

Marler, P./Hamilton, W.J. 1966. *Mechanisms of Animal Behavior.* New York, Wiley.

Miller, R.E. 1967. "Experimental Approaches to the Physiological and Behavioral Concomittants of Affective Communication in Rhesus Monkeys", pp. 125-134 in: S.A. Altmann (ed.) *Social Communication Among Primates.* University of Chicago Press.

Napier, J.R./Napier, P.H. 1967. *A Handbook of Living Primates.* London/New York, Academic Press.

Osgood, C.E. 1966. "Dimensionality of the Semantic Space for Communication via Facial Expressions", *Scandinavian Journal of Psychology* 7:1-30.

Ploog, D. 1972. "Kommunikation in Affengesellschaften und deren Bedeutung für die Verständigungsweisen des Menschen", pp. 98-178 in: H.-G. Gadamer/P. Vogler (eds.) *Neue Anthropologie* vol. 2: *Biologische Anthropologie* 2. Stuttgart, Thieme.

Plutchik, R. 1962. *The Emotions: Facts, Theories and a New Model.* New York, Random House.

Smith, W.J. 1968. "Message-Meaning Analysis", pp. 44-60 in: Th. A. Sebeok (ed.) *Animal*

Communication, Techniques of Study and Results of Research. Bloomington, Indiana University Press.

Struhsaker, Th. T. 1967. "Auditory Communication Among Vervet Monkeys", pp. 281-324 in: S.A. Altmann (ed.) *Social Communication Among Primates.* University of Chicago Press.

Tomkins, S.S./McCarter, R. 1964. "What and Where Are the Primary Affects? Some Evidence for a Theory", *Perceptual and Motor Skills* 18:119-158.

Vine, I. 1970. "Communication by Facial-Visual Signals", pp. 279-354 in: J. Crook (ed.) *Social Behaviour in Birds and Mammals.* London, Academic Press.

Woodworth, R.S. 1938. *Experimental Psychology.* New York, Holt.

PAUL LEYHAUSEN

The Use of the Comparative Method in Motivational Theory

I should like to open my comments with a reference to last night's discussion. The question was raised whether we are justified in "jumping right into the lake and trying to swim", or how cautiously we should hesitate when encountering a seemingly overwhelming and hopelessly entangled jungle of intertwining variables of which we can see neither beginning nor end. Should we boldly cut out one piece we think we can tackle, go ahead with that and see where it leads us? As a strategy I personally advocate somewhat more boldness than was suggested last night. If we were over cautious or too sophisticatedly hesitant all the time, I think we still should not have found the source of the Nile; frankly, I don't care how often along my way I fall into pitfalls as long as I or someone else can help me out again so that I can start anew. I have been asked why I fastened on one particular movement among so many, such as hitting a substrate with the palm of the hand. My spontaneous answer would have been: for no particular reason; I could have taken any one of a dozen or so others. But on second thought I *can* give a reason: it is the first clear-cut striking movement to appear in the human baby, and it is quite distinct when it first emerges out of all the more or less uncoordinated movements of the forelimb.

Now, if we want to compare motivational systems, the first question is what is a motivational system about? What criteria can we rely on, what are its properties? How can we even speak of a motivational system? Nothing we can overtly observe is itself a motivation. Motivation is, at least in the first instance, a kind of fictitious concept, similar to that of energy in physics. Employing that concept allows us to give a short description of events which conveys our assumption that there is a common factor or set of factors in this sequence of events. The second question is why employ this concept of motivation in organisms at all? If I push a pencil over the table there is an external influence that moves the pencil. Why not speak of a motivation of the moving pencil? So, clearly, the cause-and-effect relationship in processes which we consider to be motivated is something more than is contained in the cut and dried, rather naive assumption of the early behaviorists that where there is a stimulus there is a reaction, and a standard reaction at that, in the sense of Kuo's (1929) contention, for instance, that every movement is a forced movement. This is certainly true of the movement of the pencil, but is it true of the movement of an organism?

The first clue is already given by the word reaction, because, in spite of the strict behaviorist theory, in *re*action there *is* action, something which the *organism* provides from its own resources even when it reacts to a stimulus. What, then, does it provide? In the first place it provides a certain state of readiness. This could be understood at first sight in the same way as the cartridge's readiness to go off at the pull of the trigger. There is a latent power waiting to go off when a certain external event happens. Let us leave it at that for the moment. The next question is, what stimulus? And we immediately realize that the stimulus is not something *per se,* the stimulus is not exclusively defined by the physical event happening in the environment of the organism, but rather the organism, by virtue of its state of readiness, helps this physical event to *become* a stimulus. For instance, van Hooff's studies of the evolution of facial expressions in primates demonstrate quite clearly how the motivational state of the receiver animal contributes towards the physical event becoming a stimulus. Otherwise the changes in meaning which homologous signals undergo, not only in the course of the evolution of closely related species but also in the same animal according to its changing receptivity, would be quite inexplicable.

The first to outline this principle was Jakob von Uexküll (1909). According to him, the animal "subject" and the stimulus "object" are tied together into a closed "functional circle". The object carries a sign or signal (*"Merkmal"* = something noticeable or noteworthy). The sign, however, is noticeable not so much for its objective properties as for what the subject "reads" into it. There must be, von Uexküll claims, some process in the subject which creates a state of readiness or expectancy which makes the sensory organs and brain centres perceive a given object or situation differently at different stages of that process and therefore at different times; it will let them disregard one kind of object and fasten onto another with preference. The readiness to react to a particular object in a specific way will cease temporarily after the performance: by reacting to an object the animal "stamps" it with a "consummation mark", which extinguishes the signal (*"Das Wirkmal löscht das Merkmal aus")*. According to von Uexküll the object, as far as this specific reactivity is concerned, literally ceases to exist in the environment of the animal for some period of time. Thus the functional circle itself is almost entirely the doing of the subject. One need not carry the argument to such Uexküllian extremes in order to realize that all this is indeed very different from the pencil, since at any given time *its* movement is absolutely determined by the current external influence. All the same, we still could try and employ the relational symbolism of stimulus and response (S→R); there is something appearing to the perceptive apparatus of the animal and the animal *re*acts to it. However, at times no stimulus, no object may appear, it is not "available", but the internal state of the subject may change. Then, I think, we come to a point when the explanatory power of S→R breaks down, when its symbolism is no longer in keeping with actual events.

When a specific readiness has built up to a certain degree without an appropriate stimulus appearing, the animal becomes restless or goes and looks for it; under certain circumstances, given a certain level of evolution, it is even able to arrange things so as to bring about an adequate stimulus situation. We must then conclude that in the system of the subject itself there is some more or less closed circuit of events which, of its own accord and following its own course and regulating mechanisms, provides the subject with a kind of initiative. And this capacity of the subject to exhibit initiative is predominantly what we have in mind when speaking of its motivation.

Still, we normally shall not see any particular kind of activity or any special sequence of actions out of the subject's repertoire except in the presence of an adequate stimulus object or situation. Therefore, to initiate motion there are usually two requirements to be fulfilled: the presence of something which represents or which the subject may take to represent a signal, *and* the internal readiness and capacity in the subject to perform an appropriate response. Hence it follows that normally motivation equals readiness *plus* motive. To use ethological terms, there is a specific readiness, a specific tendency to perform certain activities on the one hand, and on the other a specific situation in which these can be performed most adequately, most satisfactorily, plus a stimulus object within this situation, plus experiences with this situation, the object *and* the kind of end thus reached (a "goal"), which all together form motives. Thus, motive, in my definition here, is everything releasing and supporting a behaviour sequence. Ethology, however, has always emphasized that it regards the system of specific readinesses, as well as the individual readiness, not as constants throughout. Furthermore, the system is structured. A readiness to do something is not an overall or general "drive" but has already been parcelled out to specific kinds of activities. The number of selective advantages which this obviously has I have compared to the budgeting of public institutions (Leyhausen 1965a), which also has parcelling functions; thus an official can no more spend all the available money on a single task than the animal can expend all "drive" in one effort. If they could, there would be nothing left for other necessities.

From these facts Lorenz derived his concept of reaction-specific energy, which in 1940 von Holst boldly changed into "action-specific substance". From various properties of the changes occurring in the readinesses and their mutual relationships he was convinced that a great variety of specific eliciting substances probably exists in the brain for which the different functional neuronal networks have differential thresholds. I cannot go into this here, but there is a lot of circumstantial evidence from neurology and other fields now which points in that direction (Leyhausen, 1965b).

Now, how can we use this general idea of a dualistic motivational system to compare the actual behavioural sequences or tendencies performing their species-specific tasks in related or widely differing types of animals? How can we find similarities not only between the results of these activities but also in their structural and temporal sequential order? When we find "congruities", an apparent "sameness", in two different species, for instance, is their behaviour derived from the same source, and how can we establish that it is the same? What tools do we have to investigate this question and answer it to our own satisfaction, that is, with a reasonable degree of certainty?

There are two possible ways to do this, two starting points. The first is to use an organism which, at least at first sight, looks utterly simple, as simple, for example, as the tick described by von Uexküll looks to the uninitiated. After it has reached sexual maturity and been fertilized, the tick practically reacts in only two ways: First, positive phototaxis and negative geotaxis make it climb to the top of a blade of grass or something similar, where it "freezes" and waits until the blade is shaken in such a way that it signals with a fair degree of probability that an animal has touched it. The tick then drops, and if it hits a surface with a temperature of 37° – 38 °C and smelling of butyric acid it stings and starts sucking. In all nature there is hardly any object other than the skin of a mammal which meets these two specifications. Of course, if we want to make inferences to the human situation this example is really too simple and does not provide us with much material for further comparison.

The second way is to do just the opposite: choose a rather complex situation and do exactly what I suggested at the begining: shut your eyes, pick out one part of the tangle and try to sort it out. I confess that this was the actual procedure I started with when investigating a rather complex system, namely the predatory system of a highly evolved carnivore, the cat. There was something which helped me along very much in this enterprise and that was that at the time I very naively believed in authority. A great deal of analysis had already been done on instincts in lower vertebrates.

Baerends (1941) and Tinbergen (1950) had evolved schemes of the organization of such instinctual sequences. Baerends spoke of a "hierarchy of moods" because he found out that the "mood" of a digger wasp would change when it met some particular stimulus situation without the wasp actually having to *do* anything; it had only to perceive something and need not engage in *any overt motor reaction* to undergo motivational change. This was a very solid discovery which was not sufficiently appreciated at the time and which was, I must say rather unfortunately, completely overlooked by Tinbergen in his analysis of the stickleback's reproductive behaviour system; thus, out of Baerends' hierarchy of moods Tinbergen made a sequence of internal states of readiness determining a "hierarchy of instincts", and in order to put this into an operational concept he understood it as a temporal sequence of motor patterns, successively organized on ever lower levels of complexity. In so doing he forgot something very crucial to the whole problem which Lorenz (1939) has said many years ago, namely that the essential *organizing* factor in such complex chains of activities or sequences is not the more generalized behaviour at the beginning of the sequence, but the most specialized at the end of it. But in confusing temporal sequence with hierarchical organization, Tinbergen, I am convinced, missed this organizing capacity and function of the consummatory act. Its dominating influence on all the appetitive behaviour preceding it disappeared from view. If we accept that the sequential order of events is also the order of the hierarchy, then the top ranking item in the hierarchy must be the more generalized searching behaviour at the beginning, and the consummatory act would be the most specialized and least significant with respect to the organization of the whole pattern.

Tinbergen's concept of the "hierarchy of instincts" was taken over by more or less all ethologists at the time, including myself. Hence, of course, I tried to fit my findings into this framework when I started to analyse prey-catching behaviour in cats. For some time this was quite "successful", too: when investigating a complex system, one has at first an almost natural tendency to see the overall picture rather than the variable details, and of course such a temporal sequence as postulated by Tinbergen does exist in "typical" cases. So that was how I described it, set it down on paper and published it, and I don't really know what it was which first made me realize that something was fundamentally wrong with it. I think it was a very "unscientific" liking for the cats themselves and my "stamp collector's mentality which made me collect new species. At that time I thought I knew all about cats and was planning to move on to another animal group, the viverridae, when I saw a pair of Asian golden cats—at the time a great rarity—at an animal dealer's and, on the spur of the moment, bought them. Within a fortnight of coming to my laboratory they had convinced me that I knew absolutely nothing about cats yet, that everything I thought I had seen before in predatory behaviour as fitting a strict one-way organization was wrong. The new results proved to be inconsistent with Tinbergen's main principle. But what exactly does wrong mean? Some sequential order of activities certainly exists—as a rule—as I can easily demonstrate with films. But what was wrong was the idea that this one-way system of sequential order should be fixed once and for all. There are many exceptions which need explaining. Invariably it is the exceptions which hold the key to the whole problem. Under certain conditions, which I cannot go into now, the various parts of the whole sequence started, so to speak, to lead an independent life of their own. They had a tendency to appear out of context and with no apparent goal or survival value. The survival value of the whole sequence is, of course, that the animal catches its prey and eats it. However, the animals would also catch mice and rats when satiated, as many of them, in fact, as hungry cats do. Thus hunger was ruled out as a necessary prerequisite for the performance of prey-catching. Under certain conditions the individual movements of watching, stalking, approaching, catching, throwing about, pushing with a paw, and carrying around run off out of context; sometimes an animal will even pick up a mouse, carry it to some sort of crevice, push it in and then try to hook it out again. This "angling" movement, hooking things out of holes or burrows, is very definitely an innate element in all of the cats, but it has developed more strongly in some species than in others, depending on their way of life. But when one serval, for instance, had got the mouse out, it still did not kill it but pushed it in again. Some animals play like this for quite a time, even when they are hungry. In such instances a sudden switch in the animal's whole attitude may occur. Its expression changes from "playful" to "serious business" from one second to the next, and it will snatch up prey it had seemed unable to cope with only a moment ago.

Thus a whole sequence, which in the course of evolution has apparently been designed or selected for to provide this animal with its food, begins in a way to "break down" into its smaller constituents, which start living an independent life of their own. This means that the animal now shows appetence towards performing these particular sub-activities; each can, given certain conditions, become a consummatory act just as well as any of the others. Moreover, it can be shown that some of these activities play a necessary role in other systems—functional circles (von Uexküll), major instincts (Tinbergen)—without losing their identity. Thus, eventually, the elements are defined by their own characteristics in time and space, but no longer by the function which doubtless provided the major part of the selection pressures forcing them to be evolved, just as Lorenz claimed a long time ago. They have become independent "tools" of the animal, which it can use in various contexts and which can be combined in various ways to form what we call motivational systems. If you were to compare various motivational systems in animals, you would immediately see that the same principle may prevail in functional circles other than prey-catching. This is not necessarily so, however, and we know for certain that in lower vertebrates if often is not. The fighting system of the little cichlid fish *Astatotilapia* (Seitz, 1940) is organized in a fairly rigid system, that is, in a sequential hierarchy as postulated by Tinbergen. If the fish breaks off at some point, it has to start all over again from the beginning. Here, then, the single elements do not lead an independent life. In such cases we are still justified in ascribing one mood, one readiness, one appetence or one action-specific energy to the whole sequence, with different thresholds to each element, as Lorenz has shown in his revised model of 1950. But in higher vertebrates the system seems to develop more and more differentiation and dissociation. The hypothesis now is that since mammals, and especially the higher mammals, are all of monophyletic origin, their motivational systems are likely to be structured on the *same functional principles*. That does *not* mean that the single elements in the system are identical in all species, or that each species must possess a homologon of every element, but only that the functional principle on which the system is based seems to be the same in all higher mammals, just as the functional principles and the gross structural elements of the circulatory system are basically the same throughout the whole order of mammals. Of course, in each species of mammal there are traits it acquired after its departure from common ancestral stock shared with other species, very specific developments in its motivational system as a whole as well as in some of its elements. But the general working principle, the general structure, is something that in a way characterizes *the* mammal in the same way as the circulatory system does.

Now, when we want to compare motivational systems from species to species, the first thing we do, of course, is select a similarly structured and in this sense "same" general function, a "functional circle", as our object. After investigating this in every conceivable detail within a closely related group, say a genus or family, we move on to the next group of animals most closely related to the one already studied but on a somewhat lower level of evolution. For cats this would be the family of viverridae: mongooses, civets and the like. This family offers a great advantage to the investigator of motivational systems because it contains a large number of contemporary species. These represent, so to speak, very different levels of evolution while on each level they seem to have filled all the available ecological niches. Thus in this group it is quite possible to study the interference or the interrelationship between evolutionary levels and ecological conditions. With these animals we are lucky because, though some are omnivorous or even strict vegetarians, most of them are carnivores and so we can continue with the analysis of the same functional system, namely prey-catching. This, however, is not necessarily differentiated to the same degree in all species. First, then, we have to take stock of what is there. The differentiated sequence which we observe in the cats—for example, running up and stalking, lying in wait, crouching, and the final jump—then appears to have originated from a very simple pattern which basically consists only of just running up and biting, and biting repeatedly. There is at first no precise aim in this biting. The first factor which tends to direct the bite to a particular part of the prey's body seems to emerge as a kind of by-product: a small moving object first stimulates the primitive predator to run towards it, and the predator's zeal makes for a tendency to bite at the front end of it. Hence the predator usually hits the head or neck of the prey animal. Hitting these vulnerable parts right away ensures more efficient, quicker killing and thus reduces any risk the predator may be running in overcoming the defences of the prey. This provides the selective advantage which causes the precise ways of aiming at head or neck, typical of the advanced herpestoid predator, to evolve out of this tendency. Another killing method of mammalian predators can be traced back to predatory reptiles, especially some lizards, but not crocodiles. The crocodile's way of working on a big lump of food—dead or alive—is partly due to the fact that its teeth are not fit to cut up something too big to swallow whole. A crocodile bites into its prey and twists or rotates its body until the chunk grasped comes loose. Other saurians grasp with their paws and bite and pull their heads sideways in a sawing motion. This sideways tearing is still present in cats as a movement to loosen pieces from bones, etc. In all mammals it has developed into various movements of sideways shaking, and resulted eventually in what we call a death shake, a way of killing different from the neck bite, which developed by parallel evolution in five different orders, namely in the larger didelphids and dasyurids (which are not closely related), in some insectivores, in a number of carnivores and in a pinniped, the sea-leopard. It is plain, then, that abilities like these can evolve separately and independently in widely different taxa.

The use of such comparisons for our topic is that it opens up, for example, a new approach to the often discussed idea that pre-humans, on their way to becoming humans, became predators. You know that a good deal of reasoning, or rather speculation, has been based on this hypothesis. Ardrey (1967) and others before him have argued that because early man started out as a predator he became more aggressive and more violent towards his conspecifics. From what we have seen there is absolutely no foundation for this assumption. It is rather the other way around. Lower primates which are carnivorous, like lemurs, lorisids, marmosets and the like, certainly still possess some specialized predatory movements similar to those of insectivores and other predatory animals. But these movements seem to get lost on the way in the evolution of higher primates, and as far as we know no higher primate has any specialized kind of behaviour, of motor coordination, which we could call predatory. What they do when catching things and eating them is readapt for that purpose movements which originated from completely different functional contexts. They grasp prey as they would pick fruit, and kill it by hitting its head against a stone like a coconut, or just tear it apart. For this reason I think that the similarities which man's behaviour undoubtedly shows in some respects to that of other predatory mammals, especially those organized in social groups, are pure convergencies, analogies not at all based on homology. Man, like all higher primates, certainly has no specialized prey-catching and killing instincts like true predators have, and therefore I like to call him a hunter, not a predator. This, of course, is of great significance with respect to theories like those of Ardrey.

DISCUSSION

Strutural and Functional Comparison

von Cranach: What strikes me is that we are moving from structural comparison to functional comparison.

Leyhausen: No, we are not. I am using functional comparison in order to find elements which I then use for structural comparisons, and only through these can I find out what is structurally alike or homologous, and what is not.

Cain: I fear I may have started this possible confusion when I asked in my paper, "what is a structure?" When is it complex enough to be recognized as the same, as structure in the sense that it may be composed of elements which are morphological, or behavioural, or physiological, or in fact anything else? I take it that Professor Leyhausen was using the word structure here for an ordered sequence of behaviours.

Leyhausen: A piece of behaviour ordered in space and time.

Cain: This interests me enormously, because I asked my question "what is structure?" in this sense, and we had a sequence of empirical attemps to answer it. I think this particular case represents a most important empirical attempt.

Leyhausen: To establish this, we use filming in the way a histologist uses a microtome. The single frames of the film are like slices out of a time sequence, and in analysing these we can identify the whole sequence as well as its elements; we can see variations and look into their causes. What I would like to add is that the cat is probably the best investigated animal as regards brain stimulation, and that we have a very close collaboration in this respect with John Flynn at Yale. By hypothalamic stimulation we get only the innate elements, never the learned ones which may in the course of life supersede the innate ones to a large extent. That way we can localize the mediating structures.

Cain: Behaviour and anatomy, in fact, are here so close that in some circumstances one cannot really talk separately of the two, because one *is* the other and therefore they are not independent evidence. It is always difficult to start a specialized use of a word, for obvious reasons, but would it be possible to reserve structures for ordered sequences, and when we are referring to anatomical structures to say anatomy or morphology?

Leyhausen: In a metaphorical way we refer to it as the anatomy of behaviour.

Cain: Please don't.

Leyhausen: We have borrowed our methods from comparative anatomy. Besides, morphological structures are, in principle, also processes. They are only so slow as to beat our immediate perceptual faculties.

Levels of Arousal

Chance: One of the major efforts one has to make mentally is to get rid of the paradigms, of the schemes irrelevant for the study of behaviour. But the question I want to ask really refers to the sequences or rather the elements of behaviour which could become a sequence; for instance, when exploration in primates leads to play with another animal, and in turn to clinging, which again may lead to exploration and then to clinging. Bill Mason relates these cycle-activities by what he calls changes in the level of arousal. I think this term is always terribly difficult to identify, but in a certain sense there is a relationship, because when play gets more and more active, and changes its elements faster and faster, clinging occurs. This tells you, as you go through the levels of play, what will happen if you give the animal a toy to which it will also cling at the higher intensity levels. Here I am asking you whether you can relate the elements back into some sort of sequence. To what extent is the sequence, the motivation sequence, dependent on sufficient levels of stimulus contrast? And as to this level of arousal concept, do you get the use of certain components at certain levels of activity? (Is this singling out of the elements of the separate pairs of different elements in cat behaviour due to the fact that they are not in the wild any more?)

Leyhausen: One should not confuse the level of arousal with the specific readiness, or what Lorenz called the *Aktualspiegel* or actual current level of readiness, because this is something separate. Of course the arousal, or general tonicity, puts the motor system of the animal into the state of being ready for use in action, and it does influence the intensity and also the duration of what may follow, but not the specific activity.

Chance: But doesn't it also influence the specific activity to some extent?

Leyhausen: In a situation where the overall balance of the animal's various propensities provides the appetence for some activity, the level of arousal will determine whether this activity does or does not appear. But which activity will appear depends on the specific state of readiness and the specific balance of these different specific states of readiness. There is no direct relationship—it is important to understand that. And the answer to your question is that we cannot relate different playful activities and the appearance of separate units out of whole sequences to a given level of arousal.

Chance: I believe it may be true, but I do not see how.

Leyhausen: For the animal in the wild all this has its survival value, including the independence of the single elements. We know from Schaller's investigations, and from my own, that free-ranging predators, especially those living on small prey, very often perform their hunting and stalking activities in vain; therefore the appetence to perform these elements must be very high in spite of frequent failure of the whole sequence. Besides, in playful use of these elements (which is also present in wild animals) the animals learn a lot, not only about this particular activity but also about their environment and its manipulation. So the whole system makes for greater flexibility in the directions in which learning interest can be aroused, and this of course is essential for the integration of learning and innate substructure.

Animal and Human Motivational Systems

von Cranach: I think that in this discussion we should certainly deal with the general question of whether motivational systems in man are fundamentally different. I feel that in this case I am speaking for thousands of psychologists who are convinced on good evidence that the cognitive component of human motivation makes it something different, which has the consequence that we conceive of human motivation as analogous to animal motivation. If we look at a human motivational system like achievement motivation (which is very well described), we find that looked at as a whole it has features like the motivational systems you just described. If we proceed to look at details, however, we find that it is a composite of emotions and cognitions behaving like a dynamic system in which the physiological, let us say, sources of energy play a rather minor role, as there is always enough energy for performance present in the human organism. All these details lead to the question of whether the aforesaid similarity is a superficial one after all.

Leyhausen: The similarity is not superficial, but fundamental; the superstructure in human motivational systems is of course specifically human and cannot be derived directly from systems of the kind I described. The source of disagreement between psychologists and ethologists seems to stem from the fact that psychologists are mostly concerned with the motive complex, which is vastly more elaborate and more complicated in the human than in any of the animals hitherto investigated and is represented in greater detail in consciousness, and that the ethologists are more concerned with the part I just described. Both tend to oversimplify the other's problems, to look at the other's concerns as simple things, as you did when you said that there was always enough "drive" energy in the human organism (which is not true, by the way), and like some ethologists who tend to concentrate on what they call innate releasing mechanisms and to pay little attention to subsequent learning, for instance. But my answer to you is that in order to explain why I still think we have to talk about these things here, I should have to go into my theory of what I call stabilizing systems, *Verspannungssysteme,* between the various propensities, and the greater overall stability which the systematic interrelation provides for them. But we have not enough time to go into that now. Suffice it to say that I consider the diversity and the great number of single propensities so essential to the human capacity for great diversity of interests, for appetences in so many directions, that no other theory can give us an adequate idea about what motivation initially means and where it stems from.

Classification of Complex Behaviour

Ambrose: In trying to compare the behaviour of human and subhuman species, one of the basic problems we are all concerned with here is arriving at legitimate units of comparison. Now there are many different bases for classifying complex behaviour into units suitable for comparison, motivational systems being one example. This is one instance of a more general type of classification, namely, causal classification where all behaviour due to common causal factors is grouped together. There are problems of course in causal classification. Take as an example male sexual behaviour, which is all behaviour affected by male sex hormones, that is, by an internal causal factor. But there are also external causal factors, so should we not classify male sexual behaviour as all behaviour involving the reproductive organs and directed at the partner of the opposite sex? Even this classification has serious inadequacies. With regard to the general problem of using motivational systems to unitize behaviour for purposes of comparison across species, I am not quite clear to what extent you are saying the unifying factor is the internal causal factor or the external causal factor.

Leyhausen: I should say you are right. The grouping of behaviours according to supposedly common causal factors can be a heuristic method, but not an explanation. On the other hand, I do not believe hormones are a very good example, because we know that sexual hormones affect many more things than we would like to group as sexual behaviour, and we know that sexual behaviour can persist, at least for a while, in the absence of hormones. To your last question my answer is that what I have in mind is not so much a restriction to internal processes, to the causes within the system, regarding it as a more or less closed system for that purpose, nor on the other hand a confinement to external causes. What I have in mind is rather a system of activity, reactivity, and precoded and preadapted reafference and feedback circles. The question of what a feedback is in this context must of course be taken into consideration. I think that with the concept of feedback, or what von Holst called the reafference principle, we are in fact capable, or we have at least the theoretical possibility, of combining these two internal and external causal factors in one complex system, which is what we really want to do. How far we have succeeded is another question.

Animal and Human Behavioural Systems

Wickler: My first point relates to the question whether or not motivational systems in man are fundamentally different from those in animals. Realizing that *the* animal does not exist and therefore motivational systems in different animals might be fundamentally different, we decided to investigate the motivational systems of "pair-bonding" in two very different animals. What we found is that actually the mechanism leading to pair-bonding has to do with the emancipation of the subsystems of Paul Leyhausen's system. What keeps the animals together may be quite different; in the song bird for instance, it may be the urge for singing, which seems to have become emancipated and runs off on its own, like the hiding and getting out again of food does in cats. In a duet singing bird, learning of the particular conditions under which the behaviour can be performed (*i.e.* the presence of the partner) leads to seeking and maintaining proximity to that partner and therefore heightens the possibility that both will feed, nest and breed together, too. Whether this should be called a motivational system for pair bonding depends on how motivational systems are defined, as Dr. Ambrose said. Is it the internal factors, the external situation, or both? I think, and that is what I was aiming at in the second part of my paper, that the danger lies in the use hitherto of terms like *aggression,* which actually is a functional thing like the wing. Difficulties will start as soon as one analyses the physiology or anatomy or, in fact, anything in one particular case and transfers the results to another. This is particularly true of the more abstract terms like aggression, pair-bonding, etc. There are some cases in which these confusions have not been disentangled.

My second point is that the problem of the comparisons between human and subhuman behaviour again requires definition of what we mean by these terms. Is human behaviour every behaviour performed by a human? Or is it behaviour shown only by humans, and not by any other species? In the first case, the definition will not be of much help. If any behaviour exhibited by a human is human behaviour, so what? In the second case, we have subhuman behaviour in humans, too. I think we used it in this sense, human behaviour as behaviour typical of humans, and that means that humans show subhuman behaviour as well and have subhuman motivational systems.

Masters: I should like to follow on that because I agree very much; I think that the distinction between readiness and motive is very useful for tackling Mario von Cranach's problem, because it seems very clear that with the emancipation of various components of motivational systems in higher mammals, and particularly in primates and even more particularly in humans, one has what (among others) Karl Pribram calls the multiple determination of behaviours. That is what causes the difficulties in disentangling behaviours where everything is connected with all other things. That brings us back to the problem of how to relate the uniquely human behaviours in the cognitive area to the elements that Paul Leyhausen called specific readiness and the dissociation of various elements in the motivational and subjective readiness of human organisms to interact. What we talk about as possible homology in the motivational sphere has probably to do primarily with those readinesses, and not with cognitive components. However that may be, it is the functional analogy between functional systems like pair-bonding that we really have to look into.

Ploog: In the abstract given to you before the conference I tried to say the same thing, from a different starting point. There is no purely animal behaviour in man which can be separated from human behaviour, and vice versa, there is no pure human behaviour which can be separated from animal behaviour, or even from subhuman primate behaviour. But I really want to make the following point: I think that Leyhausen's relative hierarchy of motivations is a good point to show that the principle of relatively independent actions is far more exaggerated in the primates than it is in the cats. I will try to make that clear: if one uses the monkey itself as a sort of dummy and investigates social communication in groups by means of electrical brain stimulation (which may be a good way to find out about the motivation of primates), and if one elicits electrically one given social signal hundreds of times and investigates the responses to that same signal hundreds of times (which always leads to a chain of events), one finds that there is no mathematical means to come to a deterministic view of these sequences. One has to use probabilistic means, and even if one collects much data and arranges for proper comparison of the places in the chain, one ends up with a highly improbabilistic model where the same sequence with reference to one given signal occurs almost never. I think that, in man, this is even more so than in subhuman primates; in addition, it is not only the motivational system that is more disconnected from the rest of the machinery, it is also the cognitive apparatus. There are a lot of data about cognitive differences in apes and man, and definite statements can be made about these differences. I think there are fundamental species-specific differences, for example, between the human baby and the ape baby. All this comes, of course, from the Piaget school. I think if one tried to relate the disconnected motivational system to the cognitive system, one would understand why psychologists think so differently about human motivation, and why biologists tend to find a common denominator between human and subhuman primate motivation systems.

Leyhausen: I think I quite agree with Wickler's remark, which amounts to the same as I said. I also agree with what Professor Ploog said, and for this very reason I began my work on cats, for they are on a medium level of evolution, if I may say so. With cats, some things are less complicated, but still complicated enough to show in which direction the system has gone, and I think that this relative availability of the different elements is a precondition for human evolution. If somebody wants to dissociate human motivational systems entirely from any animal motivation system, he will have to show what their evolutionary history then was.

REFERENCES

Ardrey, R. 1967. *The Territorial Imperative*. London, Collins.

Baerends, G.P. 1941. "Fortpflanzungsverhalten und Orientierung der Grabwespe *Ammophila campestris* JUR", *Tijdschrift voor Entomologie* 84:68-275.

Kuo, Z.Y. 1929. "The Net Result of the Anti-Heredity Movement in Psychology", *Psychological Review* 36:181-199.

Leyhausen, P. 1965a. "Über die Funktion der Relativen Stimmungshierarchie. (Dargestellt am Beispiel der phylogenetischen und ontogenetischen Entwicklung des Beutefangs von Raubtieren)", *Zeitschrift für Tierpsychologie* 22:412-494.

— 1965b. "Das Motivationsproblem in der Ethologie", pp. 794-816 in: C. Graumann (ed.) *Handbuch der Psychologie* vol. 2: *Motivation*. Göttingen, Hogrefe.

Lorenz, K. 1939. "Vergleichende Verhaltensforschung", *Verhandlungen der Deutschen Zoologischen Gesellschaft, Rostock; Zoologischer Anzeiger*, Supplement 12:69-102. Translation: "Comparative Study of Behaviour", pp. 239-310 in: C.H. Schiller (ed.) *Instinctive Behavior*. New York, International Universities Press.

— 1950. "Ganzheit und Teil in der tierischen und menschlichen Gesellschaft", *Studium Generale* 3:455-499.

Seitz, A. 1940. "Die Paarbildung bei einigen Cichliden. I: Die Paarbildung bei *Astatotilapia strigigena* Pfeffer", *Zeitschrift für Tierpsychologie* 4:40-84.

Tinbergen, N. 1950. "The Hierarchical Organization of Nervous Mechanisms Underlying Instinctive Behaviour", pp. 305-312 in: *Physiological Mechanisms in Animal Behaviour*, Symposia of the Society for Experimental Biology 4.

Uexküll, J. von. 1909. *Umwelt und Innenwelt der Tiere*. 2. ed., Berlin, 1921.

MICHAEL R. A. CHANCE

The Organization of Attention in Groups

Since the last war, a great deal of interest has been taken in communication between individuals. This has been approached from electronic theory and through linguistics, and much of the interest in ethology has arisen because it has uncovered forms of communication in animals and of non-verbal communication in man. But how, in the first instance, does communication become possible?

GROUP COHESION

To answer this, it is worth posing the question of why individuals in a group cohere? Is it, as is implied by those who regard communication as the essence of sociability, in order to make communication possible, or has communication arisen on the basis of an infrastructure of cohesiveness which originated in response to some other requirement and in other ways? To make communication possible, individuals have to be within earshot, range of vision, in contact or be informed by territoriality. Territoriality may relate them to other individuals either by utilizing the same resources, by the visual geometry or by territorial marking smells which facilitate serial use of the same area by individuals of a community. An infrastructure of social relations which has a primary effect of keeping animals together can therefore come about in a number of different ways; this is what we must now consider, since it is these infrastructures that form the basis of social life.

We are here concerned with the comparability of structures in behaviour systems and I shall be specifically concerned with this problem in social structures. To start with, however, the recent discovery of two distinct but essentially similar forms of primate social cohesion (constructed in different ways) demands our attention.

Every species of sub-human primate has its own unique patterns of social relations which result in a set of sub-groupings within the society as a whole, but, as pointed out in *Social Groups of Monkeys, Apes and Men* (Chance/Jolly, 1970), three types can be identified in the majority of groups so far studied. These are 1) the assemblies of adult females and their infants, 2) adult cohorts of males, and 3) clusters of juveniles: young who are no longer dependent on their mothers, but who have not yet joined either of the adult groups.

These sub-groups account for the coherence of their members over large periods of time, but in addition to them there are two distinct modes in which the society as a whole is brought together. These involve the group members' attention to a focal individual. One of these, which I have called the *hedonic mode,* based on display as a means of gaining attention, is a recent discovery. Chimpanzee and gorilla societies are constructed in this mode. The *agonistic mode,* by contrast, is based on conflict and on avoidance of attack from the dominant male. Macaque and baboon societies are constructed in this way.

To argue that these two types of society are constructed on these two modes is to suggest that, for the greater part of their active lives, individuals relate to each other in these two different ways. The possibility of being persistently aware of a dominant focus arises in the open terrain inhabited by baboons and the rhesus macaque, but it is not possible in the forest conditions usually inhabited by the great apes. Monkeys foraging in open country are potentially under attack by predators and need to keep in sight of the dominant males, whereas gorillas moving on the ground through the forest keep in touch by a chain of attention ending up on the focal male. Chimpanzees, on the other hand, moving through the trees, are often separated except by being able to call to one another, and in consequence, mutual display is used to re-unite the group after its members have come together again.

We shall now go on to consider the features typical of these two modalities as shown by chimpanzees on the one hand and macaques or baboons on the other. Then we shall show that these two systems, by which behaviour is organized, have been found in the behaviour structure of two species of macaque. The great apes are exceptional in that they appear not to be able to cohere by agonistic means.

Hedonic Cohesion

First let us consider how the cohesion of the whole group is achieved, in an hedonic society, through display. Schaller (1963) in his study of the wild gorilla, was the first to point out that when the dominant male was ready to move off after a rest period, he stood in a prominent position and in a characteristic posture: legs fully extended, vertical and rigid, looking ahead of him. After he had adopted this posture for some time, the rest of the group gathered round him and he then moved off. Schaller also pointed out that when an adult male beats his chest and advances towards an intruder, this is not only acting as an intimidating display towards the intruder, but also indicates his whereabouts to the rest of the group.

Reynolds and Reynolds (1965) in their study of the Budongo forest chimpanzees, early on found themselves witnessing the so-called "carnivals" in which all the chimpanzees hoot, jump about, beat the resonant bases of trees and wave sticks in the air, as they display towards each other in this manner. They noticed that this happened particularly when two groups of chimpanzees met in the forest and that then this mutual displaying was followed by much contact behaviour including touching, kissing and some pseudo-sexual behaviour. Reynolds and Luscombe (1969) found that it was much easier to understand the nature of this mutual display when they studied chimpanzees in a 30 acre enclosure in New Mexico at the time when the American Air Force had a research unit there. On my own visit to this place, I noticed that chimpanzees would wander over the enclosure in the early morning, often separated by 200 yards as they explored in different directions; then perhaps, as a group clambered into the old trees, they would see another group. Immediately, they moved over towards them and made contact, showing some excitement when they met. On one occasion this led to some old sacking and sticks being thrown into the air. Briefly, what Luscombe and Reynolds found was that at times when the fruit was provided early in the morning a lot of mutual displaying took place. At the same time, the animals gathered into a group.

They also found that the amount of attention that other individuals paid to any particular chimpanzee depended on his displaying ability, but that the amount of attention did not correlate in any way with the amount of aggressive behaviour shown by any individual. Moreover, an individual's ability to become the focus of attention through displays led him to become the centre of a group of individuals when fruit-sharing took place, and in this way a successful displayer was also successful in getting a large share of the fruit. Subsequently, the group broke up into two's and three's; these and other small groups then moved off in pursuit of different activities.

As I pointed out earlier, the great apes are peculiar in that they do not show any persistent agonism; this means that, apart from the coherence of small groups, the whole group is brought together only by calling over large distances followed by mutual display binding the individuals' attention from time to time to one another.

Agonistic Cohesion

The mechanism that holds agonistic-type societies together is of a different kind. As pointed out earlier, because baboons and macaques live largely in open country, it is possible for them to be aware of the dominant male at all times, and for him thereby to be the focus of the group's attention. Although they may be largely preoccupied by foraging and thereby be spread out over quite a large area, they are nevertheless ready at any moment to coordinate their own behaviour with that of the dominant male in defence of the group as a whole. As individuals, they do not move out of sight of the centre, so we must assume that they are always attentive to it. This has been demonstrated in the way the hamadryas baboon and the rhesus macaque behave as the troops move away from their sleeping sites in the morning. Here the direction of the movement of a troop as a whole results from the initiative of sub-adult males moving out in different directions, and from the eventual coordination of this initiative of theirs with the direction the dominant male ultimately decides to take.

Kummer's study (1968) of the one-male groups of the hamadryas baboon, moreover, makes it clear that this unit is the result of the confinement of the individual's attention within his own group, focussed on the male. In this way, baboons move about as a very tightly packed group. Their attention is coordinated by the infliction of neck bites by the male on the members of his harem if they stray any distance away.

From time to time in semi-captive colonies—and during fairly prolonged episodes in the wild—threatening, chasing, fleeing and various forms of submission are shown in vigorous agonistic encounters between adults of the group, mainly the males. At all times the individual in an agonistic community must be ready to defend his status or avoid being the object of severe attack, and this can only be done by persistent awareness of the dominant focus. The dominant males, having a superfluity of aggressiveness, may be provoked either accidentally or by transgression of certain spatial and behavioural regularities. Hence an agonistic society is held together by a persistent awareness of the centre, based on a readiness to behave agonistically at any moment. It is not surprising, therefore, that spacing out and status are persistent preoccupations of the individuals of such a group. Most interactions within the group (*e.g.* submission, flight) are designed to terminate the conflict inherent in the situation. In this respect, therefore, agonistic behaviour contrasts markedly with hedonic behaviour as it tends to terminate active interactions, whereas hedonic behaviour tends to promote social intercourse.

COMPARISON OF HEDONIC AND AGONISTIC BEHAVIOUR

In order to make a comparison between the agonistic and hedonic modes, it is easier to describe one of them and to contrast this description with the corresponding features of the other. In doing so I shall attempt to abstract from the information available the characteristic features of the hedonic mode. I shall do this by contrasting the behaviour of the chimpanzee, which exhibits the hedonic mode in its most characteristic form, with the agonistic behaviour shown by baboons and macaques. Until we have as clear an idea as possible about the characteristics of the modes of behaviour in the sub-human primates, we cannot hope to interpret human behaviour.

Undoubtedly the feature that has the most far-reaching implications, is the flexibility of behaviour in the hedonic mode. This leads to both an increase in the range of behaviour and an increase in the readiness with which a switch in behaviour can take place. Thus the chimpanzee extends the variety of behaviour which it exhibits in the wild, and also displays a readiness, in semi-captive situations, to acquire entirely new forms. This may well reflect an enhanced ability to receive rewards for achieving a new skill. This, in turn, undoubtedly reflects a greater flexibility in the control of attention and the individual control of excitement.

One of the outstanding features of chimpanzee society is the prominence of contact behaviour, as illustrated by Vernon Reynolds (1972). In the first place, greetings involve contact behaviour in a number of ways; for example, touching hands, touching another's body, kissing, various forms of hugging and frequently, when greatly excited, some form of sexual behaviour. This is often followed by the group's breaking up into pairs of chimpanzees who move off with their arms round each other, leading to the separation of the various sub-groups of the society and providing the opportunity for a later re-union.

Since the members of an agonistic society are constantly aware of each other and in each other's presence, the opportunity for meeting in the way chimpanzees do does not arise. The meeting between two individuals in an agonistic society always involves a heightening of the latent conflict present, and the individual meets this situation by postural and facial appeasement gestures such as presentation, lip-smacking and "appeasement face". If any physical contact is established, this leads to ritual grooming, in which one individual grooms the other; occasionally mutual grooming takes place among the group. Ritual grooming is very stereotyped and consists of a sharp downward and sideways combing movement of the fingers of one hand, while the other hand is used for examining and picking at the exposed skin. In the forms of contact shown by the chimpanzee, however, there is a great fluidity. While pairs of (or even three) sub-adult chimpanzees will be seen moving over fairly large areas in an exploratory fashion with their arms round each other, the sub-adult macaques or baboons have no contact, except that the younger sub-adults may jump on the backs of elder males in periods of social excitement.

From the work of Mason (1965) it is possible to infer that contact in all these instances is a form of reassurance. Undoubtedly, the widespread occurrence of contact in the behaviour of the chimpanzee reduces tension and enables the individual rapidly to control his excitement.

Chimpanzees in the wild are offered a great variety of ways of using sticks as tools. Sticks or branches are used to hit the ground or to throw at potential predators, and also for poking into termite hills to extract the termites; in these circumstances awkward projections are broken off. Suitable twigs may be found by the discarding of many until a usable one is found or until one is fashioned in the way described.

Macaques and baboons have not been seen using tools often in the wild, whereas chimpanzees frequently make and use tools though only for the single purpose of whittling ants from ant hills. Hence the whole range of problem-solving behaviour which is open to the chimpanzee—studied by Köhler (1927) (see also Chance, 1960; Schiller, 1957) in semi-captivity—is not readily available to the agonistic species in the wild. A startling recent discovery by Menzel (1970) has shown that a group of young chimpanzees who had grown up together in semi-captivity were able to develop and jointly exploit a new method of using a pole as a ladder. In all these studies the ability of the chimpanzee to adjust the use of the instrument to the potentiality of the situation—in order not only to reach a set objective but to use an invented skill for more than one purpose—indicates clearly that the chimpanzee is capable of extending its repertoire further into the human range than was hitherto known and probably more than any other sub-human primate. Groups of chimpanzees can considerably expand their repertoire by invention, whereas groups of baboons and macaques are much less likely to.

It is a logical requirement that an enhanced power of inventiveness will depend in the first instance on an increased flexibility in the *behavioural repertoire*. Apart from "a larger and more diverse *initial behavioural repertoire* (IBR)", Christopher Parker (1974) makes clear that this flexibility can be increased in three distinct ways: 1) less rigid relationships between elements of the IBR and eliciting stimuli, 2) a more efficient mechanism for making changes in stimulus-response relationships, and 3) an increase in the complexity of relationships among stimuli that can be integrated into a common determinant of behaviour. Agonistic arousal increases stereotypy and therefore prevents the operation of 1 and 2 whereas hedonic relations, because of the greater control of the agonistic arousal through contact, increases it. Changes in stimulus response relationships are made more likely as well by the flexibility of the hedonic mode.

A notable feature of this expansion of repertoire is the switch of attention from social to non-social aspects of the environment, clearly seen in the mutual inspection of an object by two chimpanzees. Two chimpanzees will sit next to each other and one will watch the other trace the line of a crack in the floor with his finger or a stick, and will momentarily indicate the shared attention between them by running his finger along the crack.

A greater control of attention through an awareness of external features is particularly well illustrated by a mother's awareness of the situation of her young. Both gorilla and chimpanzee mothers are able to anticipate and protect their infants from potentially dangerous situations (see Chance/Jolly, 1970). A macaque or baboon mother protects her infant from potential danger only by restricting its activities or responding to its calls.

It seems that, from studies of other species of monkey and baboon, provided they are withdrawn from their social context they are potentially capable of exhibiting some, if not many, of the problem-solving capabilities of chimpanzees, but only as individuals (Köhler, 1927; Chance, 1960; Bolwig, 1963).

However, many rudimentary elements of problem-solving and inventiveness can be found in some species with basically agonistic societies; for example, long-tailed macaques (Angst, 1971). But the limitations on individual behaviour inherent in the mechanisms designed to reduce conflict rigidify the behaviour of predominantly agonistic species (such as the rhesus macaque and the baboons) to an extent which precludes possibilities based on variability. Agonistic societies contain a very large element of conflict, both in the relationships between individuals and the state of the individual's motivation. As a result, "spacing out" is a way of reducing the impact between two individuals. So also is a "cut-off" act. This is a deflection, in the simplest instance, of the visual awareness of another individual by deflecting the gaze. Very often, this means paying attention to something else and not simply closing the eyes, for example. This brings about a continual interruption of attention, so preventing ongoing attention to a single feature of the environment, or the mutual inspection of the same object by two individuals, so typical of the hedonic mode.

In the agonistic mode, displacement of behaviour may take the form of scratching in the ground, for example, but if so, this is rigid and repetitive, not controlled by an awareness of the environment—as when a chimpanzee moves its fingers along a crack in the ground. The fixity and rigidity of these displacement activities and deflections of attention originate more from the value of these actions in curtailing the conflict between individuals than from an intrinsic interest in the alternative object towards which attention is directed.

Moreover, in this way attention is frequently deflected onto another individual of lower status. This may then be used to displace aggression by threats or actual chasing, as in the following example concerning rhesus monkeys (Sade, 1965):

> On April 10th, 1962, for no apparent reason, 1956-Male-l attacked an unrelated adult female while she was drinking. Immediately, the dominant male of the group, Old-Male-A, attacked 1956-Male-l, who fled. Old-Male-A chased him round for about half a minute or more; from a bush, up a palm tree and down the other side, round a building and down a trail. Suddenly, out from under another bush 1956-Male-l's parent, Old-Female-l, came running on her hind legs carrying her five day old infant. She ran to stand at her son's side, and together they made violent threat gestures and vocalizations at a part of the area empty of all monkeys and observers, and away from the dominant male. Old-Male-A stopped chasing to look at what they seemed to be threatening, then chased 1956-Male-l again, who again threatened loudly away from Old-Male-A. Old-Male-A sat down three feet away peering again in the direction in which 1956-Male-l was threatening. Old-Female-l climbed to sit between them and her son immediately sidled up to her, sat touching her and groomed her. She walked away and 1956-Male-l followed her closely, grooming her whenever she stopped, leaving Old-Male-A peering at nothing. Parents often defend their offspring from the attacks of monkeys more dominant than themselves by diverting the attention of the attacker in just the manner described.

BI-MODAL POTENTIAL IN THE MACAQUE REPERTOIRE

So far, we have described the difference between the hedonic and the agonistic mode in terms of the behaviour of species which are organized predominantly in one or other of these two different modes, and we have abstracted the characteristics of each mode. Following my suggestion (Chance, 1967) that rank-ordered behaviour was best understood in terms of the structure of attention, Virgo and Waterhouse (1969) have found, from their study of the Bristol Zoo macaque colony, that there were two structures of attention, each focussed on a different individual in the colony. They showed that the structure of attention based on agonistic behaviour was focussed on an adult female, and the other, based on grooming and affiliative relations (sitting next to, etc.), was focussed on an adult male. This immediately drew attention to earlier observations of my own at Bristol

Zoo, where I observed that changes of the leadership of the colony followed a period of vigorous display by the future dominant male. Hence it became clear that the features of both modalities are present and employed by members of a single species, the rhesus macaque, and therefore the propensities for these two behavioural features were present in the individual rhesus macaque. Current studies under way with Tom Pitcairn and Walter Angst are providing evidence of the same potentials in the behaviour structure of the long-tailed macaque. We are nearing the point, therefore, when we are able to say of many sub-human primates that an individual possesses a potential for developing these two modes.

What, then, are the essential consequences for the understanding of behaviour structure? Since the hedonic mode is alone capable of enabling behaviour to be constructed entirely on the basis of external information, and the stereotype of the agonistic mode is in such marked contrast to it, we are forced to conclude that the agonistic mode is constructed from fixed action patterns and is therefore programmed from internal sources.

HUMAN COUNTERPARTS

In our discussion of the infrastructure of social cohesion and control we have suggested that, depending on the basic modality in which individuals are operating, the type of communication between them will differ. In human relationships, communication of the type we have been discussing is considered "non-verbal". By its nature, this type of communication is concerned with the relationship of the communicator with other individuals in a group, or, as I shall call this attribute of an individual, his primary-group relations.

The infrastructure of social communication is posture, gesture and facial expression as well as tone of voice, and constitutes the framework of control within which linguistic intercourse takes place.

In an important sense the relationship of individual to individual in a group is tacit, and the way the individual becomes competent in this sphere is through practice in social relations. These are the implications of Bernstein's work, about which I am going to speak, for he has discovered that language can be subordinated to, and used only as an extension of, this tacit relationship.

Michael Argyle (1972), Adam Kendon (1967) and Paul Ekman (1972) on the other hand have shown that, in the conversation of literate individuals, eye contact, head and hand gestures are used to control the attention between speaker and listener. This occurs in what we colloquially know as "talking to one another." Therefore, language is used in two quite different ways, depending on the way it is related to the infrastructure of interpersonal relations.

Both Bernstein (1965) and Jensen (1968), in different ways which are incidental to their main interests, have uncovered the fact that through differentiation in development (ontogeny), different mental faculties can be subordinated to, and come to be part of, this infrastructure controlling relationships within a group.

Both have used class differentiations as a means of studying the operations of "intelligence", and Bernstein has been specifically interested in uncovering the fact that differences in intelligence of pupils are related to the patterns of social behaviour found in the family and their general social background. But this is only incidental in Jensen's work.

Bernstein suggests that the limitation of intelligence in some classes of pupil arises from the fact that they possess a restricted code of language, which, through its own structure, indicates that it is used as an extension of the non-verbal social control system which the individual has grown up to consider the essential framework within which language itself is to be used. As an example, children commenting on a strip cartoon of children playing ball and accidentally kicking it through a window pane, used the pronouns "you", "they" and "we". "They" and "we" indicate out-group and in-group identities while "you" indicates the speaker's awareness of a single individual within the group.

Nouns describing the objective existence of the components of the cartoon, the ball and the people involved, the window, etc., are not used, and, because of this lack of specification, the language operates only because there is acceptance by the viewer of what is referred to, so that elaboration is redundant. It leaves so much unsaid that it is necessary to see the picture in order to make intelligible the meaning of what is said. The picture, in this instance, stands for the shared social situation.

Significantly enough, in this context, the pronoun "I" is not used, since the person does not see himself as an entity in the total situation. The situation and the action taking place within it is merely an experience *of his*. He does not see himself as part of an objective situation.

An elaborated code, on the other hand, renders explicit what is implicit: through the use of nouns that can be modified by adjectives, it increases the flexibility of the verbal repertoire and leads eventually to the development of a fully syntactic language.

Restricted codes are rigid by virtue of their being clamped on to a social situation. There are hints throughout Bernstein's work that individuals who grow to adolescence with a restricted code form social groups in which strong control is the primary concern of members; that is to say, a strong social identity among those controlled, and control of the group by a dominant leader. Persons in such groups respond to events predominantly through the way these affect their personal relations within the group and less in ways which could show that they are aware of a wider framework of reference. In personal discussion with Bernstein, he agreed that rank-ordered social structure was typical of those using restricted codes. Elaborated codes—because of their use of nouns describing objects and persons in terms of their identity rather than their relationship to the speaker—are based on a form of language which is concerned with the assessment of external reality as opposed to the social reality of experience within a group. An elaborated code is therefore appropriate to behaviour supported by the hedonic mode and presumably arises out of it. This is consistent with the suggestion in what Bernstein says about the use of the child's imagination as the basis for the construction of an elaborated code. An example of an agonistic human social group would appear to be provided by a social group using restricted codes, since these codes are an extension of the system of social relationship, which is rank-ordered.

The social relationships of individuals using a restricted code are, therefore, essentially similar to the agonistic control structure of a sub-human primate group. I think, therefore, this demonstrates that such a system can develop into a controlling mechanism in human society.

Evidence that an individual's capacity to learn is also related to whether or not he is operating primarily in the context of a social group comes from the work of the psychologist Jensen, (1968). He was able to distinguish between individuals who are primarily associative learners and others who are able to develop powers of classification and abstraction.

Jensen is primarily a psychologist, and therefore only really interested in the fact that these two faculties can appear in different individuals, and in what educational situations they occur. Having demonstrated this distinction, however, he has noticed that powers of simple deduction (associative learning) go together with competence on the playground and concern with social relations, as opposed to the handling of abstract information and classification. Because he is not primarily concerned with the patterns of behaviour that go with these different powers of learning, the behavioural distinction is not as clear as we would like it to be, but the fact that there are individuals primarily concerned with social relations, and others concerned with these only as a part of their social awareness, is a distinction which I confirmed through personal correspondence with him.

It looks very much as if we have here the discovery of different types of learning which may be linked to restricted codes, because they are both linked to the development of patterns of behaviour which also restrict the individual to operating within a small or primary group.

The use of language in a restricted code is therefore the subordination of language to an agonistic social structure, in which the information content of the linguistic system is limited, the potential quality of language left undeveloped, and the mental faculties not fully developed. If this argument is correct, then it leaves still to be defined the social structure which permits the full realization of the capability of language to convey objective relationships in the sense usually ascribed to scientific statements. Such a social structure is latent in the hedonic mode, in which social relations are not fixed, and, therefore, are not part of a system specifically functioning to ascribe a role to the individual's behaviour. In such circumstances the individual's attention is free to be guided by interest in external features, an essential requirement for investigating the properties of the external world and the extension of this process into man's pursuit of scientific knowledge.

The extraordinary capacity of man to combine different patterns, components of his behaviour, convention and awareness, and the difficulty of knowing how to obtain valid evidence that specific elements of these are built into a particular pattern, make it difficult to do more than point to further examples where the basic pattern suggests the operation of agonistic-type cohesion.

One thing we should note, therefore, before we go on to consider further examples, is that an agonistically structured group consists of individuals primarily concerned with their social relations, whereas the hedonic structure permits varying degrees of awareness of the physical environment and the intrinsic nature of both other individuals and the nature of the physical environment.

Moreover, the experience of the physical environment, as well as of other individuals, is, in the agonistic mode, *necessarily* an extension of the social experience. Hence if we see that the primary concern of a person in a group is with maintaining the fixity of the social relations and that the individual's awareness of objective reality arises out of his or her social relations and not out of concern for the people themselves, we are probably witnessing the operation of the agonistic mode in these individuals. This seems to me to be the basic characteristic of the Hell's Angel brotherhood.

A report in the *Times* of August 31st, 1971, was headed "Pop festival—should have called police sooner". This was when 39 Hell's Angels terrorized people at the Weeley pop festival in Essex, for love of their motorcycles, as Colchester magistrates were told. Mr. Michael Whatcott, a motorcyclist, said that the gang had not been violent and had threatened no one. "Our boys were given permission to take jobs fighting fires. *The only thing we have got in the world are our motor bikes, our fellows and our girls.* We have spent years and every spare penny building up these bikes; they are our one and only life. It hurts to see them damaged, it hurts to see our wounded, but they are used to that and so are our girls."

This statement indicates that their lives centred around their motorbikes and their girls and this is borne out by reading *Buttons: the Making of a President* (Mandelkau, 1971). The cohesion of the group is their predominant concern and therefore other people are seen as a phenomenon outside the experience of the group; this leads to frequent confrontations.

One other aspect of the Hell's Angels is their concern for badges and uniforms, which are also seen as extensions of an individual's person. Badges function as a simple method of identifying a group and those outside it as indeed uniforms do when they assist the definition of social roles. These fixed elements help to define social position.

Recently a controversy on why children of inner-city areas in the United States of America show lower academic achievement than the rest of the population was published in the *Atlantic Monthly* (229, 1972). W. Labov described the differences between the standard English of the classroom and the vernacular language used by members of the street culture. His study, like that of Basil Bernstein's, is concerned with the relationship of language, culture and social structure, and shows that the intellectually poor performance of ghetto children is related to differences of culture and social experience. He produces evidence to his main concern that shows rank-ordered social relations as the dominant structural influence in the form that the culture takes.

"We see", he writes, "many speech events which depend on the competitive exhibition of verbal skills, singing, sounding, toasts, rifting, louding—a whole range of activities in which the individual gains status through his use of language. We see the younger child trying to acquire these skills of older children hanging about the outskirts of the older peer groups and imitating their behaviour."

I am not directly concerned with the nature/nurture controversy, which is wholly wrongly conceived, but evidence is here further provided that the peer groups create rank-ordered social relations wherever the culture permits it.

At the beginning we argued that an infrastructure exists, determining the form of social cohesion in sub-human primate groups, and that this takes two forms, the agonistic and the hedonic. This is not the same as saying that, in humans, non-verbal means of communication take these two forms, but that the cohesive infrastructure, especially the agonistic, is capable of determining not only the form of the social relations but also the consequent dependent forms of communication at any level at which the individual relates himself to others. This has been demonstrated in human society because language has been shown to be subordinate to this infrastructure wherever restricted codes are in operation. We must envisage, therefore, that this same infrastructure will, in the appropriate circumstances, be capable of dominating the form of social relations expressed through convention or the structure of thought. If such occurs, man is not free, but if hedonic-type relations can prevail, the dimensions of man's freedom are as wide as we know they can be.

THE ASSESSMENT OF ATTENTION IN GROUPS

Doubts are often expressed about the competence of ethology to study human society. If we take the whole complex phenomenon known as society, no single discipline could possibly claim sole relevance, though many of the arts claim to be concerned with an overall entity known as society. So let us see what can be said about the relevance of the ethological method. It is concerned with the study of a quite specific part of social relations: namely, the face-to-face situations where observation is indeed possible and then it also becomes possible to note what human features, such as different forms of language, *are* associated with each form of social relation. Yet even here, care is needed to specify the exact way in which social relations take place and the exact composition of the group. In human social gatherings, size is a factor. Some groups are small enough to consist of the same individuals who come together at different times. This is what we encounter in Hell's Angels, street corner groups and Glasgow gangs (Whyte, 1969; Patrick, 1973).

Unlike the re-assembly of chimpanzees, however, they exhibit a form which is agonistic in intrinsic structure. Larger gatherings of humans such as formalized meetings exhibit constant forms with more or less conscious and often explicit regulations governing behaviour, though the individuals involved change. These are ongoing groups composed of varying degrees of familiarity between the individuals involved. When these break down they do so very often into inter-individual fights, except where co-ordinated subgroups exist within the original gathering. So, with these provisos in mind, let us turn to the forms of attention within groups.

The original concept of attention structure was derived from an attempt to discover the mechanism underlying rank order in sub-human primates. Having identified the nature of social cohesion in the structure of attention, the problem now consists of assessing this element in its own right. One of the first direct pieces of work on this is that of Kummer and Kurt (1965). They showed that, in the hamadryas baboon, the interaction of the members of two adjacent one-male groups (next to each other on a sleeping niche) was largely confined to other individuals of their own one-male group, despite the fact that many of the individuals were as close to members of the other group as to members of their own. This showed clearly that their attention was primarily confined within the one-male group, and that this also had a large component directed to the male, whom they followed as soon as he got up and moved off. (Personal communication.) Subsequently, Kummer has carried out a series of observations on pairs of males of different species confined in a large enclosure in their natural habitat. When two members of the same sex of any species are held together in a large enclosure and the distance separating them is assessed at regular intervals, the results can be plotted on a distribution curve showing mean distance and variation on either side of this. Savannah baboon males and females show normal distribution curves with a mean of two to three feet separating the females, indicating that the tendencies attracting the males to each other, and those tending to separate them, are in balance. This is consistent with the way we know their attention is fixed on the dominant individual from whom they keep a mean distance in wild groups. The curves of the distances separating male Patas monkeys on the other hand, are of quite a different kind. Not only is the mean distance about 15 feet, but the curve of the distribution is not smooth or even normally distributed; it shows a series of peaks separated by troughs which can be duplicated if a second set of observations is made on the same pair or a different pair in the same enclosure. If, however, the objects in the enclosure are distributed in a different way, the peaks and troughs appear in different parts of the distribution curve, showing that the attention of the male Patas monkeys is divided between their companions and prominent objects in the enclosure. When they move closer to their companions, they do so only in steps which are determined by the distances separating prominent objects in the enclosure. This is consistent with their known behaviour in the wild where a Patas male, being separated from the females by a long distance, keeps close to a prominent object in the environment where he displays on the approach of a predator.

One of the subdivisions of attention structure is, therefore, between prominent or dominant individuals on the one hand, and features of the environment on the other. This same division of attention has been demonstrated in Gary Emory's study (1975) of a gelada baboon colony and a mandrill colony housed in identical enclosures in the San Diego Zoo. He used head position and body position in relation to the direction towards or away from the dominant male, and subdivided the segments of a circle into four quadrants: that of 45° on either side of the location of the dominant male; 45° on either side of a line at right angles to this and to latter direction on the left or right of it; and 45° on either side of a line directly away from the dominant male. He then showed that the frequency of head positions and body positions in these segments distinguished the pattern of attention structure in these two species. Members of a gelada baboon colony showed a preponderance of head and body positions sideways to, and looking towards the dominant male, showing that a preponderance of their attention was towards the social centre. In the mandrill, on the other hand, the peripheral members of the groups spent a large proportion of their time looking towards the environment and away from the dominant male. Head and body position is therefore one metameter by which attention structure can be studied, and within which greater detail and elaboration of attention should be made possible in the future.

This being so it could be inferred that head position does give an indication to another observer of whether or not an individual is looking in his direction. An experiment on human communication by von Cranach (1971) reports that head position and head movement play a significant part in a receiver's or an observer's estimate of whether or not an individual is involved in a face reaction, as it is called, with another individual. If this can be further substantiated, then head movements constitute a metameter for both man and monkey. In a further study he found that the beginning of verbal communication is usually preceded by an orientation of the body, head and gaze; so we assume that these behaviours are hierarchically ordered and express the intensity of readiness to communicate. Gaze, head and body are turned towards the partner, and in inhibited, depressed patients an increase in orientation reaction seems to announce improvement, even before verbal interaction is resumed. Further evidence that a correlation exists between gaze movement and total motoric (that is to say, movements of the rest of the body) is found in the work of R. Jung (1963). He summarizes his findings by saying that, in men, voluntary gaze movements are almost never restricted to the muscles of the eye. Neck and body muscles are involved in auxiliary innervation. Eye movement and nystagmus are but sections of a general orienting movement of the total motoric. Further, since we generally tend to look markedly at a partner (or markedly not to look at him) horizontal gaze movements are practically always accompanied by movement of at least the head when interrupting longer periods of looking at the partner or looking away from him.

THE MANIPULATION OF GROUP ATTENTION BY INDIVIDUALS: ADVERTENCE

Although the concept of attention structure was initially formulated in terms of the preponderance of attention paid by subordinate individuals to those higher in rank—especially to a single dominant individual in an agonistic society—features of the hedonic mode showed quite clearly—from the ways chimpanzees behaved in particular—that each individual to some extent competes for the attention of others in the group by forms of display.

A period of mutual display involving a somewhat competitive form of attention *seeking* rather than attention *paying*, was the way in which the infrastructure of social cohesion was built up in a chimpanzee society. Studies of our own colony of *Macaca fascicularis* (long-tailed macaques) have shown us quite clearly (Chance/Jones, in prep.) that subordinate individuals solicit the attention of the dominant male, and in so doing enter into a relationship with him which is not previously determined in its character. The act of soliciting therefore appears to be attractive in its own right. From time to time the oldest juvenile male, Arlo, looks in the direction of Percy, the adult male, and, raising an arm above his head, brings it down with a sharp striking movement. He does this at some distance—several feet—from Percy, who often responds with a lunge threat, and may chase him, during which time Arlo retreats. He may repeat the gesture but give a facial expression we call "mime", which is a signal of appeasement in the circumstances of aroused threat, and hence indicates his willingness, or wish, to approach. This may eventually appease Percy to the extent that he will reciprocate with a "mime", and then a play session may follow in which Arlo leaps across Percy—just out of his reach—and eventually approaches him, "miming" vigorously, and enters into a wrestling match. They may hug each other ventroventrally. Arlo's success in initiating a play session, however, is not one hundred percent and he may end up having to escape from Percy's threats. This soliciting of attention is not quite the same thing as calling attention to oneself, as monkeys do when they jolt and shake the branches of a tree or jump up and down in a prominent place. This simply brings the attention of everyone to them and does not involve them with a particular individual. Nevertheless, the relation of Arlo to Percy shows quite clearly that one individual attempts to manipulate the attention of another individual in the group and that, therefore, the study of the structure of attention must take this element into account.

Von Cranach (1971) observes that the assumption that "looking behaviour" functions as a signal can be found in many studies of the subject as an explicit or implicit supposition that determines the design of the investigation and the interpretation of the results. In discussions of this kind, the assumption is that the signal is providing information, whereas in the case of Arlo it appears that the signal is merely attempting to establish a channel of communication. This is the prior underlying condition for establishing states of communication. Yet Argyle (1967) and Kendon (1967) start from the other end when they say: "We may distinguish three ways in which visual orientation functions in interactions. One, to look at another is a social act in itself; two, to meet the gaze of another is a significant event and may often be an important part of the goal sought in interaction; and three, in seeing another, much important information about him may be gathered in addition to his direction of gaze." Von Cranach (1971) places the same emphasis when he says that his own paper will deal with the first two of these statements, which maintain that the gaze of one partner is significant for the behaviour systems of both partners, thus inferring a communicative function. In another place he says, however, that gaze direction and eye movement are used only in combination with other orienting behaviours, emphasizing a point we have already seen demonstrated in the position of the head and body in relationship to attention. Analysis of conversation between individuals undoubtedly leads to the view that many of the changes in head position, eye contact, and other components of a non-verbal interchange, are modulating the communication channels between the speaker and listener. So one of the differences between human and non-human communication is that the non-verbal channels in the non-human primate are modulating as well as carrying the message, whereas in humans the non-verbal is modulating an interchange or a system of communication which is largely verbal.

The question now arises: what is the real relationship between these two systems? This was the object of a collaborative study between Hilary Callan, an anthropologist, and myself and Tom Pitcairn, biologists (Chance/Callan/Pitcairn, 1973). In order to implement the study, we realized that we would have to go back a stage in order to avoid being clamped into the framework of thought which governs existing procedures about the study of this relationship. We soon realized that we had to study the social microstructure, which we defined very loosely (for the present) as the total of all that is patterned and systematic about social interaction in the immediate and quasi-immediate presence of others. At the start we took *presence* to mean straightforwardly *physical location,* and by small-scale social activity we meant social behaviour involving small numbers of individuals rather than broad categories or groups of our society. Nevertheless we were dealing with the study of human groups, so we allowed ourselves to be guided by Goffman (1969) who defined the study of social microstructure as the bringing into one framework concepts and findings derived from three areas of enquiry: the individual personality, social interaction and society. By attempting to assess, or be aware of, the influence of society we referred to everything a socially functioning human being does in face-to-face interaction which is done partly with reference to a submerged series of ought-statements derived from culture, learning, the person's role, the nature of the situation, and so forth. We studied two of the three components outlined by Goffman, namely social interaction and society. We chose to study the inmates of a psychiatric hospital for two reasons: the rules of the institution were relatively simple and easy to identify, and mental patients had exhibited patterns of social behaviour strikingly similar to those of sub-human primates (Grant, 1969). Our observations were made on a schizophrenic group and a psychotherapeutic group. Before we began we found it necessary to distinguish a response in the psychological sense of an elicited piece of behaviour and also in the social sense as an acknowledgment. These were, so to say, at two levels, and yet closely inter-related in their action. The fact that an acknowledgment was possible in a group, or need not be accorded by the group, made it clear that at least part of the behaviour of the patients was an attempt to become part of a group by being accepted into it. In fact, social cohesion had to become social acceptance in a human group in order for the group to operate. An acknowledgment means acceptance by the group. In terms of attention this means manipulating the attention of the rest of the group in such a way that the individual becomes recognized and takes his place as a participant in the ongoing activity of the group, whatever that is. We have seen from Kendon (1967) and Argyle (1967) that non-verbal components are used in these circumstances to modulate the conduct of the conversation. What became evident from our study of the psychotherapeutic group was that many individuals had not yet reached this position of acceptance and were constantly striving, apparently in a partially familiar situation, to gain acceptance. *Advertence* was the process by which they advertised their presence and their willingness to become part of the social activity. (Activity is here defined as the purpose of the group as well as the actual ongoing interaction between members.) In our observations of the psychotherapeutic group the aim was to identify the level of behavioural and social integration at which advertence operated. It eventually became clear that there were in fact four rough types of self-presentation termed *alert, away, closed* and *huddled*. To various degrees, and in this order, these forms of advertence constituted more and more exaggerated forms of self-presentation, representing, curiously enough, the greater degree of withdrawal from participation in group activities as the patients moved away from *alert* through *closed* to *huddled*. In fact it was only those patients showing *alert* and/or *away* from time to time who participated in what was the then activity of the group, namely, discussing the problems presented by an individual who was at that moment the centre of interest and concern. Hence participation in this sphere of social activity indicated an ability to handle information which was not directly concerned with the individual's problems. Patients who failed to present themselves and their own problems in a sufficiently acceptable way to the group were driven into, or acquiesced in, the role displayed by those in *closed* and *huddled*, two rather more fixed categories of the advertence. This general pattern is repeated in the forms of advertence identified in the schizophrenic group, with different elements. Here the location of patients at marked distances from one another, the patrolling of these intervals of physical space, and the use of certain idiosyncratic displays—usually involving the whole body—which were specific to the individuals concerned, were prominent advertive techniques. In a less pronounced way, so was the use of physical objects and features of the environment such as tables, chairs, newspapers, cigarettes, etc. Finally, institutional factors such as hospital routine were combined with the latter to achieve, avoid and modify social contact.

In this group there were usually two or three individuals occupied in conversation and continuing activities of one kind or another, usually centering round the routine of the hospital, such as serving of tea and meals and clearing up after other activities. They kept up a continuous interest focussed on the nature of the whole proceedings. Away at the other end were individuals who were spatially isolated and who had to adopt various exaggerated forms of display in order to obtain attention for a brief interval during which contact was made and then rapidly lost with other members of the group. (This reminds one of the exaggerated displays of chimpanzees indulged in when two groups meet after having been separated from one another for some time.) Mutual display and mutual recognition involve gross movements of the whole body. What appears to be characteristic of advertence is that it is used as a means of establishing group acceptance of an individual by manipulating the attention of the group. Hence the initiation of a new group, or the realignement of an old group so that the individual can play a socially acceptable role in it or be personally accepted into the activities of a group when these develop, are two forms of such intervention. Advertence, therefore, is a form of interaction of group members involving the re-creation of group relations, and is typical of at least semi-unfamiliar conditions.

Another term is needed for the part played by non-verbal elements in the communication system of groups where coherence has been established at the level of operational activities. In such groups, the speaker, for example, announces by an eye flash that he is going to start a new point, or emphasizes one in the same way. The audience, by paying attention and being in *alert,* recognizes his attempts and urges on his participation. When individuals are seen to be "sleeping" in the audience it is another way of saying that quite a few of them have gone into *huddle,* and that the speaker's activities no longer matter to them or can be taken for granted by the group acting as a whole. In groups where technical discussions are under way there is a matrix of common understanding of a reality principle which allows an individual to come in with his contribution at the point where the technical requirements involve the presentation of the knowledge which he has in his specialized role. If he participates in another form it may be an attempt to direct the activity of the group into different channels, but this can only be done with a certain degree of advertised opposition; that is to say, withdrawing the attention of the group as a whole from the current form of ongoing activity in an attempt to transfer the attention of the group first of all onto oneself, but ultimately to get a realignment of the group's discussion.

One can now see that there are various degrees of coherence possible in types of discussions, dependent partly on the compatibility of the information flow from different individuals towards the common pool. These may be graded from negotiation, through forms of bargaining to various types of international confrontation. It is proposed that "annunciation" or the "annunciative role" should be used for the way personal advertising is used or attention manipulated in a group with fully operational designation. "Advertence" could then be reserved for defining the manipulation of group attention when it is used mainly for the acceptance of an individual within an existing group. Charisma then becomes a special form of advertence used by individuals to form a new group.

The essence of understanding here lies in being able to detect whether or not the advertive activity of an individual is directed towards gaining from the group an acceptance of a principle or mental concept of his social role, rather than merely establishing a new relationship of the members of a group to the advertiser. Advertence may place the advertiser at the focus of attention of the members of the group or merely establish the individual amongst the accepted communicants.

The distrust of charismatic activity by bureaucrats is derived as much from their resistance to the simple realignment of group activities as from the dangers that arise in an individual achieving a focal position in the attention of the group merely to bring about disruption of its activities without a constructive alternative. This latter is a real possibility, for it is a phylogenetically old development of such a state of affairs that such an individual can fix his focal position by the use of threat whether consciously or because of phyletic compulsions. When this happens, revolt is the ultimate phyletic sanction, providing the only way of opening up the group to future change. As yet the conscious formulation of a counterbalance to agonistic charisma, let alone tyranny, has not yet been found and is urgently required.

In agonistic forms of social cohesion the whole group is involved in the common activity of paying attention to a focus to avoid being aggressed against, but attention in the hedonic mode does not *ipso facto* provide social cohesion for the group as a whole. When individuals attempt to gain entrance into a group, or when an individual creates a group by drawing the attention of a number of individuals to himself, advertence is the method of achieving the attachment of the individual to the rest of the group.

HOMOMORPHIC LEVELS OF ATTENTION

The foregoing discussion of proxemics reveals a geometry of attention of individuals participating in a group activity towards the group's social centre. This centre is at any one moment the attention-gathering activity of a single individual. This individual may in fact be manipulating the attention of others in order to gain acceptance of himself as a member; such was the behaviour of the neurotics in therapeutic groups (when it was termed advertence), and such will be part of the activity of all those who manipulate the attention of others in groups. Advertive activity is part of what individuals do in groups. At a different (and higher) level of complexity, advertence is part of *annunciation* when it is used to deflect the group's attention to new aspects of a problem under discussion. In important respects the attention structure of advertence and annunciation is homomorphic in that both are directed at the focus of the group's attention, although in the case of advertence this is only a single individual and his place in the group, whereas in annunciation the focus is on the subject matter. The individual becomes a focus of attention only in his relation to that subject matter. The correspondence is in the framework of the organized behaviour of a single individual.

NOTE

This article first appeared in a modified form in Robin Fox (ed.), *Biosocial Anthropology*, London: Mandaby Press, 1975. The essential points of the discussion have been incorporated into the body of the paper.

REFERENCES

Angst, W. 1971. Personal communication.

Argyle, M. 1967. *The Psychology of Interpersonal Behaviour*. Harmondsworth, Penguin.

— 1972. "Non-Verbal Communication in Human Social Interaction", pp. 243-269 in: R.A. Hinde (ed.) *Non-verbal Communication*. Cambridge University Press.

Bernstein, B. 1965. "A Socio-Linguistic Approach to Social Learning", pp. 144-168 in: Julius Gould (ed.) *Penguin Survey of the Social Sciences*. Harmondsworth, Penguin.

Bolwig, M. 1963. "Observations on the Mental and Manipulative Abilities of a Captive Baboon *(Papio droguera)*", *Behaviour* 22 (1) 24-40.

Chance, M.R.A. 1960. "Köhler's Chimpanzees—How Did They Perform?", *Man* 60 (179) 130-135.

— 1967. "Attention Structure as the Basis of Primate Rank Orders", *Man* 2 (4) 503-518.

Chance, M.R.A./Callan, H.M.W./Pitcairn, T.K. 1973. "Attention and Advertence in Human Groups", *Social Science Information* 12 (2) 27-41.

Chance, M.R.A./Jolly, C.J. 1970. *Social Groups of Monkeys, Apes and Men*. London, Jonathan Cape.

Chance, M.R.A./Jones, E. In preparation.
Cranach, M. von. 1971. "The Role of Orienting Behavior in Human Interaction", pp. 217-237 in: A.H. Esser (ed.) *Behavior and Environment, the Use of Space by Animals and Men*. New York/London, Plenum Press.
Ekman, P. 1972. "Universal and Cultural Differences in Facial Expressions of Emotion", in: J. Cole (ed.) *Nebraska Symposium on Motivation*. Lincoln, University of Nebraska Press.
Emory, G.R. 1975. "Comparison of Spatial and Orientational Relationships as Manifestations of Diverse Modes of Social Organisation in Captive Groups of *Mandrillus sphinx* and *Thercopithecus gelada*", *Folia Primatologica* 24 (4) 293-314.
Goffman, E. 1969. *The Presentation of Self in Everyday Life*. London, Allen Lane.
Grant, E.C. 1969. "Human Facial Expression", *Man* (4) 525-536.
Jensen, A.R. 1968. "Patterns of Mental Ability and Socio-Economic Status", *Proceedings of the National Academy of Science* 60:1330-1337.
Jung, R. 1963. "Neurophysiologische Untersuchungsmethoden: Nystagmographie", in: Bergmann (ed.) *Handbuch der inneren Medizin*. Vol. 1. Berlin, Springer.
Kendon, A. 1967. "Some Functions of Gaze Direction in Human Social Interaction", *Acta Psychologica* 26:22-63.
Köhler, W. 1927. *The Mentality of Apes* (translated from the German by Ella Winter). 2nd ed. revised and reset. London, Kegan, Paul & Co.
Kummer, H. 1968. *Social Organization of Hamadryas Baboons—A Field Study*. Bibliotheca Primatologica 6. Basel/New York, Karger.
Kummer, H./Kurt, F. 1965. "A Comparison of Social Behaviour in Captive and Wild Hamadryas Baboons", pp. 65-80 in: H. Vagtborg (ed.) *The Baboon in Medical Research* vol. 1. New York, Head Press.
Labov, W. 1972. *Atlantic Monthly* 229:59-67.
Mandelkau, J. 1971. *Buttons: the Making of a President*. London, Sphere Books.
Mason, W.A. 1965. "Determinants of Social Behaviour in Young Chimpanzees", in: A.H. Schrier/H.F. Harlow/G. Stollnitz (eds.) *Behaviour of Non-Human Primates*, vol. 2. New York/London, Academic Press.
Menzel, E.W. 1970. "Menzel Reporting on Spontaneous Use of Poles as Ladders", *Delta Primate Report*. Tulane University Press.
Parker, C. 1974. "Behavioural Diversity in Ten Species of Nonhuman Primates", *Journal of Comparative and Physiological Psychology* 87 (5) 930-937.
Patrick, J. 1973. *A Glasgow Gang Observed*. London, Eyre Methuen.
Reynolds, V. 1972. "Man Also Behaves", pp. 143-158 in: Jonathan Benthall (ed.) *The Limits of Human Nature*. Harmondsworth, Penguin/New York, Dutton.
Reynolds, V./Luscombe, G. 1969. "Chimpanzee Rank Order and the Function of Display", in: C.R. Carpenter (ed.) *2nd Conference of the International Primatological Society: Behaviour*, vol. 1. Basel/New York, Karger.
Reynolds, V./Reynolds, F. 1965. "Chimpanzees of the Budongo Forest", pp. 368-424 in: I. DeVore (ed.) *Primate Behaviour*. New York/London, Holt, Rinehart and Winston.
Sade, D.S. 1965. "Some Aspects of Parent Offspring and Sibling Relations in a Group of Rhesus Monkeys", *American Journal of Physical Anthropology* 23: 1-17.
Schaller, G.B. 1963. *The Mountain Gorilla: Ecology and Behaviour*. University of Chicago Press.
Schiller, P.H. 1957. "Innate Motor Action as a Basis of Learning: Manipulative Patterns in the Chimpanzee", pp. 264-287 in: C.H. Schiller (part trans./ed.) *Instinctive Behavior*. New York, International Universities Press.
Virgo, H.B./Waterhouse, M.J. 1969. "The Emergence of Attention Structure Amongst Rhesus Macaques", *Man* 4 (1) 85-93.
Weber, M. 1968. *On Charisma and Institution Building*. Selected papers ed./intro. S.N. Eisenstadt. Chicago/London, University of Chicago Press.
Whyte, W.F. 1969. *Street Corner Society*. University of Chicago Press.

JOHN H. CROOK

Problems of Inference in the Comparison of Animal and Human Social Organizations

The attempt to reconstruct the most probable origins of human society depends upon the validity of inferences concerning the evolutionary descent of behavioural characteristics. Many authors have treated single aspects of human social behaviour, facial expression and gesture, aggression, violence, territorial behaviour and the like and, by reference to the literature in animal social ethology, have come to a diversity of conclusions regarding the biological structure underlying the character concerned. Such authors usually dichotomize between biological (*i.e.* genetic) sources of information involved in the determination of the behaviour on the one hand and acquired, traditional, societal or cultural sources on the other. This dichotomy, the modern version of the nature/nurture controversy, does little service to contemporary discussion as it neglects the complexity of the systemic relations between levels of determination responsible for the social context in which the behaviour happens and of which it is a part.

The problems of comparing the advanced social organizations of higher vertebrates are of a quite different order from those involved in the comparison of morphological characteristics. In the latter case, the ontogenesis of a character can usually be meaningfully discussed in terms of genetics and the epigenetic processes. The same has usually been assumed to be true of fixed action patterns of behaviour such as those common in the courtship of birds. Indeed, the stereotypy of ritualized displays has often been used in taxonomic studies of families such as the ducks (Lorenz, 1941), gulls (Tinbergen, 1959) or weaver birds (Crook, 1964). In considering social organization, we are concerned not only with the communicational aspects of sociability but, in addition, the spatial dispersion of populations, group composition and variability in size and membership, the structuring of relationships and the dynamics by which individuals move from one social position to another.

As long ago as 1878, Espinas had realized that there were systematic principles governing animal social organizations and that these allowed for varying degrees of adaptability. The sources of information determining contrasted levels of organization, say population dispersion and pair bond formation, often differ greatly so that evolutionary accounts of social organizations treated as if they were unitary characters simply will not work. Social organizations of species are cybernetic systems with varying degrees of stability through time. Their comparison at both the inter- and the intraspecific levels thus necessitates a comparison of whole sets of systemic relationships between measurable variables (Wilson, 1971). The implications of this essential fact for inferences concerning social evolution forms the background for this short paper.

From the evolutionists' viewpoint, a number of problems are involved in comparing non-human social organizations both with one another and with that of man. I propose to treat these in sequence.

PROBLEM 1. THE ABSENCE OF A MORPHOLOGICAL BASIS IN THE COMPARISON OF SOCIAL ORGANIZATIONS

In the comparison of the displays and other behavioural activities of an individual, description tells how different parts of the body move relative to one another. The behaviour is contingent upon a morphological basis and, in the case of fixed action patterns universal to a species, it may often be treated as an innate trait and discussed in the traditional manner of comparative anatomists. Some authors have used the term "social organization" as if it were logically equivalent in status to the concept of the fixed-action pattern. This, however, is to use the term to blanket unthinkingly a wide diversity of intercalating processes commonly determined by widely differing sources of information. Thus, while a full account of the communication system based on displays might perhaps be treated as equivalent to a morphological system, this cannot be true of dispersion patterns which perhaps may be largely a consequence of the location of resources.

In studying social organization then there is no close association between morphology and behaviour such as has formed the basis for taxonomic and evolutionary inference in the ethology of individuals (see Petrucci, 1906; Crook, 1970a). Furthermore, fossil evidence for societal organization is scanty so that inferences from palaeontology to dispersion patterns or modes of relating in the reproduction of a species are usually extremely shaky. The causal processes accounting for what is termed social organization are "multiplex" as Hinde (1975) puts it. From whatever perspective one approaches the issue, a systemic interrelationship between factors of differing informational provenance is commonly the rule. Inference from description to the evolution of such systems is difficult and the meaning in this context of terms such as homology and analogy, convergence and parallelism much called in question.

PROBLEM 2. HOW SHOULD WE DESCRIBE SOCIAL ORGANIZATIONS?

The use of the term social organization has undergone changes in recent ethology. To Lorenz as to Niko Tinbergen (1953) the term mainly denoted the interactions between individuals functioning with respect to mating, fighting and group life. Prime emphasis was placed upon the fixed action patterns or social rituals mediating reproductive or protective strategies. Hinde's recent structural approach (1975) to the organization of primate social groups is similar. He conceives of three levels of complexity, interactions, relationships and group structure. He sees relationships as "the content, quality and patterning of interactions and structure as the content quality and patterning of relationships." These approaches allow valuable descriptive accounts of organization seen as it were from within. This "internalist" view is fundamentally interactionist and social organization is conceived primarily as the patterning of interactions.

A different, more functional approach is to treat social behaviour with respect to the prime strategies of biological significance to individuals—mating and rearing, protection from predators, foraging. Thus, for example, in Crook's monograph on social organization in weaver birds *(Ploceinae)* (1964) social behaviour is treated in terms of two aspects: population dispersion in relation to foraging resources (gregariousness, flock congregation, dispersal systems, territorial dispersion, etc.) and pair formation in relation to breeding success in contrasted circumstances (polygamy, monogamy, colonial versus territorial breeding, displays and rearing practices). Crook found it possible to associate four grades of social organization with differing types of resource exploitation in different ecologies. This approach, foreshadowed many years earlier in Sir Julian Huxleys attempt (1923) to explain the diversity of mating systems in birds, has proved valuable as a tool to comparative understanding of social organization in the classes *Aves* (Crook, 1965; Lack, 1968) and *Mammalia* (Eisenberg, 1966; Crook, 1975a, b; Crook/Ellis/Goss-Custard, in press) and in smaller taxonomic units such as the *Carnivora* (Kleiman/Eisenberg, 1973), ungulates, especially *Bovidae* (Jarman, 1974) and primates (Crook/Gartlan, 1966; Crook 1970b, 1972; Eisenberg/Muckenhirn/Rudran, 1972; Goss-Custard/Dunbar/Aldrich-Blake, 1972). These studies are primarily correlational so that inference to the actual causal factors involved are not explicitly tested. Furthermore, the early stages of such analyses rely perforce on relatively superficial knowledge of the ecological factors involved. The value of such work is that it has stimulated increasingly critical research on 1) direct experimental studies of the relations between social dispersion and resources (*e.g.* Zahavi 1971); 2) monitoring studies in which relations between ecological factors and aspects of social organization are researched longitudinally or the differences between contrasted research locations analysed quantitatively (*e.g.* many Japanese monkey researches and Richards' sifaka study, (1973), etc.); 3) careful theoretical studies deepening existing hypotheses (*e.g.* Ward/Zahavi, 1973; Downhower/Armitage, 1971; Dimond/Lazarus, 1974); and 4) broad surveys of a Darwinian stature providing wider canvasses for speculative thinking (*e.g.* Lack, 1968; R.B. Martin, 1972).

Detailed examination of the social processes determining the biological success of individuals within a population reveals that in addition to the main life strategies, other more subtle behavioural processes may be involved. Primate males, for example, manipulate others through social subterfuge to increase rank and/or probabilities of mating (Crook, 1970; Deag, 1974). Many of these processes are probably of local occurrence and thus part of the traditional behaviour of particular populations. In a sense we have here the primordial roots of cultural change and history.

We need to be aware that the social organizations of advanced vertebrates such as communal babblers, hyaenas, elephants, lemurs, baboons or apes can be approached from several perspectives, no one of which fully delineates the whole and all of which need to be placed in juxtaposition. The table shows three approaches to comparative description of primate social organizations. A complete account of the social system of any particular species would require examination from at least each of these three perspectives, a task that is clearly neither simple, cheap nor achieved in a short time.

Perspective A concerns the spatial dispersion of individuals in a given population or deme. Theoretically, individuals may be randomly or evenly dispersed or clumped into congregations of varying size and composition. Conditions of random or even dispersion are rarely attained so that populations are commonly distributed in some pattern in which more individuals occur in some places (at certain times) than at others. This dispersion pattern is, however, quite usually predictable through time once sufficient information concerning it has been collected.

The pattern is described in terms of the temporal cohesion of groups, their membership in terms of age-sex categories, the way in which membership changes according to circumstances and through time, the relationships between individuals in terms of who stays close to whom and the degree to which whole groups come together or stay apart. A society may show great

A. *Structural Perspective*	B. *Interactional Perspective*	C. *Social Dynamics Perspective*
1. Spatial dispersion, individuals, groups	1. Communication-channels, signals, responses	1. Group growth, cohesion and division
2. Temporal cohesion of groups, variability in membership, throughput of constituents	2. Bond forming processes	2. Individual social position changes contingent on weaning
3. Composition of groups as stable units	3. Hierarchy formation and maintenance	3. Social position changes contingent on puberty
4. Relational associations between group constituents	4. Parent-child interactions throughout growth	4. Social position changes contingent on aging
5. Relational associations between groups in deme	5. Social subterfuge	5. Individual exclusion in sexual competition
		6. Patterning of consortship interchange
		7. Relationship formation, rules of alliance and control
		8. Intergenerational replacement in social positions

consistency in patterning through time or may undergo changes dependent upon density, the seasonal occurrence of mating and giving birth, or climate. These "phase" changes (McBride, 1964) may amount to quite major reconstructions of the way individuals space themselves and spend their time.

The determinants of spatial dispersion include both extrinsic and intrinsic factors. Extrinsic factors include the distribution of foraging sites, the availability of food for capture, nest site or refuge availabilty, the occurrence of cover and so on. Intrinsic factors are behavioural, such as the extent of gregariousness or territory creating mutual intolerance or again the behavioural responses to endocrine changes as a result of seasonably induced breeding. Spatial dispersion is thus the topographical expression of the interaction of these factors in space and time.

Perspective B concerns the processes that mediate both the cohesion of the spatial pattern and the changes that occur within it. Here we have the communicational network, the processes of courtship, bond formation and mating, the rearing practices of the species and the roles of the two sexes in it, the circumstances in which aggression occurs and the ways in which aggression is contained. Finally there are the complex processes by which old individuals are replaced by younger ones. Lorenz, in several papers, explored the use of signals to reduce aggression in an opponent, an important factor in the evolution of mechanisms mediating spatial and sexual rivalry through threat display rather than damaging violence. It is clear now that at least in certain primates even more complex practices of social subterfuge buffer the antagonism between rivals. Thus Deag and Crook (1971) report on the complex use of mutual baby presentation between males in the regulation of agonistic interactions. The local character of some of these complex practices suggests they can reasonably be described as proto-cultural.

Perspective C treats of the way in which individuals occupy differing social positions through time. When groups grow in size their cohesion weakens and they split into separately ranging units. When individuals are weaned, reach puberty or assert mating rights against older rivals, they shift "roles" or social position. Aging animals move likewise. Formation of alliances between male members in a hierarchy or of females in a one-male reproductive group produce power-sharing caucuses whereby the behaviour of other individuals may be constrained or manipulated. Counter alliances or group shifting may be part of a compensatory process whereby young reproductives gain mates. In many species young males are excluded from reproductive groups and often suffer higher mortality rates in sub-optimal habitats. Re-entry into the reproductive community is often associated with elaborate bond forming behaviours, social guile or violence.

Social dynamics are the result of 1) demographic processes contingent on changes in numbers in relation to resources including the availability of mates and, 2) behavioural processes whereby sexual needs are satisfied and/or bonds formed, commonly against the will of an established, perhaps collaborative, power sharing structure enforced by the existing reproductives—usually males. Thirdly, 3) social dynamics are concerned with the intergenerational replacement of older individuals by younger ones in existing structures and/or the dispersion and creation of new reproductive social units moving in new ranges. Both processes are likely to involve competition as foraging resources and mate availability are commonly limited.

In recent papers, Hans Kummer (1975) has begun to explore the rules which govern dyadic interaction and change in social position. In a captive gelada baboon population, males operated social strategies in relating that tended to maximize their compatibilities with others in both reducing antagonism with other males and in gaining access to females. Females aim to maximize their access to chosen males. In realizing the goals of these strategies both sexes get involved in intervening in the established relations of dyadic pairs. The result is an exceedingly complex competitive process eventuating in relative degrees of success or failure for the players. Size, age, social skills, "confidence" are all attributes influencing the process. Social skills in the use of allies by subordinate individuals and the promotion by dominant males of supporting relationships with agonistic inferiors are all elements in the process. Kummer's work relates well to Dunbar's findings in the field (1974). Taken together these studies greatly extend our comprehension of primate social dynamics.

PROBLEM 3. HOW ARE WE TO COMPARE SOCIAL ORGANIZATIONS?

Comparisons of specific patterns of population dispersion and mating organization often correlate well to form a complex of characteristics clearly associated with a way of life fundamentally dictated by a foraging economy. In the weaver birds, for example, forest species are usually insectivorous solitary and territorial whereas savannah birds feed on seeds and are gregarious and colonial. Dispersed, stealthy hunting seems related to foraging for insects in forests while flocking and colonialism may facilitate finding patchy seed resources through an informational exchange at sites of congregation (Ward/Zahavi, 1973). The polygamy of the savannah birds correlates with the great surge of food abundance in the savannah rains allowing females to rear young largely unaided while males maximize time and energy spent in mating. Furthermore the displays in courtship, while comprising comparable innate elements, are constructed differently in each species according to nest visibility, site and structure (see Crook, 1964, 1969, and discussion in Lack, 1968). B. Nelson's studies (1970) of the *Sulidae* show how the social behaviour of the boobies and gannets is likewise related to contrasts in foraging economy in different types of marine environment. Such studies suggest that selection of a complex set of co-adaptations has beautifully moulded foraging and reproductive behaviour into patterns of clear survival value and that the foraging strategies are prime determinants of the system. All features can nonetheless be treated as products of natural selection.

Difficulties arise when the same species shows intraspecific variation in social organization according to contrasts in local ecology. Here one may choose to focus on the more stereotyped aspects of social life and neglect the more flexible; but to make this choice is an arbitrary decision resulting from personal bias.

It may be argued that social structures comprise the interactions, relationships and higher order associations of individuals and that only these socio-behavioural aspects of group organization are the proper subject for comparison between species. If this were possible it would certainly greatly ease our task. To a limited degree we can do this, for example in comparing the mating systems of ungulates. However, we soon run into problems since the behavioural manifestations of the ways in which individuals relate occur within an environmental context which may impose constraints on the behaviour concerned. For example, the type of herding behaviour shown by male wildebeeste in attempting to group together females for mating occurs in a variety of different types of population dispersion. In the most grassy part of the Ngorongoro crater, Tanzania, the female herds move little and males establish territorial areas in which they herd those females that enter. Elsewhere the females graze more widely and the males' behaviour lacks topographical reference. Here it may be argued that the mating system is a universal character of the species while the dispersion of the animals in space is a response to environmental conditions and of less importance in comparative study. Things are, however, rarely that easy.

In Alison Richards' beautiful study (1973) of the sifaka *(Propithecus)* in the forests of Madagascar, it is clear that important aspects of social behaviour depend very much on local conditions for their manifestation. Between populations of this species, whether or not collaborative group territorial defense occurs is a function of local ecology. Where food dispersion is seasonally affected in relatively arid forest, group areas are defended. In less arid forest with similar food abundance but less seasonality the animals overlap in ranging extensively and group conflict does not occur. Many recent primate studies point in similar directions.

If then we take a broad canvas and make correlational studies of social organization in relation to ecology, marked associations are commonly found. For example, the range of avian and mammalian social organizations is not enormous and general descriptions relating this range to ecology has provided valuable indications of the way in which social form may have evolved and radiated in differing ecologies through time (Eisenberg, 1966; Lack, 1968; Crook, 1965; Crook/Gartlan, 1966; Crook/Ellis/Goss-Custard, in press). This too is true in animals close to our own species. Thus recent studies of the orang-utan (Rodman, 1973; MacKinnon, 1974) the gorilla (Schaller, 1963) and the chimpanzee (van Lawick-Goodall, 1968, 1971; Sugiyama, 1973; Wrangham, 1975) show how their differing behaviour is related to contrasts in their foraging ecologies.

In a survey of mammalian social systems it appeared to me that reproductive and rearing practices were generally the more conservative aspect of social organization. By contrast, social strategies in relation to foraging are commonly more flexible and open to direct environmental modification. Food resource exploitation is affected not only by the density and abundance of food items but also by the density, abundance and mobility of the foragers. Gartlan and Struhsaker (1972) for example, suggest that mixed species congregations of guenons in West African tropical forests occur when food abundance is in excess of demand but that, in other seasons, when this condition does not obtain, the different species partition the habitat into more clearly distinct foraging niches and do not consociate. The formation of these groups under ecologically permissive conditions is considered possibly to be part of an anti-predation response to monkey eating eagles.

In another example, Indian langurs are found in multimale or one-male reproductive groups apparently in association with contrasts in habitat and the abundance of the monkeys in relation to resource availability. In a review of the literature, Crook (1972) shows with simple theoretical models how transitions between multimale and one-male groups may depend upon changes in the abundance, dispersion, availability and seasonality of food resources. The detailed investigation of the processes and factors involved has now started, as the work of Clutton-Brock (1972) for example, shows.

The broad comparison of species' social organizations in relation to resources suggested that they could be well explained as ecological adaptations. Comparisons of key attributes thus led to hypotheses regarding origin, descent and radiation in evolutionary terms. The organization was seen to be the collective result of the inherited behaviour of individuals causing them to pattern their spacing and relationships in ways that tended to maximize their reproductive success. The complexity of rearing systems can also be approached in this way treating parental investment in ensuring survival of offspring to reproductive age as a prime characteristic. In some conditions both parents need to invest, in others males invest much less than females and maximize impregnations rather than assisting in rearing. These strategies can be interpreted as adaptations which in no way contravene neo-darwinian principles (Trivers, 1972).

Flexibility of response to ecological diversity implies behavioural adaptability so that individuals may consociate in differing ways in different parts of their ranges. Here it becomes of key interest to find out exactly how the relations with environment are achieved. Comparison here tends to treat both intrinsic and extrinsic variables and to show how changes in their values bring about systemic shifts, thus focusing on the differing ways in which dispersion systems are controlled. Identification of relevant environmental variables is one aspect of this work. The mechanisms whereby animals respond individually and societally are another aspect so far but little studied.

PROBLEM 4. HOW DO WE COMPARE SIMILAR SOCIAL STRUCTURES THAT ARE MAINTAINED BY DIFFERING SOCIAL DYNAMICS?

Early classifications of cercopithecoid primate social organizations included two types of group within which reproduction and rearing occurred. These were the multimale group (several males, their relationships commonly organized according to a status hierarchy, with a large number of females and young) and the one-male reproductive unit in which closely bonded females associated with a single reproductive male. In comparative studies species or local populations were often allocated to one of these two categories. Research on the patterning of interactions that comprise the social relations within these groups shows however that even when structural organization in terms of dispersion, party composition and sex bonding is descriptively very similar, the dynamics maintaining the organization may be strikingly different. Perhaps the clearest example of this lies in the comparison of *Papio hamadryas* societies (Kummer, 1968) and those of *Theropithecus gelada* (Crook, 1966; Crook/Aldrich-Blake, 1968; Dunbar, 1974).

Both animals may be found in large groups made up of distinct one-male units together with a surplus population of non-reproductive males. In both cases, animals sleep on rocks or cliffs and disperse outwards to forage daily. At first sight then, there is a structural similarity. The Patas monkey (Hall, 1966) also has one-male units and has been associated with geladas and hamadryas in the past. But here the units move in very large separate home ranges and do not coalesce in herds. In fact further consideration shows that what these three species have in common most is their terrestrial (non-arboreal) mode of life in open country. Yet other species also have one-male reproductive units. In forest terrestrial animals such as the drill (Gartlan, 1970) and probably the mandrill very similar social behaviour is found. Again in more distantly related cercopithecoids such as langurs and some guenons comparable structures occur. In this case, the taxonomic distance is apparent because the signal systems, body form and cries of these arboreal animals differ greatly from those of the terrestrial species which are more closely related taxonomically. Here convergence of social organization in relation to ecological contingencies rather than habitat characteristics seems likely (Crook, 1972).

In the case of the gelada – hamadryas comparison there are rather striking ecological similarities, both live in areas with seasonal aridity although the montane gelada also experiences a long wet season with lush vegetation never experienced by the desert living hamadryas.

The differences between these two species are as follows:

— The non-reproductive males of the gelada occur in coherent groups while those of hamadryas form a relatively unstructured periphery to the reproductive units.

— Gelada herds split up into their basic social units especially when food resources are sparse. Thus in the dry season herd sizes are smaller and party frequency higher. Hamadryas move in more coherent troops that fragment for dispersed feeding by social units which forage but which come together again daily to form the coherent troop.

— Members of a gelada reproductive unit may disperse quite widely (up to 200 yards or so) when foraging and, although young reproductive males are dominant leaders, males in their prime are largely peripheral to the unit which is maintained by bonds between females. By contrast, male hamadryas are domineering and females are prevented from moving far from them.

— In correlation with the third point, the association of a young hamadryas male with a reproductive unit does not result in the development of sexual relating between him and the young females in it as is the case among geladas. Rather the hamadryas assumes a mothering protective role which establishes a relationship between him and very young sexually immature females. Thus in the early stages of the formation of reproductive units, male geladas relate sexually but this is not the case among hamadryas.

— Dunbar (1974) shows that these contrasts relate to more subtle differences in relations between individuals which show up when sociometric analyses are compared.

It becomes clear that apparent similarity in social structure results from considering primarily the population dispersion aspect of the system and that this masks important behavioural differences in the way the animals relate. Clearly, in trying to understand the gelada, hamadryas and Patas societies, it is in the unravelling of the social process itself that the most valuable answers lie. Important contrasts in the innate basis of individual temperament must be in play. Thus the intolerance of male hamadryas to female wandering must be based on thresholds that trigger at different levels than in the gelada. These contrasts probably have roots in phylogenetic history.

Hamadryas and anubis baboons, both species of the genus *Papio,* doubtless share many behavioural attributes missing from the gelada repertoire. One may expect that the derivation of one-male units from multimale units (or vice versa) among *Papio* baboons is based upon changes in response thresholds within the same overall repertoire whereas the one-male *Theropithecus gelada* unit operates on the basis of a contrasting motivational system and behavioural repertoire. It is then at this level that phylogenetic analysis resumes its importance.

It seems reasonable to expect also that the development of local social traditions is likewise influenced by the innate basis of motivational thresholds and behavioural expression. For example, male care of babies and their utilization in regulating male-male interactions occurs in both Japanese and Barbary macaques but only in the latter has it developed into a really complex ritual apparently functioning in agonistic buffering and the maintenance of group cohesion. We do not know yet how local these rituals are but we do know that the feral animals on Gibraltar differ from the wild ones in the Atlas mountains. The extent of the contrasts between Japanese and Barbary macaques suggests that their roots may lie in genetic differences. Yet, we cannot be sure. We know that complex behavioural traditions in macaques can be built up from reinforcement contingencies that are themselves the result of pure happenstance. Potato washing and swimming developed in a particular island population of Japanese macaques as a consequence of shore line provisioning. This was an observer's device for obtaining information and has little to do with the natural history of the monkey's island. The same may be true of the origins of local traditions elsewhere. Accidents of inter-individual relating may set off social reinforcement contingencies that are novel and unique to a particular group. This in turn may affect other individuals so that a patterning of responses may develop and spread through a deme. Given time, one would predict that individual responses to such an established tradition may vary and that some individuals would succeed in reproduction under the changed societal circumstances better than others. A social selection pressure different from those operating prior to the establishment of the tradition may thus come into operation so that motivational traits associated with reproductive success are differentially selected. This means that the frequencies with which certain motivational expressions, inter-individual interactions and patterns of relationship occur, will diverge between populations. The extent to which genetic changes persist will then become a function of the degree of genetic interchange between demes as well as a function of the survival value of the changed behavioural traits in different societal traditions. Kummer's work on introducing *Papio anubis* to *Papio hamadryas* groups indicates the kinds of behavioural threshold and interactional hiatuses that may be involved here.

We are back in an area where the nature/nurture problem once more rears its head in a peculiarly complex form. It seems that without detailed experimentation on inheritance and genetics we can only guess at the extent to which phylogenetically divergent traits are involved in the determination of such complex interaction sequences. That they are involved seems certain, but the extent to which they determine the patterning of behaviour remains imponderable from field data of the available kind. Is this important? It does make any decision about the evolutionary homology or analogy of a given interaction pattern uncommonly difficult. To pretend otherwise through a misguided use of Lloyd Morgan's Canon or to make guesses which have important consequences in theory seems to me to be quite irresponsible. Given our present state of knowledge about the informational sources determining the ontogeny of relationships and social structure, the answer to questions about their homology can only be an appropriately informed "don't know".

PROBLEM 5. WHAT IS THE RELATIONSHIP BETWEEN EVOLUTIONARY HISTORY AND CULTURAL HISTORY IN THE ORIGIN OF ADVANCED MAMMALIAN SOCIAL SYSTEMS?

In discussing the way in which social organization changes through time it is now clear that several different processes are at work. Furthermore while all are going on at the same time, they operate at different rates. In an earlier discussion I attempted to tease out some of these intercalating processes (Crook, 1970b). Restating them here they include:

— Changes in individual activity rates and time energy budgeting contingent upon gradual shifts in the availability of environmental commodities. These changes affect the time energy expenditure on social relations as contrasted with object relations. Changes in the reinforcement schedules provided by nature in relation to individuals' needs impose differing patterns of learnt behaviour.

— Changes in dispersion of resources may impose shifts in the spatial arrangements of groups and individuals. Competition and the acquisition of means for coping with competition brings about contrasts in the spacing and relating of individuals. These may become relatively fixed as traditions. That which an individual learns will be related to the satisfaction of needs in ways that closely resemble the behavioural adaptations that arise due to the natural selection of genetic traits.
— The long term stabilization of spatial grouping and relational traditions provides a societal environment in which animals mate and rear their young. We may expect that those genetic traits underlying the temperamental basis of biologically successful behaviour in given societal conditions will be positively selected. Naturally for selection to occur in this way the societal norms will need to be stabilized over many generations, and this in turn necessitates long term environmental stability. Such conditions are, however, not rare in certain habitats, such as the Amazonian basin rain forest for example, and both gradual change and stabilization of traits in this manner seems a highly probable mechanism in evolutionary history.
— Changes in the population dispersion aspect of social organization are unlikely to occur independently of changes in social dynamics. Thus the contingencies of both the environmental and societal conditions provide the selection pressures described under the various categories of sexual selection (see Crook, 1972). Likewise the development of genetic traits basic to the intelligence underlying the social skills used in hierarchy modification or maintenance, agonistic buffering and the like is likely to have occurred in ways that are not independent of environmental conditions.

In general then, animals respond directly to environmental changes by behavioural shifts. They also set up patterns of acquired behaviour in relation to changes in natural reinforcement schedules and these become the basis of elaborate traditions of interaction. These in turn provide the societal environment which, together with the physical environment, yields the contingencies that select shifts in the genetic constitution. These sets of causal factors are thus "nested" within one another in a highly systemic manner. There is a marked interdependency between their effects on behaviour so that without extremely detailed analysis the weighting to be attributed to any one of them in explanation is difficult to determine. No analyses of social systems in mammals can yet approach a resolution of this problem although valuable small-scale attempts on particular species and the creation of useful models are opening up the way (*e.g.* Downhower/Armitage, 1971; Trivers, 1972).

Historical change in social organization requires much more extended explanation than is involved in the neo-darwinian explanation of evolution of morphological characteristics. In the latter, environmental selection operates progressively on relatively passive genetic material through the transgenerational elimination of certain unsuccessful traits. Change in social organization involves the programming of behaviour rather than simply its selection or elimination. This programming itself sets up further dependent programmes through the establishment of acquired tradition. These programmes comprise the social environment within which the selection of behaviour occurs. The argument thus needs extension to encompass this complexity.

With respect to man, Talcott Parsons (1966) argues that the information controlling social processes arises at four levels that can be usefully abstracted for analytical purposes. These levels are the organismic, the personal (*i.e.* temperamental, personality, character level), the societal and the cultural. These differ in their contribution to the functioning of the system as a whole. Organism and personality interact to mediate individual adaptation and the acquisition of appropriately goal-directed behaviour while the social level consists of factors integrating patterns of inter-individual interaction and relationship. Parsons considers the cultural level to be of prime importance in man since it operates through the "legitimation" of living patterns, acquired values and rationalizations of behaviour in relation usually to ultimate or absolute values. These are often organized as a system of religious or philosophical beliefs relating man to fate and the universe. Changes in belief systems can thus have major effects on a social system and may operate without direct reference to ecology.

Maitland Bradfield (1973) shows how cultural, economic and ecological factors interact systemically in the evolution of nomadic, settled and agricultural community styles of tribal life. Thus the development of religious beliefs, matrilinearity in kinship and ownership of land by Pueblo women in certain Amerindian societies of North America is derived from female cultivation of grass tracts near temporarily flowing streams in a seasonally stabilized residence pattern. This in turn is the consequence of the effects of food-rich locations upon an otherwise nomadic ranging pattern of a hunting-gathering people.

It must be clear by now that in examining social evolution we need the concept of stages. Each stage comprises a systemic interaction of sets of factors that provide information from different organizational levels. In birds and lower mammals these are primarily environmental, phylogenetic and innate behavioural levels. In advanced mammalian societies such as those of many primates a new level, that of traditional behaviour involving social skills adaptive to success primarily in intergenerational competition, comes into play. In man this level develops as a result of language and education into systems of value and legitimation. Yet the boundaries between these abstracted levels are often not clear and the interaction between them is obviously complex.

Using an analogy with the gelada baboon as his first stage, Clifford Jolly (1970) derives protohominid hunting society from it through a careful consideration of palaeoecological history, analogies in morphology and modern accounts of baboon and chimpanzee society. His paper shows it is clearly possible to provide a plausible and coherent account of the probable derivation of human societal organization and the emergence of culture from primordial elements.

We now know that chimpanzees, like men, are commonly hunters and carnivorous. Furthermore, males are the animals most involved in this activity (Teleki 1973; Wrangham, personal comm.). If we are to seek the patterns by which human hunting evolved, it may be useful to consider not only its probable origin and development in our ape relatives but also its convergent or parallel evolution in the true carnivores. Because of the great taxonomic distance between primates and carnivores the argument here is clearly by analogy. Nonetheless, it leads in an interesting direction. Schaller (1972) points out that the formation of hunting groups in the *Carnivora* is associated with the following effects:

— Group hunting results in greater success than individual hunting and groups take large prey animals.
— Group size is flexible in relation to the character of the prey hunted.
— Group hunting often involves cooperative circling, short cuts and relaying procedures which reduces the energy expenditure of the group. These tactical manœuvres thus develop prior to language.
— Division of labour occurs in some species, some individuals hunt, some track, some kill, some carry while others care for the young. Yet others may maintain vigilance. These functional roles are also interchangeable.
— In dogs and wolves greeting ceremonies prior to hunts seem to strengthen bonds and synchronize behaviour.
— A tendency for intraspecific aggression to out-groups tends to produce separate group ranging or group territorialism, as in the hyaena (Kruuk, 1970).
— Food storage in trees, in water or under a stone may occur within the range.

If we superimpose these characteristics upon a chimpanzee derived picture of possible hominid hunting techniques we get a vision of a process remarkably resembling that of some primitive peoples (see also Wash-Burn/Lancaster, 1968). The origins of inter-group aggression and warfare are clearly deducible from such an account through considering the effects of food shortage on the relationship between groups in neighbouring territories.

PROBLEM 6. HOW FAR ARE HUMAN PSYCHOLOGICAL ATTRIBUTES UNIQUE THUS RENDERING COMPARISONS BETWEEN HUMAN AND ANIMAL SOCIAL ORGANIZATION OF PURELY ACADEMIC INTEREST?

The literature on human social evolution is now full of accounts tracing the way in which non-human primate grades slowly changed into those of early men. Apart from Jolly's paper, the discussions of Washburn and Lancaster (1968), Fox (1967; 1972), Reynolds (1968) and Tiger (1969) are guides here. The theorizing involved draws on many sources, many analogies and several theoretical bases. We can in fact picture, not clearly perhaps but at least in outline, the way in which human society emerged from that of animals. It is however at this point that an anxiety strikes home. Many ethologists consider their work of value in aiding psychology towards a comprehension of our species yet, when all is said and done, an enormous gap suddenly yawns. For what goes on right now in the streets, offices and institutions of Manhattan, Whitehall, or the Kremlin has seemingly little relationship to these just-so stories of long ago. After the canvas has been painted as best we can, we are brought up short by the abrupt question, so what?

If we look at the gallant attempts of ethologists such as Lorenz (1966) and Morris (1967) to scale the heights of sociological relevance, the result is a great deal of sound and fury signifying very little. The substantive contribution isolated from the rhetoric and the special pleading amounts roughly to this:

— Many aspects of human non-verbal communication by facial expression and gesture resemble those of non-human primates. They appear universal to our species and are therefore likely to depend upon innate bases. The same appears true of the components of certain communicational rituals, the greeting interaction for example (see van Hooff in this volume; Vine, 1970; Eibl-Eibesfeldt, 1970; Kendon/Ferber, 1970). Yet these same elements in the communication repertoire mingle with culturally induced ways of expression to produce complex codes of manners that carry the communication of affect in a given society.

— Psychogenetic research with rats, dogs and birds shows that determinants of behaviour such as intelligence in maze running, tameness or wildness and other aspects of temperament are inheritable and can be selected in breeding experiments. It is likely that human motivational characteristics involved in violence, cruelty and aggression have been selected during centuries of hunting-gathering existence. Yet these are superimposed upon an older vegetarian mode of life and patterned emotionalism not unlike that of chimpanzees. The mix is unique and subject again to a wide range of cultural controls expressed in value systems of behavioural legitimation.

— Ethological motivation theory of the 1930's has been used to suggest that the inheritance of temperamental traits implies acceptance of particular "instinct theory" interpretations of behavioural control. In fact Lorenz's account (1966) of the motivational control of aggression is widely doubted (*i.e.* Montagu, ed., 1973) and he continues to repeat ideas (1974) the basic elements of which were forged in another period (1940). There is in fact no need to confuse this issue with the inheritance from animal forebears of a diversity of behavioural traits controlled by mechanisms that may be of different kinds.

— It is now possible, as we have already shown, to draw up a picture of the interacting factors that determine complex social organizations in advanced mammals. We can also suggest pathways of historical change whereby these early stages progressed to the human level.

Yet the essence of man in this type of discussion is somehow lost. So long as the continuity of animal and human traits is the focus of attention, the uniqueness of man remains somehow unseen. It is this invisibility that accounts for the shock; these accounts of human origins appear to leave something vital out. Until ethologists think far more deeply, this trivial but vital blockage will remain.

Timbergen has pointed out that some of the most valuable contributions of ethology to psychology are methodological, a point taken up by Blurton Jones (1972) in a book in which study designs largely developed with animals are applied to humans. Tinbergen (1972) also shows that functional analyses of single behaviours in animals offer analogies for human studies in spite of the added complexities at this level. He emphasizes that the unique feature of human adaptability lies in cultural transmission of novel information to succeeding generations. It is this that has allowed the astonishingly rapid conquest of the terrestrial environment and the anthropogenic changes in habitat that are the results of steadily accelerating rates of population increase and the extension of urban and suburban habitats. He points out several areas in the mental health and educational fields where human adaptations to an older way of life may be failing to cope with these artifacts of cultural "success." Yet human peculiarities seem to lie deeper than this and for information here we need to turn to those gropingly tentative sciences seeking to explore the psychodynamics of the human mind (*e.g.* see Berger/Luckmann 1967).

I would like to argue that man's linguistic ability to refer to objects and their relations using a verbal code of nouns and verbs has had psychological effects that raise man to a distinct grade of psycho-social complexity. These effects are so wide reaching in their implications that, although thin lines of continuity cross the boundaries between this and animal grades, any comparison is of little value unless the distinctiveness of the human level is adequately examined. The effects of language capacity in man include the following:

— The grammar of language usually divides space and time according to a here/there dimension and into past, present and future tenses. Ashley Montagu (1965) has pointed out that this linguistic enshrinement of space and time conceptualization is a crucial element in developing control over events. Clearly in hunting this improved ability in timing and in precise symbolic reference would be of great adaptive significance.

— A consequence of language for human beings in society is that an individual is referred to as an element in the space-time object relations of other individuals. In hearing of himself in this way, a human being comes to see how others see him. In language this takes the form of the distinction between the nominative and the accussative cases, I and me.

— The course of individual development is prolonged in the human species with an especially long period of dependency on the mother. The attributions of values to the child's behaviour begin during this time. A child is good or bad in relation to his mother's tolerance of his demands. Mother's behaviour, furthermore, is largely patterned by the social traditions of her background and group expectations about how the mother-child relationship should be. The child obtains a picture of the relations between what it wants, what it does, and what mother (and "others") expects, and internalizes this as a self-referring judgment. Here lies the origin of repression when the child's frustrations and consequent aggression and resentment cannot be expressed for fear of loss or punishment. To cope with these internal conflicts, the child elaborates defences against the powerful and frightening feelings the conflicts evoke and these emerge in behaviour as emotional masking and skills in social manipulation.

— Children come to "introject" the parental value systems and thus conform with the social expectations of their milieu. Yet this is at the cost of maintaining active defences against repressed tendencies that are constantly re-activated. As the child becomes adult and relates sexually, a situation of dependency resembling that which he has on the mother recurs. This then becomes the prime arena in which the archaic fears and defences of childhood are once more aroused. The phenomenon of transference is thus established.

— In man, there is thus usually a store of habitual responses involving self-frustrations. While on the one hand these are coping skills, on the other they reveal a need to work through repressed material to relieve the fossilized conflicts of childhood. The "unconscious" in this form must be an unique human phenomenon and its power to make and mould lives is little short of fabulous.

The human individual is thus a highly sentient being that refers to himself, evaluates his or her behaviour in relation to others and modifies or masks many motivational impulses in order to maintain a societal position. Of course, once the rules are known, subterfuge of many kinds is used to improve self-respect, status with regard to others, control over commodities and reproductive and economic success in the intergenerational turnover. Here men, baboons and chimpanzees parallel one another, yet at the human level, behaviour depends on relations between "persons"—individuals defining themselves in various ways in terms of childhood introjections and the defences these have involved.

Men and women then are "persons" in this sense and without consideration of the psychodynamics and psychosocial development of the person any inference from animals to man is likely to be incomplete or, worse, naive.

And there are yet more complications. The societal value system of legitimation which governs the acceptability of an individual's behaviour is in itself the collective defence mechanism against behaviour which at the individual level evokes fears of overwhelming loss or punishment. As Talcot Parsons (1968) puts it, the social and personal systems are articulated through the functioning of the "ego" while the "superego" is the point of articulation between culture and societal organization.

In the same way that the linguistic code is both the basis of self-reference and eventually of individual identity so, at the societal and cultural levels, the legal-political and religious-moral codes regulate the social structure and value systems defining the world in which that identity grows and to which it belongs. Indeed the total system is often a closed circuit with the ultimate authority figures, the "gods" that sanction both societal organization and individual behaviour, being the collectively accepted and projected arbiters of childhood emotional security. Even in the world of modern science our *Weltanschauung* remains essentially paradigmatic—the seemingly appropriate view of the universe to which current history in ideas has brought us (Kühn, 1962). As for Poincaré, such a view may be no more than one among many alternative geometries of existence.

Our visions of the nature of men and women are commonly partial and ill formulated. The white rat model of Hull and Skinner, the anseriform image of Lorenz or the sexy monkey story of Desmond Morris do disservice precisely because they neglect the complexity of man's distinctive, characteristic personal psycho-social existence. Yet this complaint may be levelled with equal force at most behavioural sciences. Thus so long as psychology treats of factors controlling individual behaviour, sociology of social behaviour, and evolutionary biology and physiology of the organismic structuring of life, we can treat man *alternatively* as *psyche,* as *socius* or as *organism* depending on the perspective we adopt. A few authors have had the breadth of vision to examine, haltingly maybe, the cybernetic interplay and dynamism of the whole. G.H. Mead, L. von Bertalanffy, Talcot Parsons, J. Piaget and Herbert Marcuse are among those who have attempted such examinations. Inferences from animals' behaviour to man can only become "relevant" to contemporary human problems if approached in this light. The rest is academic, good material for understanding history and essential reading in universities but the cutting edge is blunt.

SUMMARY

— The understanding of animal social organization and the inferences from this to human society depend upon the comprehension of processes involved in the social structuring of populations and the causal levels determining behaviour shown within social structures.
— Social organization is approached by way of three perspectives; that concerned with the spatial dispersion of individuals, that treating the dynamic interplay between individuals involved in intergenerational change, and finally the study of communication processes mediating both spacing and interaction. Both social structure and inter-individual behaviour may be analysed with respect to prime survival strategies of individuals in relation to reproduction and rearing, protection from predators and resource acquisition. Correlations between social organization and ecological factors suggest the evolution of behavioural complexes or grades adapted to specific habitats.
— Study of intra-specific variation in social organization and the nature of the dynamics maintaining group cohesion and change show that the organization is responsive to direct ecological influence, the acquisition of traditional behaviour, and the selection of innate patterns of interaction. These types of change operate at different rates. In particular, social selection of temperamental traits occurs within an organization itself expressing a direct response to ecology and social patterning. The genetic basis of temperament defines thresholds in behavioural interactions that may constrain the adaptability of a social structure in shifting environments.
— Similar social structures may be mediated by differing social dynamics so that in discussion of analogy and homology between social organizations, a broad knowledge of both is required. These differences may occur, for example, because of a general convergence in spatial disper-

sion and grouping patterns in relation to resource dispersion developing on top of large differences at the interaction and relationship level. These last may be the result of differing phyletic inheritance.

— A synthesis of primate ethology and social anthropology provides today a fairly broad picture of the way in which human social organization evolved. Yet the involvement of the cultural level in man introduces a complexity which few have attempted to treat in relation to ethology. This raises problems as to how far relatively simplistic inferences from animals to man are of any more than academic interest.

— This issue is explored in the final section where we focus on the uniqueness of the human person rather than on lines of derivational continuity in behaviour from animals to men. The person comprises a unique introjected identity and human interaction concerns relationships between such identities. Identity formation depends on the elaboration of linguistic, legal and moral codes and the slow development of an individual's behaviour within a social context structured by such codes. Yet these codes are neither independent in relation to one another nor arbitrary. They represent collective mechanisms of societal and behavioural regulation based intimately upon the personal needs and defences of individuals.

— Social ethology, if it is to include man within its scope, needs to examine the implications of man's uniqueness as well as the evidence for his animal heritage. The complexity of the issues at stake is clear, yet if inferences from animals to man are ever to be more than partial, we need to widen the horizons of contemporary approaches considerably. One way of doing this is to reconstruct most behavioural science courses in universities around the issues we have raised so that the interplay and implications of present day ethology, sociology, anthropology, physiology and genetics can be more clearly examined by students. Better education would precede deeper thinking in this area and probably facilitate a rapid expansion of the subject.

DISCUSSION

Chance: A long time ago we had a meeting on the social life of early man; I don't think anybody knew anything about the subject at that time. Nevertheless, the meeting revealed one feature of man's early life, which I still find fascinating and which, as far as I know, has not yet been challenged. If we assume that man emerged in the area about the Great Rift Valley between the Congo forests and the Eastern seaboard, this provides an area of immense environmental diversity, so it is likely that there were groups of men adapting to all these various environments and nevertheless breeding back into a common gene pool which in turn became immensely variable, and so may be the source of the diversity of man. If this is the case, can one be quite so definite about man being savannah or woodland or whatever, and how long have we been separated from each other into spaces as we are now?

Crook: Surely this depends on what period in time you are considering? I accept your point that there probably was a great diversity in early hominid populations and a great range of adaptive radiations at each developing stage. The question is whether we are right in locating a centralizing tendency in the savannah. Certainly many studies do suggest that the savannah played a key role especially in the development of foraging and hunting procedures.

Cain: I would like to raise two small points in regard to what Dr. Chance said. First, the savannah *is* a variable environment. It varies both spatially and temporally whereas the tropical rain forest, where I have lived for quite a long time, is solid and invariable beyond belief. The second is a more far-fetched point. The English aristocracy of the 18th century who were highly cultured people, what environment did they create for themselves? They built themselves mansions and surrounded them with lawns, flower beds, and small groups of trees, but they never created a rain forest, they never made a seashore and the sea! But the vital point is that the environment of the savannah does seem to offer lots of reinforcement. The savannah environment is immensely variable, and this again is stimulating, as a variable type of environment tends to reinforce the suggestion which indeed is supported by a lot one knows.

Phylogenetic Conservatism

Cain: As for your reference to phylogenetic conservatism, I must ask you to abandon the term in the sense that nobody ever observed this phylogenetic conservatism.

van Hooff: I do think that phylogenetic conservatism is nevertheless a good thing to keep in mind. If you have a certain adaptive modification—for instance, a certain social organization—in an environment that is rather stable with respect to those aspects that determine the adaptive significance of the modification, you can imagine that such a modification itself again contributes to a genetic selection of characters that are more adapted to the particular way of life. In this way, even differences in social organization—for instance differences in social behaviour—may become based on genetic differences. I am thinking of the work of Kummer (1971) and Nagel (1973). They showed that the social organization of the hamadryas baboon, which differs from that of other baboons by its harem system, is largely determined by the peculiar "herding of females" by the male; they, moreover, presented strong evidence that this "herding" is a genetic characteristic of this species. So, in this case, the typical aspect of the social organization may be based on an innate trait.

Crook: There is another point. It is quite striking how the macaques have a quite limited range of social structures, whereas the langurs have a wider one. The question is why one genus should be more limited in its range of social formation than another one.

Cain: Yes, I quite agree, but not until you have shown that this conservatism is actually based upon phylogeny. As soon as there is some evidence for this, I will agree that it is a useful term. All I say is that you are putting a label on it which is also an interpretation.

Crook: Yes, that is correct.

Hécaen: You mentioned the different degrees of rigidity of social organization in certain species. Do you have some reference about the cognitive capacities of those species, for instance, of the macaque and the langur?

Crook: I can answer that only very indirectly. The work on cognitive capacities focuses mostly on rhesus monkeys and on chimpanzees. I do not think that sufficient work has been done, let us say, on the langur or the macaque, to enable effective comparisons to be made on that level. It would be very interesting, but I don't think it has been pursued very far. There are a number of macaque species studied in considerable detail from a developmental point of view; for example, compare Kaufman's article with that of Hinde in "Ethology and Psychiatry" (1974).

Comparison of Social Organization

Ploog: I was quite intrigued by Dr. Crook's way of comparing social organizations, it reminded me in some ways of the techniques used by people working on the roots of language. Humboldt came up with seven or eight roots, others listed as many as twelve or fourteen, and placed them in a rather similar way into the system of social organization, in which it was necessary to communicate in one way or another. I could think of a way of abstracting from your way of comparison, in which you compared the various social organizations as to differences in the mechanisms of communication. Then one could probably tie this into a theory of the biological foundation of language, and that could be relevant to questions about why language came to man so late.

I was also intrigued by this comparison in relation to our present subject, concerning brain structure and behaviour. I think it is obvious that it is much easier to relate brain structures and behaviour in simple animals. Already that is rather difficult. But I think everybody would also agree that the thousands of features of primate brains versus the features of non-primate brains are very important and outstanding. If one would not try the impossible, namely to relate specific aspects of social organization to specific brain structures, but would start from the question "what is the purpose of everything that has been added during the mammalian development," one would probably come up with a very general statement that there must have been a great need for all these various social organizations you have pointed to. Then the brain might become more important and be included in our comparison.

von Cranach: The first idea one gets when looking at your list is that it is difficult to make generalizations, because you have put down the dimensions within which such a generalization could be made. But we can, of course, try to compare various societies on one level. We could then say that the best results of the investigation have been lost because those other dimensions are not in it, and we could try to make comparisons on all dimensions at once. This seems to be quite possible by means of the methods Dr. Rössler has presented; that means very abstract generalizations, in which much of the essential content of the study would be lost. So your "so what" really presents a dilemma, and I do not see a way out of it, but I hope Dr. Cain can suggest one. Dr. Chance has suggested the dilemma in seeking out agonistic and hedonic behaviour. He has more or less stated implicitly that some dimensions are more important than others, so that you do not lose much if you compare on the important levels. Is that really the case? And should we try, as Roger Masters suggested, to sort out important levels on which comparisons can be made?

Aims of Animal–Human Comparison

Cain: Let me respond to John Crook's "so what" by saying "not at all what you expect". This is a very serious question indeed; why are we doing this? Broadly speaking, the purpose of these comparisons is obviously to stimulate better generalizations, but generalizations about what? If we are interested, as I am, primarily in the course of evolution, and there is such and such proof to look at, then of course everything that is being done is purposeful. The distinction between homology and analogy is immensely important because genetic transfer and adaptation to environment are parts of the course of evolution. Nor is there any problem for the evolutionist in comparing man and animal. Man indeed has evolved from other animals, and the comparison in principle is therefore correct. But what do you expect to find in such a comparison? This question is particularly appropriate concerning comparison of man to other animals, more so than for comparison between other sorts of animals, and for two reasons.

First, referring to the 19th century, nearly all the anatomical evidence (and there is a large amount of this, particularly regarding detailed anatomical elements), which was relied upon by the comparative anatomists for showing the affinity between man and the great apes, shows nothing of the sort. They were looking at functional features relative to the verticality of the trunk, and the fact is that man stands up and apes hang down. In other respects, however, we are looking at two groups that are already known to have many parallels, whether homologous or analogous.

Secondly, Julian Huxley put the matter extremely well when he introduced the terms grades and clades. Clades are phyletic lines, and grades are stages of organization. He said that as far as clades were concerned, man is one twig on the hominid tree. Having reduced man anatomically nearly to one of the other animals, he then introduced a new grade for man to indicate that, though anatomically so close, they are behaviourally different. The difference is in fact so great that one needs to offset man against the rest of the animal kingdom. If that is really true, then the purpose of comparison between man and the great apes is almost pointless.

Crook: That is why I said my "so what".

Cain: The next point I want to raise is that man is also in the absolutely unique position of the self-domesticated animal. He has changed his own environment to his own specification so rapidly—and not very well, admittedly, from an evolutionary point of view—that he is the classic example of an unbalanced animal. He need not be in any sort of effective relation to the environment in order to survive. Many wild animals, however, even if in environments still unspoiled by man, have to specialize: they tread a narrow path between those worn by other species, and it is a terrible consequence that a great deal of natural selection has been taken away from them by man himself. At the present time we are probably preserving more genetically impaired variables than we could imagine to have occurred successfully in the palaeolithic. I grant you that this is rather hypothetical since we have little direct evidence of the palaeolithic.

Crook: I do not want to go into details about what Dr. Cain said, but they concern the background of questions about the distinctiveness of man as opposed to his biological origin. I am in a rather peculiar position, being a biologist working in a psychology department. I sometimes wonder why my psychological colleagues spend time with animals when they could do the same work on our own species, but then I am faced by my colleagues with the Lorenz aggression book, Desmond Morris' writings and Robert Ardrey.

The questions these authors raise are concerned with our image of ourselves. This does lead to political issues in the same way as divergent views on the bio-psychology of the two sexes affects thinking in the area Kate Millet calls "sexual politics." The image which we have of ourselves affects our thinking of ourselves. The way in which we think of our origin may be crucial in determining what we do with ourselves. Undoubtedly in the last few years one of the trends has been the development of what I would call a zoomorphic image of man, and this turns the "so what" issue into a political question.

Masters: My immediate response to John Crook's question, "so what," was to think of the function of play in many primates, specifically in terms of the exploration of the environment and the possibility of developed adapted responses. In large part of course, science is exploration. But it could also be play. Since one cannot limit one's play to the practical things of society perhaps the "so what"is part of the playfulness of science. The last couple of comments however do point toward something I said earlier that may have been misunderstood, about working on a new paradigm. This word may cause trouble because it is differently understood in German, or Greek, and English. What I mean by paradigm is the frame of reference or model, even the hunches one has in some sense, and how one goes about understanding and explaining problems. Here it seems to me that the "so what" is very important, because when one looks at the comparison of species that include *Homo sapiens,* one looks at different hypotheses as one looks at one. One of the difficulties is that of levels. When we think of yesterday's talks, what strikes me is that there is one connection between structure and behaviour we have not yet talked about. That is the complexity of the central nervous system which permits individuals to recombine so many different dimensions of elements of behaviour that one can almost say that for *Homo sapiens* the entire primate repertoire plus the entire carnivorous repertoire have become functional elements in a super-repertoire by which human populations can adapt to all different kinds of things. The difficulty is that it is so general that we have to determine how to go at it. That is why I suggested that "levels of analysis" business.

Now let me bring up an example which is extremely interesting, the process by which the young male in the Japanese monkey population seems to respond to social pressure by becoming a solitary male. This seems to have some ecological preconditions, for example, the absence of predators. But where that can happen, there is a very specific function which is thereby satisfied.

What is so interesting about it is that the solitary male leaves the group and later comes back to become dominant. One could describe these supplementary dominant males as a social insurance. There was a time when the existence of solitary males was treated as indicating the absence of sociability. This may have been a cultural strategy, but I immediately thought of the function of that type of sequence in human organization, it works out like De Gaulle. It can be very useful in modern society to have some potential dominant males and keep them solitary for a while; and that is a very specific example of a function wherein one gets a functionalized explanation of the descriptive part of the hypothesis.

Crook: May I just make a short comment on that last point, namely, that the exclusion of young males is common in many higher vertebrate societies and not only in primates. In differential exclusion, the animals excluded are usually aged and juvenile males. The process forms part of the population dynamics concerned with intergenerational replacement without discontinuity in social structure.

Ambrose: John Crook is asking the question of what happens next, and at this particular stage of science we have got to. He has reviewed descriptive work and he has reviewed classificatory and correlational work, and he says that the next phase has to be explanation and prediction. I really think his "so what" question is about going into that next phase, and it may well be an expression of trepidation at the problems involved in entering it because it is concerned with arriving at and testing causal generalizations.

The kind of material John Crook produced concerns two logically distinct types of generalizations; it is about two classes of things. One has to do with history. How does one go about making generalizations about history? You mentioned evolution, and evolution is in a sense a historical concept. There are various ways that people have attempted to make generalizations about human history, such as Toynbee's with reference to the growth, maturity and decline of civilizations. The second class of generalization has to do with mechanisms. You have been talking about the relations between social structures and the organisms or individuals that compose them and the environments in which they operate. How to we arrive at generalizations about the relations between these structural, individual and environmental factors? Inevitably this requires getting into experimental work such that one tests predictions of what will happen if you systematically alter one or two classes of variables. It may be a matter of taking one group deliberately out of its environment and putting it in another one, or removing the dominant male from a group and predicting what will happen.

Chance: It seems to me that we are at a very crucial point in the whole way in which we handle what goes on. I think the point is that we must try to get information that tells us how far we can do what we like without actually running into impasses because of the restraints and constraints our nature brings about.

Ploog: I think the more troubles we have in our societies, the more scientists have to take a stand on these issues; and they will be asked for the answer to your "so what." Maybe one way to discuss this is not to ask "so what," and to give a positive answer, but to pull together all you said and to ask what we should *not* do. The answer might be easier, because many things that are done at the moment we all know are against our biological and our cultural heritage, but are done nevertheless. There might be a series of yes-answers, which may limit the field of what we should not do.

NOTE

This article first appeared in *Social Science Information,* 14 (3/4) 89-112.

REFERENCES

Berger, P.L./Luckmann, T. 1967. *The Social Construction of Reality.* Harmondsworth, Penguin. Blurton-Jones, N. (ed.)

— 1972. *Ethological Studies of Child Behaviour.* Cambridge University Press.

Bradfield, R.M. 1973. *A Natural History of Associations, a Study in the Meaning of Community.* London, Duckworth.

Clutton-Brock, T. 1972. "Feeding and Ranging Behaviour of the Red Colobus Monkey". Unpublished doctoral dissertation. Cambridge University.

Crook, J.H. 1964. *The Evolution of Social Organisation and Visual Communication in the Weaver Birds (Ploceinae).* Behaviour, supplement 10. Leiden, Brill.

— 1965. "The Adaptive Significance of Avian Social Organisations", *Symposia of the Zoological Society of London* 14:181-218.

— 1966. "Gelada Baboon Herd Structure and Movement: A Comparative Report", *Symposia of the Zoological Society of London* 18:237-258.

— 1969. "Functional and Ecological Aspects of Vocalisation in Weaver Birds", pp. 265-289 in: R.A. Hinde (ed.) *Bird Vocalisations.* Cambridge University Press.

— 1970a. "Social Behaviour and Ethology", pp. XXI-XI in: J.H. Crook (ed.) *Social Behaviour in Birds and Mammals.* London, Academic Press.

— 1970b. "The Socio-Ecology of Primates", pp. 103-166 in: J.H. Crook (ed.) *Social Behaviour in Birds and Mammals.* London, Academic Press.

— 1971. "Sources of Cooperation in Animals and Man", pp. 237-260 in J.F. Eisenberg/W.S. Dillon (eds.) *Man and Beast: Comparative Social Behavior.* Washington, DC, Smithsonian Institution.

— 1972. "Sexual Selection, Dimorphism and Social Organization in the Primates", pp. 231-281 in: B. Campbell (ed.): *Sexual Selection and the Descent of Man* 1871-1971. Chicago, Aldine.

— In press. "On the Integration of Gender Strategies in Mammalian Social Systems", in: J. Rosenblat (ed.) *D. Lehrman Memorial Symposium. Rutgers University. Newark, NJ, 1974.*

— 1975. "Primate Social Structure and Dynamics – Conspectus 1974", in: S. Kondo/M. Kawai/A. Ehara (eds.) *Contemporary Primatology,* Proceedings of 5th International Primatology Congress, Nagoya 1974. Basel, Karger.

Crook, J.H./Aldrich-Blake, P. 1968. "Ecological and Behavioural Contrasts between Sympatric Ground-Dwelling Primates in Ethiopia", *Folia Primatologica* 8:192-227.

Crook, J.H./Ellis, J./Goss-Custard, J. In press. "Mammalian Social Systems: Structure and Function", *Animal Behaviour.*

Crook, J.H./Gartlan, J.S. 1966. "The Evolution of Primate Societies", *Nature* (London) 210:1200-1203.

Deag, J. 1974. Doctoral dissertation. Bristol University.

Deag, J./Crook, J.H. 1971. "Social Behaviour and 'Agonistic Buffering' in the Wild Barbary Macaque, *Macaca sylvana*", *Folia Primatologica* 15:183-200.

Dimond, S./Lazarus, J. 1974. "The Problem of Vigilance in Animal Life", in: W. Riss (ed.) *Brain Behaviour and Evolution.* Basel, Karger.

Downhower, J.F./Armitage, K.B. 1971. "The Yellow-Bellied Marmot and the Evolution of Polygamy", *American Naturalist* 105:355-370.

Dunbar, R.M. 1974. Doctoral dissertation. Bristol University.

Dunbar, R.M./Dunbar, E.P. 1974. "Ecological Relations and Niche Separation Between Sympatric Terrestrial Primates in Ethiopia", *Folia Primatologica* 21:36-60.

Eibl-Eibesfeldt, I. 1970. *Ethology: the Biology of Behavior.* New York, Holt Rinehart and Winston.

Eisenberg, J.F. 1966. "The Social Organisation of Mammals", *Handbuch der Zoologie* Band 8, Lieferung 39, 10 (7) 1-92.

Eisenberg, J.F./Muckinhern, N.A./Rudran, R. 1972. "The Relations Between Ecology and Social Structure in Primates", *Science* 176:863-874.

Espinas, A. 1878. *Des sociétés animales.* Paris, Baillière.

Fox, R. 1967. "In the Beginning: Aspects of Hominid Behavioural Evolution", *Man* 2:415-433.

— 1972. "Alliance and Constraint: Sexual Selection and the Evolution of Human Kinship Systems", pp. 282-331 in: B. Campbell (ed.) *Sexual Selection and the Descent of Man 1871-1971.* Chicago, Aldine.

Gartlan, J.S. 1970. "Preliminary Notes on the Ecology and Behaviour of the Drill, *Mandrillus leucophaeus*", (Ritgen, 1824) in: J.R. Napier/P.H. Napier (eds.) *The Old World Monkeys.* New York, Academic Press.

Gartlan, J.S./Struhsaker, T. 1972. "Polyspecific Associations and Niche Separation of Rain-Forest Anthropoids in Cameroon, West Africa", *Journal of Zoology* 168:200-221.

Goss-Custard, J./Dunbar, R./Aldrich-Blake, P. 1972. "Survival, Mating and Rearing Strategies in the Evolution of Primate Social Structure", *Folia Primatologica* 17:1-19.

Hall, K.R.L. 1966. "Behaviour and Ecology of the Wild Patas Monkey *(Erythrocebus patas)* in Uganda", *Journal of Zoology* 148:15-87.

Hinde, R.A. 1974. "Mother/Infant Relations in Rhesus Monkeys", in: N.F. White (ed.) *Ethology and Psychiatry*. Toronto, University of Toronto Press.

— 1975. "Interactions, Relationships and Social Structure in Non-Human Primates", in: S. Kondo/M. Kawai/A. Ehara (eds.) *Contemporary Primatology*, Proceedings of the 5th International Primatology Congress, Nagoya 1974. Basel, Karger.

Huxley, J. 1923. "Courtship Activities in the Red-Throated Diver (*Colymbus stellatus* Pontopp) Together with a Discussion of the Evolution of Courtship in Birds", *Journal of the Linnean Society* 35:253-292.

Jarman, P. 1974. "The Social Organisation of Antelope in Relation to Their Ecology", *Behaviour* 48:215-267.

Jolly, C. 1970. "The Seed-Eaters: a New Model of Hominid Differentiation Based on a Baboon Analogy", *Man* 5:5-26.

Kaufman, 1974. "Mother/Infant Relations in Monkeys and Humans", in: N.F. White (ed.) *Ethology and Psychiatry*. Toronto, University of Toronto Press.

Kendon, A./Ferber, A. 1973. "A Description of Some Human Greetings", pp. 592-668 in: R.P. Michael/J.H. Crook (eds.) *Comparative Ecology and Behaviour of Primates*. London, Academic Press.

Kleiman, D.G./Eisenberg, J. 1973. "Comparisons of Canid and Felid Social Systems from an Evolutionary Perspective", *Animal Behaviour* 21:637-659.

Kruuk, H. 1970. *The Spotted Hyena*. University of Chicago Press.

Kühn, T.S. 1962. *The Structure of Scientific Revolutions*. University of Chicago Press.

Kummer, H. 1968. *Social Organisation of Hamadryas Baboons*. Bibliotheca Primatologica 6. Basel, Karger.

— 1971. *Primate Societies: Group Techniques of Ecological Adaptation*. Chicago, Aldrine. Atherton.

— 1975. "Rules of Dyad and Group Formation Among Captive Gelada Baboons *(Theropithecus gelada)*", in: S. Kondo/M. Kawai/A. Ehara (eds.) *Contemporary Primatology*, Proceedings of the 5th International Primatology Congress, Nagoya 1974. Basel, Karger.

Lack, D. 1968. *Ecological Adaptations for Breeding in Birds*. London, Methuen.

Lorenz, K. 1940. "Die angeborenen Formen möglicher Erfahrung", *Zeitschrift für Tierpsychologie* 5:235-409.

— 1941. "Vergleichende Bewegungstudien an Anatiden", *Journal für Ornithologie* 89:194-294.

— 1966. *On Aggression*. London, Methuen.

— 1974. *Civilized Man's Eight Deadly Sins*. London, Methuen.

MacKinnon, J. 1974. "The Behaviour and Ecology of Wild Orang-Utans *(Pongo pygmaeus)*", *Animal Behaviour* 22:3-74.

McBride, G. 1964. "A General Theory of Social Organisation and Behaviour", *University of Queensland Papers in Veterinary Science* 1:75-110.

Martin, R.B. 1972. "Adaptive Radiation and Behaviour of the Malagasy Lemurs", *Philosophical Transactions of the Royal Society of London* B 264:295-352.

Montagu, M.F.A. 1965. *The Human Revolution*. Cleveland, Cleveland World Publishing Co.

— 1973. (ed.) *Man and Aggression*. Oxford University Press.

Morris, D. 1967. *The Naked Ape*. London, Jonathan Cape.

Nagel, U. 1973. "A Comparison of Anubis Baboons, Hamadryas Baboons and their Hybrids at a Species Border in Ethiopia", *Folia Primatologica* 19 (2/3) 104-165.

Nelson, J.B. 1970. "The Relationship Between Behaviour and Ecology in the Sulidae with Reference to Other Birds", *Oceanography and Marine Biology Animal Review* 8:501-574.

Parsons, T. 1966. *Societies: Evolutionary and Comparative Perspectives*. Englewood Cliffs, NJ, Prentice Hall.

— 1968. "The Position of Identity in the General Theory of Action", in: C. Gordon/K. Gordon (eds.) *The Self in Social Interaction*, vol. 1. New York, Wiley.
Petrucci, R. 1906. *Origine polyphylétique, homotypie, et non-comparabilité directe des sociétés animales.* Instituts Solvay, *Travaux de l'Institut de Sociologie. Notes et Mémoires*, fasc. 7. Bruxelles, Misch et Thon.
Reynolds, V. 1968. "Kinship and the Family in Monkeys, Apes and Men", *Man* 3:209-223.
Richards, A. 1973. Doctoral dissertation. London University.
Rodman, P.S. 1973. "Population Composition and Adaptive Organisation Among Orang-Utans of the Kutai Reserve", pp. 171-210 in: R.P. Michael/J.H. Crook (eds.) *Comparative Ecology and Behaviour of Primates.* London, Academic Press.
Schaller, G. 1963. *The Mountain Gorilla, Ecology and Behaviour.* University of Chicago Press.
— 1972. *The Serengeti Lion.* University of Chicago Press.
Sugiyama, Y. 1973. "The Social Structure of Wild Chimpanzees, A Review of Field Studies", pp. 375-410 in: R.P. Michael/J.H. Crook (eds.) *Comparative Ecology and Behaviour of Primates.* London, Academic Press.
Teleki, G. 1973. *The Predatory Behaviour of Wild Chimpanzees.* Lewisberg, Bucknell University Press.
Tiger, L. 1969. *Men in Groups.* London, Nelson.
Tinbergen, N. 1953. *Social Behaviour in Animals.* London, Methuen.
— 1959. "Comparative Studies of the Behaviour of Gulls *(Laridae)*, a Progress Report", *Behaviour* 15:1-70.
— 1972. "Functional Ethology and the Human Sciences", *Proceedings of the Royal Society of London* B 182:385-410.
Trivers, R.L. 1972. "Parental Investment and Sexual Selection", pp. 136-179 in: B. Campbell (ed.) *Sexual Selection and the Descent of Man, 1871-1971.* Chicago, Aldine.
Van Lawick-Goodall, J. 1968. *The Behaviour of Free-Living Chimpanzees in the Gombe Stream Reserve.* Animal Behaviour Monographs 1. London, Baillière, Tindall and Cassell.
— 1971. *In the Shadow of Man.* London, Collins.
Vine, I. 1970. "Communication by Facial-Visual Signals", pp. 279-356 in: J.H. Crook (ed.) *Social Behaviour in Birds and Mammals.* London, Academic Press.
Ward, P./Zahavi, A. 1973. "The Importance of Certain Assemblages of Birds as 'Information Centres' for Food Finding", *Ibis* 115:517-534.
Washburn, S.L./Lancaster, C.S. 1968. "The Evolution of Hunting", *Perspectives on Human Evolution* 1:213-229.
Wilson, E.O. 1971. "The Prospects for a Unified Sociobiology", *American Scientist* 59:400-403.
Wrangham, R. 1975. Doctoral dissertation. Cambridge University.
Zahavi, A. 1971. "The Social Behaviour of the White Wagtail, *Motacilla a alba*, Wintering in Israel", *Ibis* 113:203-211.

ANTHONY AMBROSE

Methodological and Conceptual Problems in the Comparison of Developmental Findings Across Species

BEHAVIOURAL ONTOGENY AND THE OBJECTIVES OF CROSS-SPECIES COMPARISON

"The fact of development is a fundamental feature of all living forms of life. From the simplest plant form to man, all of them have a starting point, grow up, and, in some sense, die. Attention to this basic commonality can be expected to provide rich dividends for the progress of psychology as a whole." In making this statement Thompson and Grusec (1970) were expressing a faith which will be shared by many psychologists and ethologists, that study of ontogenetic change in a variety of organisms is a basic and most revealing part of the overall science of behaviour. Unfortunately they do not elaborate either upon why such study is likely to be so rewarding or upon the kinds of attention required to make it so. There is perhaps a hint that the most fruitful direction lies in attention to commonalities.

When it comes to trying to identify commonalities, however, and to arriving at statements about them which are both legitimate and productive for further research, the pitfalls whether of formulation or of methodology are numerous. They are, perhaps, most treacherous of all for those, like myself, whose research is primarily in the area of human development but who believe that a comparative perspective can yield important insights. It is notable, however, that many students of animal behaviour, even of that at relatively primitive levels, will admit when pressed that despite the magnitude of the evolutionary jump involved one of the basic assumptions underlying their research activity is that, somehow, some day, their work may have relevance for understanding human behaviour and development. One of the major points to be made in this paper is that such understanding is likely to be furthered as much by the evaluation of differences between species as by the attempt to discover principles that hold across several species on the basis of their similarities.

It must be recognized at the outset that, among both psychologists and ethologists, considerable differences of opinion exist concerning the importance and significance to be attached to species commonalities or similarities on the one hand and to species differences on the other. Such opinions stem in large part from differences in view as to the objectives of comparison. Those who have conceived psychology to be a natural science modelled upon physics have, on the whole, been preoccupied with a search for commonalities. The aims of discovering laws of behaviour and development, and of constructing models of the organization and modification of behaviour which have the widest possible generality, have tended to result in the playing down, if not ignoring, of species differences. Sceptics would point out that such aims have led also to the extrapolation of findings to species beyond those investigated, especially to man. It might be added, furthermore, that because of ethical and practical difficulties of certain kinds of experimental investigation into human behaviour, and the consequent need to resort to animal studies instead, there has for long been a vested interest in maintaining the assumption that, in many important respects, human beings have basic similarities to such convenient experimental animals as the rat, the cat, the dog or the monkey. The common use of the personal pronoun in referring to such animals, incidentally, betrays the further assumption that species differences within these various types of animal, not to mention differences between breeds, can be ignored as of minimal significance. Now the scientific desirability of establishing warranted generalizations across species, in particular concerning behaviour development, appears self-evident for the progress of psychology. What is open to serious doubt, however, is whether the frequently associated lack of interest in species differences is in the interests of such progress. In other words, in the broad-sweep attempt to gain basic knowledge about the organization and development of the behaviour of "the organism", is there not a real danger of throwing out the baby with the bathwater of species differences?

By contrast, those who conceive psychology to be essentially a biological science, with its roots in Darwinian theory, pay much closer attention to species differences. Because of the adaptive nature of evolutionary change, the differences between related species regarding both anatomical structure and behaviour repertoire and organization are as important as the similarities for understanding how species survival is achieved. The objectives of comparison in the field of behaviour, as seen in ethology, are broadly speaking not only to clarify systematic evolutionary relationships, but primarily to help in understanding the natural behaviour characteristic of any given species over its life-cycle. Recognizing that, over wide stretches of the phyletic tree, ontogeny poses many similar problems for widely differing species, interest is centred upon the different ways in which behaviour is organized to surmount these problems under the varying species-characteristic conditions of natural environment, behavioural structure and genetic programming. In studying a particular kind of behaviour in a given species, whether its causation, modifiability, developmental antecedents or its function, the differences from other species revealed by comparative study help to throw into relief the biological significance of the similarities in those species. From this point of view therefore, study of the differences that go along with species similarities, far from being a hindrance to psychology, fall into place as an integral part of its progress.

In the attempt to arrive at generalizations about behaviour that cover more than one species, the greatest danger is in being deceived by apparent similarities. Thus in regard to developmental findings the distinction between homology and analogy, which has already been examined in this conference, is all important. Caution in this regard is all the more necessary at this time following the appearance in recent years of numerous popular books in which the likeness of some aspects of man's behaviour to that of one or another type of animal has been either suggested or taken for granted. These books depend heavily upon the findings of ethological studies on animals ranging from fish and birds to the higher primates. Many of these findings represent break-throughs for our understanding of complex or seemingly meaningless animal behaviour, and it is natural that they should stimulate and guide new thinking about human behaviour. Indeed, the reasons why such studies can have important implications for the study of human behaviour and its development have been reviewed elsewhere (Ambrose, 1968). Nevertheless, the making of straight inferences from the behaviour or development of animals, even of the higher primates, to that of man has little justification as yet, and these popular books highlight the pressing need for much more careful comparative study of relations between human and animal behaviour and development in order to establish more precisely the nature and significance of both the similarities and the differences.

In the last resort, perhaps the strongest reason for undertaking comparative work which attaches as much importance to the differences as to the similarities that are revealed, is an interest in the phylogeny per se of behaviour and its ontogeny. After all, study of the development of behaviour over the life-cycle, of its similarities and differences at successive phases and of how the changes occur, is now generally accepted as a fundamental part of the science of psychology. The ontogenetic story is, however, incomplete without the phylogenetic story. The concepts, approaches and methods required for study of the two kinds of story are, of course, different; and transfer of interest from the one to the other requires a good deal of adjustment by the investigator. Nevertheless the fact remains that, as Blurton Jones (1972) has pointed out, the evolutionary story is the endpoint:

it is the ultimate explanation of ontogeny and its causation, it provides the last link in the causal chain.

In what follows, I shall first describe the essentials of the main types of developmental research study which form the basis for comparisons across species, and the special problems to which these give rise. I shall then review some theoretical concepts which have implications for the ways in which behavioural development is conceived and comparative work approached. Finally I shall consider the principal methodological problems involved in developmental comparisons across species. Hopefully both the interest and the limitations of this type of work will become apparent during the course of this paper.

THE SCOPE AND PROBLEMS OF DEVELOPMENTAL RESEARCH

The Study of Behavioural Development

From a biological viewpoint the significance of ontogenetic development is that it prepares the individual organism to become an effective adult member of its species, enabling it to contribute to species-survival through reproduction and, where appropriate, the rearing of young. Achievement of these functions requires, of course, that various other functions be carried out effectively as well, such as the acquisition of food, defence against predation, maintenance of territory. In social species, which includes most of the higher primates, all these functions require participation in group activity and the maintenance of group cohesion. The behavioural capacities necessary to carry out all these functions adaptively are represented by an appropriate behaviour repertoire and a huge set of more or less complex sensory and neural mechanisms for controlling it according to the demands and opportunities of the environment. The way in which an individual member of a species becomes equipped with these over a period of transformation from a fœtus to a mature, biologically significant adult, is central to the concerns of both developmental psychology and ethology.

The developmental study of behaviour is directed, in the main, to the changes with age that take place in the nature of the behaviours that go to make up the repertoire, in the mode of organization of specific response mechanisms, and in the conditions under which such changes do or do not occur adaptively. Because of the logistics of research, as well as of the requirements of scientific method, most developmental investigations have been highly selective, directed to specific questions about some limited aspect of behaviour or its control, usually over a restricted age span. Furthermore the species used for study has usually been chosen simply as the most suitable for the task in hand. For example, the effects of genes on development have been most easily studied in the fruit-fly *Drosophila melanogaster;* the great body of experimental studies of behaviour development has been carried out on rats, mice, birds, dogs and cats, because of their excellence as laboratory animals, their relatively short longevity, and the practicability of achieving strain control; monkeys and apes have been used in many cases because they are the nearest available relatives to man. The crucial question of how far principles of development discovered in one species can be generalized to other species should become clearer as this paper proceeds. It will be helpful first to become clear about the nature of developmental findings, and to point up the methodological problems of comparison across age within any given species before going on to consider the problems of comparison across species.

The Nature of Developmental Findings

Developmental findings in behavioural research are the result of attempts to answer one or other of two basic classes of question, namely, "*what* changes take place with age", and "*how* do the changes occur". Research within each of these classes can be further differentiated according to the extent and nature of the control imposed experimentally upon the conditions under which the changes are observed. Studies asking what changes take place with age refer, to one or another aspect of the whole range of behavioural equipment characteristic of a species which, throughout ontogeny normally enables the individual to live adaptively. Depending upon the type of change being studied, its identification may be made qualitatively by description or quantitatively by measurement. Studies of the development of this equipment can be conveniently divided into three kinds. First are those which describe the emergence of the behaviour repertoire, or aspects of it. These range from the ethogram type of study which incorporates a wide variety of kinds of behaviour, with special interest in the order and age of appearance of specific sorts of movement pattern, to the microscopic type of study which investigates how a particular pattern is built up from its components. Because interest is centred in the naturally occurring behaviour characteristic of the species such studies usually entail observation in the natural environment; study of the build-up of component parts, however, can involve experimental intervention. Second are the studies of the development of those aspects of equipment that are necessary for the control of behaviour; namely, perceptual abilities, learning and cognitive abilities, and motivation. While the early stages of research into the development of these capacities are often directed to obtaining guide-lines from naturalistic observation, demonstration of the nature and extent of any particular ability at one or more ages requires careful control of the conditions of stimulation, whether eliciting, reinforcing or internal, as well as suitable facilities for obtaining measurements of performance. Much of this work is therefore carried out experimentally, whether in the laboratory or in the field. Third are studies of development of the organization of behaviour, that is of the changes with age in the way in which the occurrence of a particular response system is controlled by external and internal causal factors: for example, changes in the specificity of the eliciting or reinforcing stimuli, in drive strength, or in the role of feed-back. Because the elucidation of stimulus-response relationships requires careful control of all variables, such studies are usually carried out in the laboratory.

Research into how developmental changes occur is concerned with the roles of genetic and environmental factors. With the recognition both that the realization of genetic potential requires adequate environmental conditions, and also that environmental factors can affect development in ways other than just through learning, studies of how development occurs have become much less preoccupied with questions of whether a given response is innate or learned. Analyses of the logical and theoretical weaknesses of the maturation/learning dichotomy by, for example, Schneirla (1952), Hebb (1953), Lehrman (1953), Beach (1955), and Hinde (1959), have led to the formulation of two different types of question (Hinde, 1970). First, given differences in the behaviour of two organisms, are these differences due to genetic or to environmental factors operating in their development? Studies of this sort of problem entail keeping the one type of factor constant while ensuring variation in the other. If organisms of the same genetic make-up are reared under differing environmental conditions, any differences in behavioural outcome are attributable to the differences in the environment. Contrariwise, if organisms known to differ genetically are reared in similar environmental conditions, any differences in outcome are attributable to the genetic differences. Once this type of question has been answered a second one arises, namely, how do the genetic or environmental factors produce their effects? Studies of the effects of genes on behaviour have shown, for example, that the effects can be subserved by changes in a variety of media such as sense organs, effectors, body proportions, general excitability or neural organization. Similarly studies of the effects of environmental factors on behavioural development have revealed wide diversity in their modes of action such as, for example, by altering arousal level, producing a change in hormone balance, eliciting responses that inhibit others, as well as by providing learning opportunities. The principal methodology employed in such environmental research has been the experiment using stimulus deprivation or enrichment or motor restriction, as used, for example, in studies of the effects of early experience on the later development of behaviour.

It is beyond the scope of this paper to give illustrative examples of all the types of study referred to above. Collections of representative examples, however, have been assembled in the case of animal studies by Hinde (1970) and in the case of human studies by Mussen (1970).

Developmental Research Designs

All of the types of research study referred to above are to be regarded as developmental insofar as they require the comparison of behaviour or capacity at one age with that at another. Essentially four different research designs have been employed for this purpose, the advantages and disadvantages of which have been discussed at length by, for example, Rheingold (1967) and Nunnally (1973).

First is the longitudinal method by which the same organism is studied at different ages, no matter whether these ages be at different parts of the life-span or separated only by a few days or weeks as in the short-longitudinal study. The special advantages are that genetic factors are held absolutely constant and the history of the organism between the two ages compared can be known with precision. Such history is particularly relevant when mechanisms of change between the different ages are being studied. The principal limitation of this method is that any differences across age are always confounded with whatever effects may result from repeated measurement of the same individuals.

Second is the cross-sectional method by which comparison is made between different organisms at different ages. In this case differences in performance can be related to differences in age only if the two samples are similar in genetic strain and only if no change has occurred in the environment to cause differences between the samples at the different ages. In the case of successive age groups of the same strain in a laboratory environment this difficulty is likely to be minimal. With different age groups selected from the natural environment, however, the wider the age-span covered, and the more heterogeneous the stock from which the organisms come and the environmental stimuli to which they have been exposed, the more caution is required in attributing any difference in performance to age alone.

Third is the mixed cross-sectional and longitudinal method. This method goes some way towards counterbalancing both the shortcomings of cross-sectional studies for the estimation of longitudinal trends and the practical difficulties of performing true longitudinal investigations of long duration. Different organisms are studied in different age groups, but within each age group the same organism is studied on two or more occasions. Thus some of the economy of the cross-sectional design is preserved together with the conceptual advantages of the longitudinal design. The partially cross-sectional data yielded are likely to provide a better approximation to genuine longitudinal data than if the data from the different age groups were solely cross-sectional.

Fourth is the method of implicit developmental comparison where the organism is studied at one age or age-level only. Although no explicit comparison is made between findings from this age and those from other ages, the investigation is planned either as a prelude to similar investigations at other ages or so that comparison with findings already available at other ages is implied. Many studies of the immature organism are of this kind, being carried out essentially with a developmental perspective.

Problems of Developmental Data Comparability

The comparison of behaviour across age poses two kinds of problems which, in other fields of research into ontogeny, either do not exist or are much less prominent. The first concerns any study of behaviour per se, whether developmental or not. Unlike physical characters which are there to observe all the time, behaviour occurs in stream of patterns and responses, and any particular piece or type of behaviour is transient and may be variable in form and intensity, occurring only under specific conditions of internal and external stimulation and varying with them. Therefore comparison of "the same" type of behaviour occurring at different occasions or different ages, whether in the same or different individuals, presupposes adequate definition of the criteria used to identify such behaviour as equivalent. The logical and procedural problems of achieving this have been discussed by Eckensburger (1973) in the context of cross-cultural research, and the many different ways in which behaviour has in fact been classified have been reviewed by Hinde (1959). Furthermore, since a given piece of behaviour can vary in meaning or significance with the context both of the other behaviour and of the stimulus conditions under which it occurs, comparison must always take these contexts into account. This problem is especially acute in the case of naturalistic observation; laboratory studies on the other hand provide opportunities for reducing it to greater or lesser extent. The second problem is that unlike most visible physical characters which, through growth, merely change in size, much behaviour changes qualitatively with age, as also do the sensory, motivational, learning and cognitive capacities which subserve the control of behaviour. Indeed, so extensive are these changes in human ontogeny that the behaviour of the human neonate has much more in common with neonates in subhuman primate species than it does with the human adult. Naturally, the smaller the range of the different ages over which behavioural changes are studied, as in the short-longitudinal study, the more easily comparable will be the data.

It is therefore very desirable that any researcher proposing to compare developmental findings across species should first of all check the validity of the findings within each species from a number of viewpoints.

- What criteria were used to justify the assumption that the types of behaviour, perception or learning observed at the different ages were in fact equivalent?
- Were the stimulus conditions or the test procedures employed to provide the behavioural data at the different ages appropriate for eliciting the relevant responses at those ages?
- If the conditions and procedures employed differed across the ages, was this fact taken into account in interpreting the results?
- Similarly, were the conditions of motivation across the ages reasonably constant, and if it was not practicable to ensure this, was due allowance made for the differences in interpreting the results?
- In the case of negative results, was the granularity of the units of behaviour to be observed and of the measures used, appropriate to the detection of differences?
- Were the measures employed comparable and if not, were any correlations between them justifiable?

The problems of applying measurement techniques in investigations of development have been well discussed by Nunnally (1973). Many developmental findings used for comparisons across species, however, are not of a quantifiable type but are primarily qualitative. That quantification of any units or variables under developmental investigation is, where possible, desirable goes without saying. It is the only way of achieving real precision in operational definition, of arriving at functional relations between independent and dependent variables or between correlated variables, and of ensuring replicability in experiments. It must be recognized, however, that in the progress of any life science, premature preoccupation with measurement can be stultifying for research, and that there is no substitute in the early stages of investigation for thorough qualitative description of biologically relevant variables. Furthermore, because so much in behavioural development involves qualitative change, clearly the scope for using measures that are comparable across age must be limited in principle.

PERSPECTIVES ON DEVELOPMENT AND THEIR IMPLICATIONS FOR COMPARATIVE WORK

When setting out to compare developmental changes across species, the investigator's initial hopes or expectations of finding similarities on the one hand or differences on the other will heavily influence the selection of all aspects of the material to be compared. These include the scope and detail of the material, the evolutionary distance or closeness of the species, and the extent of the age-range within each species. An important determinant of the investigator's expectations is not just his own preferred theoretical position but the way this affects his whole perspective both on how ontogenetic development takes place and on what are the significant variables for comparison. As a prelude to looking at methodological problems of cross-species developmental comparison, therefore, it will be worthwhile to review a number of concepts that have implications for the ways in which development is viewed and comparative work approached.

Biological Function and the Ontogeny of Behaviour

One of the cornerstones of modern biology is the concept of biological function. This stems directly from Darwin's theory of evolution (1859) according to which, in the field of behaviour, species-characteristic behaviour is the outcome of natural selection operating on genetic variation. The form and dimensions of such behaviour, and of the processes that control it, are as they are in any given species because, in the course of evolution, they have functioned in ways that have ensured the survival of that species. Questions about the biological function of a piece of behaviour are thus questions about the part which this behaviour plays, or played, in contributing to species-survival in the type of environment in which the behaviour or capacity evolved. The answer refers not just to the immediate effect on the environment and its consequences for the behaving individual, but to the way these consequences contribute to the existence of the species. For example, Tinbergen (1963) showed that the effect of egg-shell removal by the black-headed gull is to clear the nest of stimuli which, from other experiments, are known to attract the attention of predators; the consequence is that the young have, on that account, a greater chance of survival than otherwise.

This kind of approach to understanding behaviour in terms of its consequences for the individual and for the species is, of course, fundamental to ethological thinking. Equally important, though less discussed, are its implications for understanding the course of ontogeny of behaviour characteristic of a species. Developmental psychologists, influenced by ethology, are only just beginning to ask questions about biological function. Indeed they have hardly begun to accept the realities of evolution and to examine their implications for thinking about behavioural ontogeny and developmental theory. Several such implications will be briefly considered.

One represents an extension of the application of the concept of biological function to the developmental changes that occur over the whole course of ontogeny of a given species. The concept has in the past been used principally in explaining the evolution of specific aspects of an organism such as morphological characters, behaviour patterns or learning capacities. However, species differ not only in the nature and organization of such characteristics but also in the way in which they emerge and change during ontogeny. That is, every species, at a certain level of abstraction, has its own typical and specific profile of behavioural ontogeny which, to greater or lesser extent, is different from that of every other species. Consideration of this fact in the light of Darwinian theory leads to an assumption which might be formulated thus: the course of development of behaviour that characterizes the ontogeny of any species, and the changes in the mechanisms underlying such development, are as they are because they successfully serve (or served) the biological function of ensuring survival of that species in the environment in which it lives (or in which selection took place).

Acceptance of such a statement has implications for most developmental research that is concerned with selected aspects of behavioural change over some limited part of ontogeny. It immediately focusses attention upon the question of why a given change takes place in the way it does, in the sense of what kind of selective advantage it provides for members of the species living in their natural environment. Answering this type of question entails study both of the effects of that particular change upon the individual's capacity for adapting to the environment at its particular stage of development, and also of the consequences of such a change for the individual's subsequent development. This sort of answer to the questions of why the change takes place is, of course, different and supplementary to the types of answer usually sought in developmental psychology which are of a causal or historical nature based on the action of genetic or environmental factors.

The achievement of this kind of understanding of the part played by some particular developmental change in the overall ontogeny of the individual requires evaluation of the change in the context of the individual's natural environment, the demands it makes and the opportunities it provides. Because the concept of biological function applied to developmental changes places its consideration firmly in the context of the natural environment it also has implications for the comparison of developmental findings across species. In particular it gives perspective both to the similarities and the differences which inevitably emerge from such comparisons by keeping in mind the similarities and the differences in the adaptive problems posed by the environments for each species. On the one hand some kinds of adaptive problem met with during ontogeny are common to a wide range of species and may be solved by similar sorts of behavioural development, not just by related species, but by unrelated ones as well through convergence. On the other hand only small species differences in such factors as habitat, food supply, predation, social organization, or parental characteristics can be associated with very different profiles of behavioural ontogeny. Similarly with species reared in virtually identical environments, even slight differences in genetic constitution can result in major differences in how the behaviour repertoire appears and develops, in the changing nature of perceptual sensitivities, in readiness to learn in a particular direction, or in the later effects of certain types or early experience.

Consideration of behavioural ontogeny from the viewpoint of biological function thus has the effect of sensitizing the comparative investigator at every step to the liklihood of coming across species differences. Even if he finds a similarity in behavioural development in one sphere of life, it by no means follows either that such a development is subserved by the same mechanisms in the two species, or that the species are likely to be similar in other spheres of development. Such an outlook becomes reinforced when the interactional nature of behavioural development is taken into account.

Interaction and the Development of Behaviour

Developmental researchers are now increasingly adopting an interactional view of the development of behaviour. According to this view, behaviour development, the potential and limits of which are genetically determined, is dependent up on the continuous interaction of the organism with its environment, and the outcome is determined by the operation of successive endogenous and exogenous stimulative events (*e.g.* Lehrman, 1962; Rosenblatt *et al.*, 1961). Even some so-called innate behaviour patterns, such as pecking in the newly-hatched chick, have been shown to be the outcome of complex interactions within the egg in which self-stimulation plays a significant role (Kuo, 1967). No character of sensory or behavioural functioning is the outcome solely of genetic factors in the sense of genetic predeterminism, nor is any such property the outcome solely of environmental influence in the sense of learning operating on a tabula rasa. However, differences in behaviour characters either within or across organisms may be attributable to either genetic or learning factors, and it is undeniable that the ontogeny of some behaviours is controlled predominantly by genetic information while that of others is shaped mainly by processes of learning.

The findings that have led to this view of the development of behaviour have contributed much to the breakdown of the rigid dichotomy of "innate" and "learned" as different properties of emergent behaviour. Correspondingly they have led to the formulation of new theories such as that of behaviour gradients (Kuo, 1967). Such progress, however, will probably not prevent such broad categories of behaviour as "instinctive", "appetitive", "stereotyped action patterns", "reflexes" and "learned responses" from continuing to be useful in research, especially comparative research, for a good while yet. After all, as Delius (1970) points out, diversity in the ontogeny of behaviours must have arisen during phylogeny. On the one hand a behaviour pattern may be innate because there is a survival premium on having a response available without a preceding time-consuming conditioning process. Indeed, behaviour patterns dealing with types of environmental conditions that have remained stable throughout the phylogeny of a species are likely candidates for enhanced genetic control. On the other hand, responses that cope with situations too variable to permit evolutionary adaptation must necessarily arise under environmental control during the individual's lifetime.

Nevertheless, the assumption that all behaviour characteristics are always under some degree of genetic control and always subject to some degree of shaping by the environment, whether this be biochemical, intra-uterine, other organisms or the habitat, is fully congruent with recent findings which show that behaviour develops as a result of an intricate mosaic of endogenous and exogenous stimulative events. Such findings about the role of interaction in behavioural development indicate the complexity of the processes involved, and this is a further reason for the comparative investigator to be wary. Such a cautionary note will be self-evident to the evolutionary-minded investigator whose interest is in the remarkable variety of forms that ontogenetic adaptation can take. But what of those whose interest is in searching for principles or laws of development that are common to more than one species? The caution is not meant to imply that they are on a wild goose chase, merely that they should not be too ambitious, nor too dogmatic.

Laws of Development

A scientific law usually carries the connotation of being a very basic principle or generalization, the validity of which is well established. The wider the scope of its validity the more abstract are the concepts contained in it. In the field of behavioural ontogeny it will be a long time before the principles of developmental change that are currently put forward from time to time achieve the status of developmental laws. This is because any such law, if it is to be empirical and not just part of some all-embracing a priori type of theory, must include reference to the conditions under which the generalization holds. At the present stage of research the specification of all the conditions relating to any given developmental principle is usually incomplete, and the limits to the scope of such a principle therefore remain vague, if not totally undetermined.

These conditions are of two kinds. The first concerns the class of organism for which the principle is valid, and it is usually referred to explicitly. At one extreme the class may be limited to one sex of a specific strain of a particular species or sub-species; at the other it may cover a whole range of species, genera, families, sub-orders, orders or phyla. The following are examples of principles where the range has been specified. 1) Human infants go through the sequence of lying-sitting-crawling-standing-walking-running (McGraw, 1963). 2) In rodents the behavioural effects in adulthood of stimuli applied or withheld in infancy are dependent upon the type of stimulation, the age at which it is administered or withheld and the duration of the treatment (Thompson, 1960). 3) In the early ontogeny of social behaviour in avian and mammalian species, the nature of the stimuli necessary to elicit it changes from being non-specific to specific (Hinde, 1970).

Although none of these principles has yet been tested out in all of the sub-types within the class of organism referred to, the implication is that in those that have been tested no exceptions have yet been found. In view of the limited amount of research that has been done to date, however, it may well be regarded as rash to believe that no exceptions will be found in the future. Such an attitude is reinforced when account is taken of a second kind of condition that affects the validity of a developmental principle, namely, the background conditions. These concern the state of the organism and the type of environment in which the development takes place. Such conditions are usually not referred to in a developmental generalization, because it is understood that the generalization assumes "other things being equal". In the case of generalizations about normal developmental changes the assumption is made that the background conditions refer to "normality" of the state of the organism and to "typicality" of the environment in which the developments occur. However, the higher the phyletic level to which the generalization is held to apply, the more difficult it is to know what these terms mean because "normal" covers a widening range of individual differences and the "typical" natural environment becomes increasingly variable. As the number and dimensions of relevant background conditions with which an organism interacts increase, so also does the complexity of its developmental processes. Therefore, any study carried out under one set of conditions, which purports to reveal a principle of development, will require replication under a variety of background conditions that represent the normal or typical range before the scope of validity of the principle can be properly known.

The study of behaviour and the development of its organization even in any one species, let alone across species, is still in its infancy. Consequently any principles of development that are currently propounded must be treated as highly tentative and likely to be the subject of profound revision with the progress of research. For example, even such relatively simple generalizations as that concerning the antero-posterior and cephalo-caudal directions of behavioural differentiation (Coghill, 1929) became subject to controversy. Nevertheless, the reaching of valid generalizations across species remains one of the objectives of developmental psychology. Success in this endeavour is all the more likely to be achieved if it is carried out with an awareness of the nature and extent of the differences between the species concerned, and of the factors which give rise to them. One such class of factors are genes.

The Nature of Gene Action in Behavioural Development

Although it is probably true to say that the development of all behavioural characteristics is under some degree of genetic control, in fact we know very little indeed about the underlying causal processes. It is clear that genes operate on the development of the physical organs which subserve behaviour and perception, and also largely determine the sequence of appearance of different behaviour patterns into the repertoire. It is clear also that genes affect both learning capacity and the direction of effects of external causal factors on behaviour. Thus genes act on behaviour development in at least two distinct ways: by affecting the physical morphological characters that underlie behaviour, and by affecting the nervous system which controls the rest of its properties.

As yet we do not know of any specific genes affecting the development of behaviour patterns in man. There are in fact three main reasons for this (King, 1968). First, genes do not affect behaviour directly, but act through various structural and metabolic pathways which eventually affect behaviour. Our knowledge of the mode of action in metabolism is extremely limited, but that gene action is basic to it is demonstrable in the conditions leading to phenylketonuria (Knox, 1955). Second, most behaviour characters or traits are probably controlled by multiple genetic factors (Fuller/Thompson, 1960). Single genes probably act on only one enzyme in the synthesis of metabolites, which may affect many different patterns of behaviour. Each gene acting through basic metabolic processes may then have pleiotropic effects and contribute to several different behaviour patterns. Third, genes affecting behaviour through the nervous system are difficult to identify, and in fact the mode of action of genes in neurogenesis in general is as yet poorly understood. It seems probable, however, that different genes act at different stages of development and that it is this sequential action that determines the sequence and rate of development of different behaviour patterns.

The seeming impossibility of discovering specific genes affecting particular behaviour patterns or particular sequential patterns of development is probably less important than isolating them just sufficiently to manipulate the phenotype, as a guide to the genotype. There are three methods of doing this (Denny/Ratner, 1970), namely, selective breeding, comparison between different strains or breeds of the same species, and analysis of the pleiotropic effects (or unit characters) of single genes in which single pairs of genes affect a variety of structures or behaviours.

Now the results of genetic studies employing these methods have a number of implications for the comparison of developmental findings. First, the development of behaviour and the extent of its control by environmental factors is profoundly affected by genetic factors, and these can result in major differences in development even between different strains of the same species (*e.g.* King, 1968), let alone between different species. Second, as regards the behaviour units employed for purposes of comparison, it must be clear that whatever type of behaviour is used and whatever way it is measured, no matter how much it is affected by heredity, it is never a genotypic character. As Fuller and Thompson (1960) point out, the potentialities for phenotypic variation increase with distance from the primary chemical functions of genes. For practical purposes blood-group phenotypes are a reliable guide to genotype, body size and form are less so, and behaviour is still more remotely related to gene action. This applies, in principle, just as much to a species-specific behaviour pattern which is regarded as innate, as it does to a species-characteristic slope of a learning curve. That innate behaviour patterns, such as display patterns, or reflexes, can be of enormous importance in study of the evolution of behaviour from a common phyletic origin is generally accepted. It does not follow, however, that for study of ontogeny or for comparisons of ontogeny, innate behaviours are either the only units of behaviour to be considered or even the most important ones. In behavioural ontogeny we are concerned with the whole course of development of behaviour and capacity of all kinds, some controlled predominantly by genetic information and others shaped mainly by learning and other environmental influences.

Developmental Stages

We would, of course, find it hard to get along without using such shorthand terms as "fœtus", "newborn", "post-weaning", "infancy", "juvenile", "adolescent", "sexually mature adult". Each of these stages of development, although covering widely diverse and sometimes overlapping age periods for any given species, carries an implication, however vague, that the organism in the period in question is different in important respects from what it was or from what it will become.

Theorists of psychological development, however, have used the concept of "stage" much more seriously as a fundamental term in building their frameworks. Examples are Freud (1915) and his stages of psycho-sexual development, and Piaget's many stages and sub-stages of cognitive development (1953). While the order of succession of stages is regarded as constant and genetically determined, the age of appearance of new structures and functions is related to the environment which provokes or impedes their appearance (Inhelder, 1956).

Now the concept of an operationally definable stage is, undoubtedly, extremely convenient. It is a help in identifying qualitative differences that successively appear in the visible development of specific classes of behaviour such as intellectual or sexual behaviour; it is also a convenient way of handling the theoretical implications of changes in their organization.

Unfortunately, the theorizing has led to the concept becoming reified, with the implication that a given stage, once achieved, really represents a period of equilibrium in the development of a particular structure and its function, or sometimes of the whole organism, until tendencies to change transform it into the next stage. Indeed, there is an attempt at a comprehensive developmental theory, namely Werner's comparative organismic theory (1957), which is based on the concept of stage as a real developmental phenomenon. Here, the central theoretical issue, according to Langer (1970) is "Why does an organism that is well adapted at one stage ever progress to the next one." Long-term change in the development of all organisms is considered to be composed of a sequence of universal stages in the organization of structures and functions. Each stage of organization is conceived dialectically as simultaneously directed towards maintaining continuity or stability and generating discontinuity or transformation.

Now there are two questions, of immediate relevance to comparative work, to be asked about all this. First, does development really occur in stages or is it continuous? Second, if stages do really exist, perhaps in the development of particular response systems or capacities, do any general stages occur, characterizing the simultaneous achievement of a number of anatomical, physiological and psychological developments? In searching for answers to these questions it may be fruitful to look at an area of development where the issues can be seen relatively clearly, that of physical growth, on which, of course, the much more complex area of behavioural development is partly dependent. I will quote the conclusions of Tanner (1970) in this regard.

> In physical growth, there is continuity and little evidence for discrete stages. Even the development of motor skills such as crawling and walking is more continuous than sudden... In some creatures such as insects at metamorphosis, general stages of development do occur. But in man only at adolescence do we have anything approaching a general stage—here, it is true, developments in anatomy, physiology and behaviour tend to occur synchronously. But even here the degree of synchrony is only relative. Skeletal age and dental age are largely independent of each other and the spurt in bodily development is not matched by any great rise in intellectual capacity.
>
> Physical development is best envisaged as a series of many successive processes, overlapping one another in time and linked loosely or tightly as the case may be. Out of the complexity of the linkages, under equilibratory forces, emerges an overall order with visible changes in the varying sections following one another with the regularity of a continuously changing mosaic. The process is one of continuous unfolding and movement, with speeds varying from time to time in different parts of the mosaic; it is not a series of kaleidoscopic bumps. Only in certain restricted areas do rapid reassortments of the pieces occur, as they fall into new and increasingly precise patterns.

Now it seems likely that development of the organization of behaviour and capacities proceeds upon exactly the same lines, especially if one views development interactionally as previously described. For this reason, it may well be more productive for research, especially comparative research, if the term "stage" is scrapped together with its theoretical undertones. Instead, use of the term "phase" of development would appear preferable, regarding it as a wholly descriptive term which is useful for referring to different time periods in an organism's ontogeny, namely, those which happen to interest a researcher for empirical reasons for purposes of identification or comparison.

APPROACHES AND PROBLEMS IN THE COMPARATIVE STUDY OF BEHAVIOURAL DEVELOPMENT

The Current Situation

It has been said (Blurton Jones, 1972) that comparative studies of developmental systems scarcely exist and are as yet hardly feasible. Insofar as this statement refers to systematically planned comparative empirical investigation it is true, and the reason is that the groundwork is not nearly far enough advanced. The study of behavioural ontogeny within species of any kind is still in its early stages. Moreover, the conceptual and methodological problems of making valid comparisons across species are still insufficiently resolved, and even the objectives of making developmental comparisons remain a matter of controversy.

Nevertheless, comparisons of developmental findings are becoming increasingly common. The attempts at making them have, however, been haphazard rather than orderly and have often been implicit rather than explicit. Most developmental research investigators tend to limit their empirical work to one species, being reluctant to change to another one with whose biological makeup they are less familiar. Consequently most cross-species comparisons are made by an investigator who works with one species and then compares his results with those reported in the literature by other investigators studying a similar problem in another species. This approach to comparison inevitably raises serious problems regarding the comparability of the data since the procedures, behavioural units and measures employed at the different levels will have been chosen independently by the different investigators.

In the last decade or so, however, there has been an upsurge in the number of symposia, of volumes of collected research reports, and of reviews which are both comparative in aim and cross-disciplinary in approach, each concerned with some major aspect of behavioural development, or with some aspect of behaviour viewed from a developmental standpoint*. Some contain articles that are explicitly comparative. In others the process of comparison is facilitated by the editor's comments or else left entirely to the reader. Such publications are of much importance in helping to prepare the ground for more systematic and precise comparative work in the future by familiarizing investigators both with findings in other species than those with which they work, and with the theoretical and methological problems involved in making comparisons. Furthermore, most of them either include articles about development at the human as well as non-human levels or are oriented towards interpreting the non-human findings to yield hypotheses for testing in humans. For this reason the selections of contents exclude those areas of behavioural ontogeny in which man differs most from his nearest relatives, namely his capacities for cognitive functioning and the use of language, the influence of culture, and the huge diversity of skills and habits that are the result of learning. In contrast to the role of these factors in the adult life of humans their role in infancy is minimal. Consequently the majority of serious attempts to compare developmental findings are as yet confined to studies of the immature organism.

The following topics may be cited as examples of the kinds of developmental problem that have been looked at comparatively: the emergence of the infant's behaviour repertoire (Kohts, 1935), the ontogeny of hand functions (Connolly/Elliott, 1972), the development of perceptual ability (Fantz, 1967), the development of learning abilities (Vince, 1961) the socialization process (Scott, 1963), mother-infant relations (Rosenblum/Kaufman, 1967), maternal separation effects (Bowlby, 1973), the effects of hormones on behavioural development (Hamburg, 1969), and the effects of early experience on the later development of behaviour (Thompson/Grusec, 1970).

The basic comparative problem thrown up by all such studies concerns the comparability of the developmental data from one species with that from another. Assuming that the developmental findings themselves are valid according to criteria discussed earlier, and assuming also that the purpose for making a particular comparison is clear, the validity and fruitfulness of its outcome will depend upon a number of issues. These are the range of the species selected for comparison, the conditions of observation under which the developmental findings were obtained, and the levels and dimensions of description and analysis employed. These issues will be considered with special reference to developmental comparisons between humans and non-human animals.

Criteria for Selection of Species for Comparison

The types of species chosen for the comparison of developmental findings depend principally upon the purpose for which the comparison is to be made. Whatever the purpose, however, any comparison will be based upon some assumption about the nature of the relationship either between the selected species or underlying the range of species chosen for study. Past comparative work has been implicitly or explicitly based upon assumptions which can be distinguished according to four different, though related, criteria. These are the degree of resemblance between the species in their neuroanatomy, in their developmental problems, in ecological factors, and the closeness or distance of their evolutionary relationship. The significance of these will be examined in relation to a number of different purposes for making developmental comparisons.

Testing in One Species of Hypotheses Derived from Findings in Another Species

Three different purposes for this kind of hypothesis testing can be distinguished. First, preliminary findings in one species may lead to the formulation of specific hypotheses which, however, cannot be tested in that species for either practical or ethical reasons. This refers, in the main, to findings in man, usually of a clinical or epidemiological nature, which suggest developmental hypotheses that cannot be tested either because of the long duration of his development through infancy and childhood, or because relevant variables cannot be adequately controlled. For example, Harlow's studies (1958a) of the role of contact-variables in the development of the infant's attachment to its mother in the rhesus monkey are based on hypotheses derived from observations of human infants which have long left open the question of whether the infant's attachment to his mother develops through secondary reinforcement of the primary drive of feeding, or whether it results from reinforcement of a drive for contact with the mother which is itself primary. Testing of such hypotheses required procedures that could not ethically have been applied at the human level. Again, both for this reason and because of the long time taken to reach adulthood in humans, much of the research on the later effects of early experience has been conducted on birds, mice, rats and dogs, in which the developmental span between infancy and adulthood has relatively short durations that are manageable for research purposes. A second purpose for the transfer of a particular type of research from one species to another works the other way around. Findings in one species that reveal striking new knowledge about its behavioural development raise questions as to whether similar findings will also hold in another species, especially if they are likely to have practical implications for the welfare of that species. An example of this was the questioning about the possible role of critical periods in human development (*e.g.* Caldwell, 1962) following the discovery of such periods in social development first in birds and then in dogs. A third purpose is really an extension of the second kind, namely, to test the scope of a generalization that has been achieved concerning some aspect of development in one species, by identifying the range of other species for which it will also hold truc. For example Hinde (1961) has described a number of generalizations about the development of the parent-offspring relation in birds which also hold for the development of this relation in mammals, including human beings. The same has been done by Scott (1963) for generalizations that hold as between dogs and man, and by Mason (1968) for those that hold as between non-human primates and man.

The principal criterion that has been used in comparative psychology to justify findings about similarities or differences between species has been their neuroanatomical relationship. Similarity in neuroanatomy has for long been regarded as the best criterion for judging the extent to which behavioural data in one species might generalize to others. This takes account of one of the main features of vertebrate evolution, that is, encephalization whereby control of behaviour has shifted progressively from the spinal cord and become increasingly centralized (de Beer, 1928). Harlow (1972), for example, suggests that in predicting behavioural generalization, we are properly prone to seeking similarities and dissimilarities in brain structures and particularly cortical strutures. It is on this basis that he considers all his experimental findings in the rhesus monkey regarding development in such areas as infant-mother attachment, peer-group relations, and psychotic behaviour, as applying in principle also at the human level. He admits, however, that no single or simple neuroanatomical measure adequately accounts for generalization or the lack of it between animals in relatively disparate orders, whether it be total brain weight, amount of fissuration, or cytoarchitectonic structure of the neocortex. For example, in considering the development of complex behavioural capability, neocortical cytoarchitecture works well in seriating man, apes, monkeys, carnivores, and ungulates. That criterion breaks down, however, when birds are included. Birds have no neocortex, yet they have extremely complex and differentiated striated structures, and some of them can successfully solve certain problems at a performance level equal to or beyond representative carnivores and at a level approaching the performance of the higher primates. Hence the old ethological joke that, no matter what kinds of behaviour or process you may find at other evolutionary levels, you will also find it in birds.

It comes as no surprise, therefore, that Hinde was able to demonstrate analogies between birds and mammals concerning the development of the parent-offspring relation. The similarities include proximity of the young to the mother, gradual learning of the individual characteristics of the mother, the ambivalent nature of the young's behaviour toward the parent, the fear of strangers, and the use of distress vocalization. These similarities exist because of a second criterion we now come to, namely, that certain basic developmental problems are common to the young at both levels: for example, the early dependence on the parent for protection, for food and for social learning. The apparent similarities in the mode of solution of such problems are, however, clearly the result of convergence and by no means imply similarity in the mechanisms by which particular aspects of the parent-young relation develop. Indeed, the liklihood of differences in these mechanisms is supported by many differences in the parent-young relation that go along with the similarities. These include differences in the mechanics of maintaining proximity and in the role of contact during the critical period for socialization (Ambrose, 1963), differences in maternal behaviour (Rheingold, 1963), as well as differences in degree of maturity at the time of commencement of socialization. So, when it comes to comparing species on the basis of similarities in developmental problems, Rheingold's warning will always be relevant. "Apparent similarities must be judiciously weighed against differences before generalizations from one species to another can be made. How much more caution should be exercized before generalizing from genus to genus and from order to order."

The same warning applies with even more force to comparisons that are encouraged by a type of criterion that is related to similarity in developmental problems, namely, similarity in ecology. Similarity in very broad ecological features has been used, for instance, by Scott (1963; Scott/Fuller, 1965) as a part-justification for comparing basic aspects of the socialization process in wolves and dogs on the one hand and in man on the other. Ecological similarities he draws attention to include the facts that both are highly social and normally live in groups, both have evolved the capacity to develop dominance-subordination relationships, they have each been direct ecological competititors as predators on the large herd animals, both being carnivorous, and much of their social behaviour is mutually recognizable. He does not lay too much emphasis on the significance of such features, especially in view of the distance of the phylogenetic relationship between canines and man, as well as the extensive differences in the detailed course of early ontogeny at the two levels. Nevertheless the common features in the socialization process demonstrated by him are striking. By contrast, other investigators, particularly in the primate behaviour field, employing finer analyses of ecological variables, have used species differences in ecology to justify their expectations of finding species differences in developmental trends. The intervening variable here is social organization. On the basis of ornithological studies which show clearly that the social structures of species populations correlate closely with ecology, Crook (1972) has suggested a tentative model relating differences in the organization of primate social groups to differences in their ecologies. That such differences in social structure can have important consequences for the course of development of social behaviour in the young is apparent from numerous studies. As Rheingold (1963) emphasizes, "Knowledge of the state of the organism at birth, the composition of the group into which he is born, his experiences with the members of that group, including his peers, offers the possibility of predicting the organism's later social behaviour."

This review of the various criteria that are used either to determine or to justify the range of species or phyletic levels selected for comparative purposes has so far concentrated mainly on those purposes related to the making of generalizations across species. In many, though certainly not all, cases, the liklihood that a developmental generalization, especially of a large-grained kind, will hold across species will be greater the more closely related evolutionarily these species are. On the whole, closely related species are more likely than distantly related ones to have major neuroanatomical similarities as well as marked resemblances in their behaviour repertoires, in their ecologies, in their developmental problems, and even in the mechanisms by which behaviour is controlled and develops. Even here, however, caution can never be relaxed because of further factors now to be discussed.

Study of the Similarities and Differences Brought About by Evolution
The ranges of species that have been selected for this kind of study have varied tremendously, again depending on the specific purpose of the comparison. At one extreme are studies which compare phenomena at widely disparate phyletic levels, for example Vince's review of developmental changes in learning capacity in birds, rats and human (1961), Thompson's review of the effects of early experience in rats, dogs and human (1960), Bronfenbrenner's review of the effects of early deprivation in mammals (1968), and Fantz's studies of the development of visual perception in chicks, rhesus monkeys, chimpanzees, and human infants (1967). At the other extreme are studies which make comparisons between species within the same genus, such as Lorenz's classic, though non-developmental, study (1941) of the evolution of display movements in the *Anatinae,* and Rosenblum and Kaufman's investigation of early mother-infant relations in pigtail and bonnet macaques (1967). On the whole, whereas the focus in the wide ranging comparisons has tended to be upon evolutionary continuities or convergences, that in the comparison of closely related species has tended to be upon the divergences that go along with the continuities, giving rise to trends of change.

An exemplary instance of the wide ranging type of comparison is Bronfenbrenners' careful analysis (1968) of the literature on early deprivation effects across species from mouse to man "with the aim of identifying continuities and gradients as one ascends the phylogenetic ladder". His assumptions and his methodological caution are well worth noting.

> Most of the original authors reviewed quite properly restricted themselves to data about a single species, and their generalizations are typically confined to the species in question. In contrast, our objective requires us to engage in cross-species comparison and to consider the applicability to Species A of generalizations drawn from work on Species B.
>
> This is, of course, a risky business. Some may even assert that it is unscientific on the ground that no generalization can be regarded as valid for a given species until its applicability is demonstrated for animals of that

> particular species. Such an assertion is, of course, incontestable if one is interested solely in conclusive demonstrations. But, in actual practice, science is less concerned with absolutes than with probabilities. This is particularly true for research on human behaviour, where crucial experiments, even when they can be formulated, often cannot be carried out for ethical reasons. Under these circumstances, if an investigator can demonstrate that a given phylogenetic relationship or process operates (or fails to operate) in a phylogenetic series approaching *Homo sapiens,* and the relevant characteristics or their analogues are present for the human species, we must then entertain the probability that the given process operates (or fails to operate) in man. At the very least, the burden of proof for demonstrating that man is the exception surely lies with the skeptic.
>
> Unfortunately, the role of the skeptic also falls to our lot. As we move from species to species, it becomes our obligation to look not only for evidence consistent with a prior generalization, but also for the exceptions that do not fit the rule. This becomes necessary not only as a canon of scientific method but also as our means for identifying the discontinuities which, hopefully, can provide a basis for hypotheses regarding "evolutionary" trends.

In contrast to this type of broadsweep approach are studies of evolutionary trends across closely related species. The extent to which different species can be said to be "closely related" is, of course, a relative matter, depending upon the comparative purpose and the evolutionary perspective range of the investigator. Whatever these may be, however, the fact that there is a phylogenetic relationship between the taxonomic groups concerned should be shown to have been established on the basis of characters other than the ones under investigation. These may be anatomical, physiological, morphological or behavioural, and the relationship may have been derived from study of living species or from palaeontological evidence. In most instances the close relationship will not be direct but, as in the case of man in relation to non-human primates, indirect.

The movement towards making behavioural and developmental comparisons between man and non-human primates has now gained tremendous momentum. Although man's direct evolutionary ancestors are, of course, extinct, the assumption is made that his closest living relatives are close enough for evolutionary trends to be discerned. Clarke (1960), for example, has stated:

> it is possible to arrange existing members of the primate order in a graded series which (quite apart from the evidence of the fossil record) suggests an actual scale or trend of evolutionary development within the order... It is not to be inferred, of course, that man has actually been derived from an anthropoid ape of the specialized type which exists today... What a

series of this kind does indicate is that, even without reference to palaeontological evidence, there are connecting links of an approximate kind between *Homo sapiens* and small mammals of a quite primitive type; it also suggests the general trends of evolutionary change which have characterized the order primates as a whole.

On the "closest available relative" basis, numerous comparisons of developmental findings have been made between human beings and monkeys or apes, most commonly the rhesus monkey and the chimpanzee. With the exception of case studies such as those of the Jacobsens (1932), the Kelloggs (1933), Kohts (1935) and others who observed or tested behavioural development in a single chimpanzee and compared it with that of human infants, most such comparisons have been made by reviewers drawing upon a variety of different studies. At the levels of abstraction employed marked similarities and evolutionary trends have been shown, for example, regarding the characteristics of early social development (Mason, 1968) and, in more specific areas, infant-mother separation effects (Bowlby, 1973), and the development of hand function (Connolly/Elliott, 1972).

One difficulty about confining such comparisons with man to the chimpanzee or the rhesus monkey is that one can never be sure that a given generalization common to the two will also hold for other relatively close primate genera, whether in the same or different sub-families, or even for other species within the same genus. The multiplying studies of mother-infant relations in monkeys of other genera such as the squirrel monkey (*e.g.* Kaplan/Schusterman, 1972), the Patas monkey (*e.g.* Preston *et al.*, 1970), langurs (*e.g.* Jay, 1963) and vervets (*e.g.* Struhsaker, 1971), all show how extensive the differences can be, and the more so the finer the granularity of the behaviour referred to. The same holds true for different species within the same genus. Among the macaques for example, pigtails and bonnets (Rosenblum/Kaufman, 1967) show the same basic progression of types of interaction in the development of the mother-infant bond, but there are differences in details, such as in the frequency of maternal inspection of the infant's pineal region, and in the amount of infant affiliative interactions with adults other than the mother. As regards ontogenetic processes, a response system of a similar type in different species can develop through very differing mechanisms. Among some avian species, for example, species-typical vocalizations develop without the young ever having heard them; in others, however, normal song will develop only if the young have had opportunity to hear other members of their species sing; and in yet others the young must be able to hear themselves sing to attain normal song (Nottebohm, 1970). This example reflects, further, the fact that very similar behaviours in different species can involve learning of very different kinds and degrees in the mechanisms of their ontogeny.

Discussion of this matter cannot be left without referring also to the comparative significance even of breed and strain differences. These have proved to be highly relevant factors for determination of the direction and magnitude of effects, for instance, in studies of early experience effects in dogs (Scott/Fuller, 1965), rats and mice (Erlenmeyer-Kimling, 1972). In fact, genetic diversity among closely related species, subspecies and breeds, can be expected to affect ontogenetic processes in numerous ways such as the time of onset of motor patterns and sensory responses, rates of maturation and sensitivities to environmental factors (King, 1968).

In view of the ubiquity of variation of some features within any taxonomic group, the question arises as to whether restricting comparison of human behaviour to that in one or another primate group as opposed to other types of mammal will always yield the most evolutionarily relevant information. The fact that some basic features of human behaviour and development can be seen at the level of the non-human primates does not necessarily imply that all such features can be found at that level; for example, human beings are partially carnivorous, but most non-human primates are not. The extent of radiation and selective adaptation in the course of evolution can result in quite closely related taxonomic groups diverging into a variety of ecological niches, and in unrelated groups converging towards similar ecological adaptations. The likelihood of such phylogenetic changes having an effect on ontogenetic process must always be taken into account. As Blurton Jones (1972) points out, "an animal's way of life has profound effects on many aspects of its behaviour and structure. It is therefore necessary to consider, and this is what most modern studies of the evolution of man are doing, what modification of general primate patterns this might have brought about." He goes on to enquire whether man's partly carnivorous way of life has modified the general higher primate pattern of child care with reference to whether the human infant is better adapted to being cached or carried: that is, whether it is more like higher primate infants which accompany the mother wherever she goes, usually by riding on her, or more like the infants of many carnivorous mammals which are not carried around but kept in nests. Although he concludes that the arguments against caching and for continuous contact are more convincing, such an interesting enquiry would not have been carried out at all had he not in the first place looked beyond the primates to other mammals.

In general we may say that, as a model for man, as well as an experimental substitute, clearly the non-human primates, not to mention any other grouping, have major limitations. Use of the monkey or ape as such a model works best, as Mason (1968) points out, when the process being studied is fairly well defined and the type of behaviour to be observed is limited in scope; "the question of how closely the performance of the model matches human behaviour in other ways is of small importance". On the other hand, as (indirect) links in the evolutionary chain to man the non-human primates have the greatest possible relevance for revealing both the changes that have had to take place for man to evolve from that level and the continuities that have extended to him.

The Conditions of Observation

Just as human behaviour tends to differ when observed in the natural home, in an institutional setting and in the laboratory, so also does the behaviour of animals when observed in the wild, in captivity, and in the laboratory. Not only are the conditions of stimulation different in the various settings, so also are the conditions of motivation.

One of the assumptions that is usually implicit in drawing conclusions from comparative work is that the behavioural or developmental phenomena compared are "normal" or 'typical" of the species concerned. Obviously this assumption is most valid when observation takes place in the natural environment and when the presence of the investigator is unknown to the individuals under study. The accumulating numbers of descriptive studies of species at many different phyletic levels under such conditions make a welcome advance from much of the earlier descriptive work on animals in zoos and on children in institutions where, it has since been realized, a lot of the behaviour observed was abnormal in every sense of the word. Natural behaviour and developmental changes in it have, however, to be explained. Since many of the questions that have to be asked in order to provide causal explanations can only be answered by research which exerts control over the relevant variables, it is often essential to resort to observation under laboratory conditions. While this may meet the requirements of the investigator who is thus able to study relations between variables in which *he* is interested, this can only be done at the cost of depriving the animal or person under study of normal living conditions and substituting conditions which, because either strange or unnatural, may elicit behaviours that are more or less deviant. Furthermore, insofar as the investigator designs his experiments with prime reference to theoretical problems with which he is preoccupied, to the extent of ignoring the role of variables that may be integral to the biology of the animal, he is likely to come up with distorted, if not irrelevant, answers. For example, Kortlandt's observations on chimpanzees (1962) suggest that the laboratory animal is so different from the wild animal that experimental results used in comparing laboratory chimpanzees with other species need to be interpreted with the greatest caution. In the wild, while the adult is much more alert than its laboratory counterpart, the young are more retarded than those born in the laboratory.

It is therefore important that, as comparative work advances, increasing attention should be paid to the status of the findings compared in terms of the conditions under which they were obtained. In the present early exploratory stages of comparison, reviews of developmental findings in different species often have to make the most of the limited data available from whatever source. This is alright as long as its evaluation by writer or reader retains an awareness of the ways in which the data could have been influenced by the different circumstances under which they were obtained. As systematic work becomes more possible, this will compare across species only those findings that were obtained under more or less equivalent conditions of observation, whether these be field, captivity or laboratory.

It is in the evaluation of the differential effects of these conditions, whether in planning or in interpreting comparative work, that a knowledge of the biology and ecology of the species under comparison becomes so significant. Only if the natural history of an animal is known can observations be ordered to allow for known sources of variation. Such sources may be inherent not only in deviations of an artificial environment from a natural one but also in the natural environment itself: for example, factors such as regional differences in natural habitat or in climate, in the distribution of other species, and cyclical variation due to seasonal or diurnal factors.

The limitations of studying animals outside of their normal way of life show especially in some of the cross-species comparisons that have been made of the findings concerning early experience effects. Many of these findings are derived from deprivation experiments. Such experiments require study of groups of animals reared with and without some particular aspect of normal experience, both groups living in an artificial laboratory setting. Looking at such findings comparatively, it sometimes remains unclear as to whether or not the deprivations experienced by the different species are equivalent, and how far the findings on the effects have been distorted by the animals' artificial rearing conditions.

The difficulty of obtaining findings, particularly developmental ones, under the combined conditions of minimal disturbance to the animal's way of life, minimal observer interference, continued observability of the animal and some facility of control of relevant variables, has been solved to some extent by the establishment of primate colonies and animal research stations. In these, although the animals live in a more or less confined area which is partially under the investigator's control, the area is kept as natural as possible in terms not just of the flora but of the presence of the normal range of other species as well. Observation and recording techniques may even employ telemetry. An example is the University of Chicago's Lake Cove research station in Maryland, where experimental studies of imprinting in mallards are being undertaken in their natural setting (Hess/ Petrovich, 1973).

In seeking to obtain species-typical findings for comparison, a problem closely linked to that of the observational conditions is that of ensuring representative sampling. This means making allowances for both strain-effects and for the effects of variations over the range of natural environments in which members of a particular species are reared. Species vary greatly as to the range of natural habitats in which they are found, but in none is the range greater than that of *Homo sapiens*. The problems of cross-cultural research in developmental psychology, especially that of equivalence of data, have been reviewed by Eckensburger (1973). Taking cross-species comparison seriously entails taking account of the extent to which the findings employed may in fact depart from species representativeness. As soon as one goes beyond comparison of fixed action patterns to areas where environmental factors play a major role, such as in behavioural ontogeny, this difficulty arises.

Levels, Units and Dimensions of Comparison

Nissen (1951) has pointed out that "for a phylogenetic comparison of behavior, the problem of classification, of identifying the most meaningful and useful dimensions and descriptive units of behavior, is of basic importance". He also suggests that "some orderliness in behavioral evolution will appear if, instead of considering each particular character (or trait, or mechanism) separately, we group together those characters which are behavioral equivalents or alternatives". The same may be said to apply also when the interest is not in tracing evolutionary trends but in testing the scope of generalizations. The only difference is that, whereas behavioural characters grouped together in order to show how such a character evolved through closely related species must be homologous, those grouped together in order to show a broad span evolutionary trend, or for purposes of generalization, may be either homologous or analogous.

Phylogenetic comparisons of ontogenetic findings can pose special problems of classification because such findings are by no means confined to changes in behavioural characters in the strict sense. As seen earlier they may refer also to changes in factors that affect the development of behaviour such as perceptual or learning abilities, environmental factors, and mechanisms of change. One would hesitate to exclude any kind of developmental data from comparative use as long as one makes sure that one is not compounding fallacious judgments based on experimental performance of doubtful validity by making comparative judgments in terms of insignificant variables. In fact, the developmental variables that have so far proved to be of value in comparative work differ greatly in type, level, and comprehensiveness.

Some of the achievements in and problems of identifying and classifying such variables in various areas of developmental study will be considered below. It should be borne in mind throughout that one of the most central problems it that of terminology. As Schneirla (1950) and Emlen (1958) have emphasized, observation, description and terminology are interdependent. The study of behaviour, especially social behaviour and its development, is a subject upon which investigators converge from a variety of disciplines such as social anthropology, comparative psychology and zoology. Between these there is as yet relatively little overlap of concepts. "To some extent, the discipline background of the investigator may affect his choice of levels in making his observations, because his terminological-conceptual framework is such as to attune him to the ordering and perceiving of a certain class of data" (Hall, 1963).

Development of the Behaviour Repertoire

The basic issues in describing and classifying behavioural characters have been admirably clarified by Hinde (1959, 1970). He distinguishes two methods of describing behaviour, the choice of which is likely to influence the whole nature of the subsequent analysis. One refers to patterns of limb or body movement that result from a particular range, strength and patterning of muscular contractions; for example, the "tonic neck reflex", "crawling", "smiling". The other he calls "description by consequence": terms such as "approaching", "hitting", "contact-seeking", do not refer to particular patterns of muscular contraction but include all patterns, often diverse ones, which either lead or could lead to a specified result. For each type of description the degree of precision to be used will depend on the kind of analysis or comparison intended. But, as Hinde points out, whichever type is used, "it inevitably involves the rejection of some of the data and a selection (arbitrary, intuitive, or rational) of the features to be studied—as soon as we move away from the initial cinematographic record of the behaviour we are neglecting information. Just because description and classification involve rejection of data and the selection of criteria, there are always the dangers that the essential data are rejected and that the units are irrelevant, too small or too large for the task in hand."

For practical research purposes, especially those of phyletic comparison, the types of behaviour defined by the above criteria require further classification. Hinde distinguishes three criteria that have been most valuable in ethological research. First is a causal classification, which entails a grouping together of types of behaviour according to the causal factors on which they depend, those activities that have causal factors in common being classified together. For example, all the activities of a mother that are influenced by the stimulus situation "infant" can be described as "maternal behaviour", and all the proximity-increasing behaviours of an infant that are elicited by the mother can be described as "attachment behaviour". Second is a functional classification by which activities are grouped according to the evolutionarily adaptive consequences they serve, for example "greeting response", "threat behaviour". Since functionally related activities often share causal factors, a functional classification often corresponds to a causal one; but, as Hinde points out, this is not always the case and, in order that causal and functional categories not be confused, causal analysis must precede causal classification. A third type of classification is a historical one, whether according to source such that behaviours like fixed action patterns having a common evolutionary origin are grouped together, or according to method of acquisition such as "learned" or "ritualized".

The problems of categorizing behaviour for comparative purposes tend to be proportional to the fineness of granularity of the category. At the broadest level, classifications of social behaviour according to the nature of the interacting individuals are most straightforward: for example, "peer-group behaviour", "infant-aunt behaviour". Within any given class of relationship, such as the mother-infant relationship, the various terms used to label major classes of behaviour such as "maternal", "attachement", "separation", "agonistic", "play", work well on the whole. Difficulties can arise, however, at the borderlines of overlapping classes such as play and exploratory behaviour (Hutt, 1966). More serious problems of categorization are associated with specific behaviour patterns that may occur in several different relationships. "Smiling" is a case in point. Confusion has arisen not only in discussion of the ontogeny of this response in humans, for example some investigators would not call newborn lip-retraction smiling, but also in phyletic comparisons (Ambrose, 1960). For example, what are to be regarded as the essential properties of this response, lip-retraction, a relaxed open mouth, teeth-baring, or a combination of one or other of these? Such different manifestations can be seen respectively in the human, the chimpanzee, and in monkeys of various kinds. Although these and other facial expressions at the different primate levels have to be taken into account in considering the evolution of the smiling response, "smiling" is certainly not the term to be used in classifying them together, as has sometimes been done. Van Hooff's use (1972) of different terms to label the expressions at the different levels is the correct approach.

Perhaps the greatest difficulties in categorization arise with attempts to label behaviour in terms of the type of emotion associated with it, for instance, "fear responses" and "distress reactions". As Rosenblum (1971) says about the latter, "distress vocalization may be fairly common in some animals, but heavily predated animals respond in quite different ways and do not vocalize when distressed. What justification have we for putting this diversity of responses into the same box? There are clearly dangers of adultomorphizing infant behaviour, but even greater dangers of anthropomorphizing animal behaviour." Nevertheless for some comparative purposes, what may matter most is the state of the individuals compared, not the response(s) by which it is expressed. Such is the case regarding depression in the infant following separation from the mother, whether in the human or in the rhesus monkey (Bowlby, 1973). Since classifications of this sort may incorporate, both within and across species, a variety of different behaviours, the only way to discover which ones should be included in a given category is by testing empirically their interrelationships and functions.

For studying the development of the behaviour repertoire and its organization in a particular species or group of species some investigators have explicitly worked out their own comprehensive scheme of classification which enables them to order and handle the hundreds of specific movement patterns and behaviour sequences they describe. These schemes vary greatly in usefulness for comparative purposes. The basic categories used by Gesell (1954), for instance, for his study of the principles of development of behaviour in the human infant are far too broad. At each successive age level at which he observed and tested infants, he grouped behaviour into what he called "four major fields of functional organization". These were· "motor behaviour" (postural reactions, prehension, locomotion, bodily co-ordination, motor skills), "adaptive behaviour" (those varied adjustments—perceptual, orientational, manual, and verbal—which reflect the child's capacity to initiate new experience and to profit by past experience), "language behaviour" (all behaviour to do with soliloquy, dramatic expression, communication, and comprehension), and "personal-social behaviour" (personal reactions to other persons and to the impacts of culture). Such a classification suffers not only from the problems of seriously overlapping categories but also from confusion regarding the nature and meaning of the criteria. By contrast, the scheme developed by Scott (1950) as a basis for his studies of the development of social behaviour in dogs and wolves allows much greater precision in comparative work. His detailed inventories of their behaviour repertoires were made in terms of behaviour patterns which he then classified into behavioural systems. He defines a behaviour pattern as "a unique and independent piece of behavior having a complete adaptive function". Because it has specific functional characteristics he also refers to a behaviour pattern as "a natural unit of behavior": for example, a simple reflex like the scratching reflex, or a piece of voluntary behaviour such as investigating another dog with the nose. His test for specificity of function involves reducing the behaviour to its simplest parts, such as lifting the hind paw, which may have any one of a dozen different functions as part of different patterns of behaviour: "we know that we are no longer dealing with a behavior pattern when specific function disappears". A behavioural system is defined as "a group of related behavior patterns having a common general function"; "in most species such a group is unrelated to other systems". For dogs and wolves he distinguishes nine behavioural systems, and he uses these for comparing human and dog behaviour patterns, "for it is obvious that all nine behavioral systems are also well developed in human beings". They comprise the following kinds of behaviour: investigative, epimeletic (giving of care and attention), et-epimeletic (attention getting or care soliciting), allelomimetic (doing what other animals in a group do), agonistic, sexual, eliminative, ingestive, and comfort-seeking. Although these categories do not entirely rule out the possibility of overlap, they are well defined in terms of the behaviour patterns subsumed under them, some of which are usefully classed also in subcategories. This type of approach has proved particularly useful in the comparative study of the development of social behaviour.

The Socialization Process

The extensive comparative thinking that has gone into the study of this process, especially in canines and primates including man has been concerned with "equivalences" of three kinds, namely, in behavioural systems, in behaviour patterns, and in developmental stages.

For the study of social behaviour and its development in primates Harlow (1965) referred to behavioural systems as affectional systems. He distinguished five of these: the infant-mother, the maternal, the age-mate or peer, the heterosexual, and the paternal. He considered that each system operates through a different set of "response patterns". The underlying variables that produce and control these response patterns, whether external, neural, or biochemical, are similar in some systems such as the first two, but different in others such as the third. Furthermore, each system develops through its own "maturational stages". The stages for those systems most involved in the socialization process are largely mutually complementary. The infant-mother affectional system has four: a reflex stage, a comfort and attachment stage, a security stage and a separation stage; the maternal affectional system has three: a maternal attachment and protection stage, an ambivalent transitional stage, and a stage of maternal separation and rejection; and the peer affectional system has four: the reflex, exploration, interactive play, and aggressive play stages. Analyses carried out in the rhesus monkey showed that each stage is characterized by a different set of response patterns which may, however, more or less overlap across stages. Harlow believed that this system of classification would be effective for the developmental and comparative study of social behaviour in most primates, including man.

In a similar way Scott (1963; Scott/Fuller, 1965) studying in dogs the development of those behavioural systems most involved in the socialization process, especially the epimeletic, found that the most fruitful method of analysing them was by dividing their development into "natural periods". Each natural period is based on "important changes in social relationships" and is "characterized by an important process". He distinguished four in dogs, namely the neonatal, transition, primary socialization, and juvenile periods, and these formed the basis for his comparison of human and dog socialization. For the transition period, for example, he was able to show the similarities and the differences in the transition from a neonatal to an adult type of ingestion, to an adult form of locomotion, to adult sensory capacities, and to an adult form of learning. For the primary socialization period he compared crying in babies with yelping in puppies, smiling in babies with tail wagging in puppies, and various physiological indications of emotional state in the two, in each case pointing to their functional similarity. The kinds of difference he was concerned with were not just in the behaviour patterns themselves, but also in the order of the major developmental changes; for example, the period of human primary socialization precedes the development of adult capacities of locomotion instead of following them as in the puppy. Other kinds of similarity he drew attention to at both levels were: the very early visual fixation of moving objects, the development of discrimination between familiar and unfamiliar objects, the differential emotional responses to such objects and especially the development of a fear response to completely unfamiliar objects which limits the socialization process, and the fact that socialization is not dependent upon the object of attachment reinforcing feeding behaviour.

At these levels of description, analysis and comparison, both Harlow and Scott have contributed much both to the working out of basic principles of socialization in mammalian species and to our understanding of the kinds of change in this process that have been brought about during evolution. Another investigator to do the same, in primates, is Mason (1965, 1968), only instead of making his comparisons on the basis of analogous behaviours and processes, such as Scott did, he concentrated more on homologous ones. In fact he compared three classes of these in human and non-human primate infants: first, the three behaviours through which contact and communication with the mother are established and maintained, clinging, sucking and vocalization; second, the various forms of investigatory and play behaviour used in the exploitation of the physical and social environment; third, the four main features of the social deprivation syndrome: abnormal postures and movements, motivational disturbance, poor integration of motor patterns, and deficiencies in social communication. By examining all these from both causal and functional standpoints he arrived at a way of ordering them in terms of a single higher level concept, namely, "arousal". The first group of behaviours he showed to be arousal reducing, the second to be arousal increasing, and the third represent the consequences of disturbances in these arousal functions.

Furthermore, by comparing various characteristics of social behaviour in rhesus monkey, chimpanzee and human infants, Mason (1968) has been able to indicate a number of significant evolutionary trends culminating in humans. In formulating these trends the dimensions he uses have to do not so much with units such as behaviour patterns or behaviour systems but with properties of behaviour, like "rate" of development, "specificity" of stimulation, "sharpness of definition" of developmental stages, and "number" "variety" "plasticity" "reliability" and "efficiency" of motor patterns. From his starting-point of the general neotenous trend in primate evolution he indicates the following evolutionary changes: 1) There is a slowing down of the rate of ontogenetic development of behaviour. 2) This gives more time for learning, perhaps because more learning is required. 3) There is a "loosening" of motor patterns, such that the number and variety of motor acts and the tendency to combine them spontaneously into new and more complex patterns increases progressively. 4) The human infant's needs for stimulation are less specific than those of non-human primates. 5) Once a developmental sequence has been set in motion, the human infant will show a greater tendency to elaborate on it than will the chimpanzee, so will the chimpanzee in relation to the rhesus monkey. 6) Developmental stages are less sharply defined in humans than in the other primates: sensitive periods in development are more difficult to establish, there is less liklihood that the withholding of any specific experience will result in developmental arrest, and there is a much stronger tendency for behaviour to reflect an intermingling of developmental stages, different response patterns, and different motivational systems. "Of all the primates, man has gone farthest in an evolutionary venture in which the reliability and efficiency of instinctive patterns have been sacrificed to achieve the behavioural plasticity and the liberation of psychic energies that are so much a part of the human condition."

From these and other studies it may be said with some justification that, of all the areas in which developmental findings have been compared across species, comparisons in the area of the socialization process have so far yielded the richest rewards.

Development of Abilities

The study of ontogenetic changes in ability is dependent upon the use of testing procedures which require both the provision of selected stimuli and the measurement of selected responses. In cross-species comparisons of the results of developmental findings of this type it is much more important to ensure that the stimuli and responses selected and the test situation employed are appropriate to the biology of each species rather than that the tests are precisely similar.

In the case of the development of perceptual abilities, it has been clear at least since the time of von Uexküll's classic publication (1934) that the perceptual worlds of different kinds of animals can be very different from each other, let alone from that of man. The selective nature of response to stimulation, and the effectiveness of the physical properties of a stimulus in controlling behaviour are areas that have been studied in much depth, and Hinde (1970) has reviewed a number of features of the findings that create problems for comparative work. One is that even quantitative variations in a stimulus along a simple physical dimension may produce complex changes in its effect on the organism. A second concerns the danger of making premature generalizations regarding the patterns of stimulation which an animal can or cannot distinguish. A third is that, in the perception of movement, the mechanisms that determine the effectiveness of a given stimulus may be considerably more complicated than may appear at first sight.

Despite such difficulties, revealing comparisons of findings on perceptual development have been made with reference to such broad issues as the role of experience in the development of form perception, pattern perception and distance perception. Such findings have been based on the use of procedures which can be applied in similar ways across a wide phyletic range, such as rearing in darkness (Riesen, 1961), pattern preference tests (Fantz, 1967) and the visual cliff experiment (Walk/Gibson, 1961).

In the case of the development of learning abilities, cross-species comparisons are equally, if not more, fraught with difficulties. In the first place, even before comparison is commenced, the within-species developmental findings may be suspect due to lack of control over developmental changes in factors that affect learning as distinct from changes in learning ability itself. Campbell (1967) has pointed out several of these that should always be taken into account at different ages, namely, the conditions of motivation and reinforcement, the unlearned behaviour that is elicited by the test apparatus, and the extent to which the previous experience of an older animal is influencing its learning success, whether by direct transfer or by influencing its emotionality in the test situation. In the second place, when it comes to cross-species comparison, as Hinde (1970) points out, "few tests are equally suitable over a wide range of species, for differences in test scores may reflect differences in sensory, perceptual or motor functioning, rather than in learning ability itself. Furthermore, learning processes are themselves diverse, so that different species may solve the same problem in different ways."

In contrast to the huge literature that demonstrates the generality of all the traditional types of learning across the vertebrate range and beyond, there are relatively few studies of the changes in learning ability that occur during the ontogeny of a given species such as that by Zimmerman and Torrey, for the rhesus monkey (1965) and Campbell (1967) for the rat. Although there have been numerous comparative reviews of the nature and extent of differences in learning ability at different phyletic levels (*e.g.* Warren, 1965) and of evolutionary trends in factors accounting for such differences (*e.g.* Harlow, 1958b), these reviews are non-developmental in the ontogenetic sense. Nevertheless the scattered studies so far available of changes with age in learning ability in various species give a lead to classes of developmental phenomena that can be fruitfully compared. These include the age of onset of different types of learning; the ways in which this is influenced by growth and differentiation in the nervous system, by the development of perceptual abilities or by changes in the behaviour repertoire; variations with age in rate of learning, in the nature and extent of transfer, and in the retention of learned responses; the role of sensitive periods in learning and the effects of rearing conditions on later learning ability. As regards the last of these, Beach and Jaynes (1954) have pointed out special difficulties: "anatomical differences, nutritional requirements, sensory sensibility, motor development, and previous experience are variables clearly interwoven with age and cannot usually be controlled independently of each other."

Bearing in mind the interactional nature of all behavioural development, behavioural capacities are the outcome both of the exogenous capacities of the environment to provide stimuli and reinforcers and of the endogenous capacities of the organism to respond. Insofar as the latter are the outcome of processes of maturation, as distinguished through deprivation or restriction experiments, their developmental course requires study and comparison just as much as that of learning processes. Not only does each response pattern in an individual mature at different rates, different species may also develop the same response at different rates. Furthermore, within ontogeny, maturation rate changes are typical during development, some patterns developing linearly and others developing in a curvilinear or sigmoid fashion (King, 1968). Such consequences of species differences in genetic programming throw further light on comparisons of developmental findings of all kinds.

Effects of Early Experience

Comparative work in this huge area of developmental research has been concerned not so much with evolutionary trends as with efforts to establish basic principles by drawing on experimental findings at different phyletic levels. The principles that have emerged vary greatly in scope and precision from broad statements covering a wide range of mammalian and avian species such as that early experience is important, to highly specific statements limited to one strain of one species concerning the effect on a measurable variable in adulthood of a defined measurable treatment administered under controlled conditions at a given early age assessed from the time of conception.

The importance of early experience lies in the facts that many kinds of treatment have different effects on an organism according to whether they are imposed during immaturity or maturity, and that the effects brought about during immaturity not only tend to be greater in scope and magnitude but can drastically affect the nature of behaviour and capacity at maturity. For assessing the degree of importance of any particular early experience Thompson and Schaefer (1961) suggested that account must be taken of four dimensions: the extent of change in a particular function, the potential reversibility of the change, the permanence or duration of the change, and the generality of the change, that is, the number of discrete functions or structures affected by the treatment.

For comparisons of findings in different kinds of animals the problem is not so much in qualitatively equating the various kinds of early treatment used as the independent variable. Examples of the variety of those employed in many studies are: prenatally, stressors such as x-irradiation, chemical manipulations, or maternal conditioned anxiety; and postnatally, general massive stimulation, for example electric shock, handling, auditory stimulation, temperature change, or general or specific environmental enrichment or restriction, or alterations in the normal conditions of social interaction during rearing. The main problem lies in the quantitative aspects of these treatments and in the measurement of their effects. These effects are mostly nonspecific and differ in kind, magnitude and direction according to fundamental parameters of the early treatment such as stimulus intensity, amount of enrichment or restriction, and age at which it is imposed. A given stimulus intensity for one species may be inappropriately small or large for another. Furthermore, because of problems of age-scaling between species having characteristically different growth rates, equivalence of age may be difficult to assess. As with all other cross-species developmental comparisons, resort has to be made to assessment in terms of developmental phase. In this case phases such as preweaning and postweaning are commonly used, but although change in mode of feeding may be associated with other important changes it is not necessarily a crucial dividing point.

The kinds of dependent variable employed for studying effects have varied between very broad categories such as "emotionality", "social attachment", "intelligence level" and "general growth rate" to very specific ones such as "defecation level", "open field activity," "adrenal size" and "water maze learning rate". Since the broader categories are usually operationally defined in terms of specific measures appropriate to particular species, the kinds of measure used to indicate, say, emotionality in rats, dogs and monkeys are not likely to be the same. Nevertheless the issue of precisely what is affected becomes significant for any comparative enquiry concerned with the mechanisms by which the early treatment takes effect. This is especially so where the effects may not be direct, as for example in the case of intellectual functioning: most tests of intelligence in animals, such as tests of exploration or maze learning, involve components of motivation and emotionality which may be affected directly.

Indeed, the problem of unconfounding variables, whether independent or dependent, and of defining these with precision has been with this area of research from the start. This becomes a serious problem in any attempt to compare findings in animals with those in man. Human studies in this area are beset by formidable difficulties of control, having to make use either of retrospective studies in which the relevant independent variable is likely to be vague and mixed in with many others, or prospective studies in available natural situations such as institutions. For example, although contrasts between the later effects of early rearing in the family-home setting and in an institutional setting can be apparent enough, it always remains unclear how much the institutional effects are due to such different factors as amount of general stimulation, perceptual richness or complexity, the number and permanence of the caretakers, or the emotional quality of the caretaking. Consequently much of the outcome of research at the human level has gained credence and impact only because of the general congruence of its findings with the more clearcut findings from the much better controlled animal work.

Although comparisons of findings of the early experience studies at different phyletic levels certainly show that early experience is important, they have not yet provided the means for a breakthrough in the theoretical treatment of how the effects are brought about. While this may be due in part to the fact that many studies concentrate on the relations between the early treatment and the later effects without giving due attention to the nature of the immediate effects, it may well be due to a more fundamental difficulty. As Thompson and Grusec (1970) describe it, "this appears to be that too little attention has been paid to analysis and definition of the basic dimension involved—development."

DISCUSSION

Stages and Phases

Chance: If you eliminate the idea of stages, how far are you prepared to say anything about any sort of behaviour being related to the requirements of survival value in a particular social environment? To some extent, your own work suggests that the smiling response, which is a species-specific behaviour, is very closely related to the requirements of the infant at that stage.

Ambrose: In giving a complete account of how a particular behaviour pattern has survival value, especially during early ontogeny, it is important to take full account of the stage or stages of development involved, because the changing organism relates very differently to the same social environment at different stages. What I am making a plea for is that instead of just talking about, say, *the* neonatal stage, or *the* stage when this or that happens, as if this was universally understood and accepted, investigators should give a clear definition of what constitutes the stage in question. This reduces the danger that the term "stage" may be confused with what may just be the age-range of interest to a particular investigator. Furthermore, clear definition becomes particularly necessary in comparative developmental work.

von Cranach: Isn't the concept of stages based on the fact that phases are correlated, the correlation comprising several developmental traits towards something like a level, after which other developments will lead to another level?

Ambrose: Yes, the terms stage and level of development are really synonymous. The difficulty is to find the correlations. Besides, within the concept of a general stage, one could still talk about a phase in the development of a particular response. It does not have to correlate.

Chance: How about the sexual behaviour of immature creatures and Freud's ideas of sexual development?

Ambrose: I did not go into that because there isn't time. Of course, many of the major forms of adult sexual behaviour occur in component form during infancy and childhood, before the whole sexual response system is fully developed.

von Cranach: It is interesting to consider the problem of stages in relation to bodily growth. In German, we have a word, *Gestaltwandel,* which designates something more than bodily growth, namely, a kind of change of proportions, for the change of figure that children undergo between the age of five and six or thereabouts. Whereas the figure of a small child shows a relatively big head, protruding belly and short arms and legs, that of a six-year-old is more like that of an adult. Among school psychologists it is said that a child is ready for school if he can touch the left ear with his right hand over his head; I don't know whether this is true.

Ambrose: But the point is, how general is the readiness for school? The implication is surely that the anatomical changes are highly correlated with other physiological and psychological changes.

von Cranach: I would like to know, but I do not. What I am interested in are the social effects, and as our society is providing institutions for several different age-levels, it is hard to say whether it is the intellectual and affective changes that originally gave rise to these different institutions, or the other way around.

Chance: The concept of the latency period, as I understand it, has some physiological or developmental correspondence, doesn't it? Does the existence of this period between five and six correlate with anything?

Ambrose: This concept is one derived by psychoanalysts from the study of adults and is used by them in their attempts to reconstruct their patients' early history. It is not derived from direct observation of children. That is why it has fallen into disrepute among non-analysts.

Imprinting

Crook: Lately, one hears less about the imprinting concept and Bowlby's ideas based largely upon it. This is partly, I think, because the imprinting field in animal behaviour study has run into complexities which obscure the original simplicity of the idea. I am wondering whether among some psychologists there was the feeling that the imprinting work represented a new paradigm, some new way of thinking?

Ambrose: Well, I know one or two psychologists who were very taken with this critical period work and started looking for critical periods in early infancy. They were disappointed when they did not find any. I think the principal difficulty here has been over the question of irreversibility. It seems that in humans there is relatively little irreversibility, the human capacity for recuperation being quite amazing. Furthermore, there are all the studies of individual differences in infancy, which show that babies even within one culture, not to mention different cultures, can go through particular developmental processes at somewhat differing ages, and one knows that these processes can be sped up or slowed down. It is therefore very difficult to specify something like a critical period. We talk about critical periods in language development, because language is much more difficult to learn if you do not learn it within a limited age-range, but the fact remains that it is still possible to learn it perfectly well later on.

Masters: I know there are only very few carefully documented cases of humans who had no access to language before the age of say 20, but I thought that every case showed a limited vocabulary of some 40 words and a very rudimentary grammar. My question is, is it not possible that the critical periods of *Homo sapiens* are much longer and more flexible precisely because of the complexity of human behaviour, and that this complexity does not necessarily mean that there are not certain periods?

Ambrose: That may well be. I cannot answer your question because I don't know much about language development, especially later on. What happens mostly is that psychologists continue to talk about critical periods, but when pressed tend to define them rather differently. King, for example, defines a critical period in terms of the receptivity of the organism to environmental influences when rates of change are maximum, but that is a very different definition from the ethological one.

von Cranach: Isn't the idea of the critical period lost if there is no relatively fixed period anymore? If extended too far, it will become something substantially different that cannot be distinguished from learning processes, so that one could only talk about different learning capabilities in different phases.

Masters: I raised the question because of the whole problem of neoteny in *Homo sapiens.* If one assumed neoteny does produce these rather long periods, a critical period that stops only at 21 years would be possible. That would certainly be important to know.

Cain: There is no necessary connection between neoteny and critical periods. The critical period in the very sharp form in which it was first found by Lorenz is a very peculiar phenomenon in a very selected form where you need to have very rapid orientation which in itself cannot be preselected towards some definite individual. This is a highly specialized type of behaviour, and I am slightly puzzled why one should necessarily propose critical periods for the human race. That is a purely historical point.

Ambrose: In human psychology, we are interested, among other things, in the optimal conditions and opportunities for development, and if it were discovered that we were not providing the right opportunity at the right time, at a certain phase of development, this would be important for promoting human welfare. In other words, it seems to me entirely legitimate that we should engage in research into the general question of whether or not there are critical periods in human development.

Smiling and the Orienting Response

van Hooff: When you talked about the selection of units, you said that some responses may change considerably in form during development, and you referred to the response of fear. I should like you to enlarge upon that a little, and also I want to ask you whether you would say that smiling and laughter, too, change in form during development. I may as well say why I ask this: I have been trying, in connection with the comparative aspects of smiling and laughter, to find accurate descriptions in the literature. These are not easily found, and usually the descriptions are quite rough. Nevertheless, it seems that the earliest smiles of an infant tend to be not so much retraction of the mouth corners, but rather an opening of the mouth, which later on is accompanied by gurgling sounds. Now, if I were to interpret this very naively, I would say it is not a smile at all but the initial emergence of a playface. I would say the same with regard to the retractive response.

Ambrose: Infant smiling becomes a functional response at around four weeks of age. You can however see components of it, such as unilateral or bilateral lip retraction, well before this time. This is clearly lip retraction, without any mouth opening at all.

van Hooff: Is that early lip retraction shown in anything that approaches the later social context?

Ambrose: No, it tends to occur from the first week to a variety of kinds of nonsocial stimulation, whether internal or external, and it is my impression that this has to be of intermediate intensity. Then it begins to occur in response to auditory stimulation, usually the voice of the mother, and finally at four weeks or so it occurs in response to the sight of the human face. Of course young babies smile with the mouth open as well, but I think this is very much an intensity variable. That is, lip retraction only in low intensity smiling, lip retraction plus mouth opening in medium intensity smiling. At high intensity the mouth opening seems to predominate over the lip retraction and gradually the response goes over into laughing. Laughing, however, though related to smiling, is a qualitatively different response with a different causation, and this does not begin to appear until the fourth month.

There remains, of course, quite a problem in defining smiling. Should one define it merely in terms of the motor pattern of lip retraction, or should one also take into account the eyes? Many people agree that somehow the eyes of the baby seem to brighten when it smiles. I ask you to define a measure for the brightening of a baby's eyes: nobody can put their finger on it, but it seems to be there.

von Cranach: What about the orienting response; would that not belong with the early infant smile?

Ambrose: Yes, the baby's face, and usually the trunk and arms as well, are oriented towards the person who elicits the smiling. The whole appearance is that the smile is being directed towards that person: smiling is a response towards the social stimulus. At the other extreme is the fear response. In early infancy a baby who cannot crawl or run away does the next best thing; that is, to look away, and this may be accompanied by crying. The same happens in the older child, but here the locomotory component may come in as well. In still older children the fear response may be confined merely to a physiological reaction without either crying or looking away. This depends on the development of a capacity to cope with a fearful situation by trying to alter it rather than by running away.

NOTE

Examples of the range of developmental topics examined from a comparative perspective are the following titles:
Determinants of Infant Behaviour (Foss, 1961)
The Development of Motor Skill (Connolly, 1970)
Early Behaviour (Stevenson *et al.*, 1967)
Environmental Influences (Gloss, 1968)
The Roots of Behaviour (Bliss, 1962)
Stimulation in Early Infancy (Ambrose, 1969)
Maternal Behaviour in Mammals (Rheingold, 1963)
Early Experience and Behaviour (Newton/Levine, 1968).

REFERENCES

Ambrose, J.A. 1960. "The Smiling and Related Responses in Early Human Infancy: An Experimental and Theoretical Study of their Course and Significance." Unpublished doctoral dissertation, University of London.

— 1963. "The Concept of a Critical Period for the Development of Social Responsiveness in Early Human Infancy", pp. 201-225 in: B.M. Foss (ed.) *Determinants of Infant Behaviour*, vol. 2. London, Methuen.

— 1968. "The Comparative Approach to Early Child Development: the Data of Ethology", pp. 183-232 in: E. Miller (ed.) *Foundations of Child Psychiatry*. Oxford, Pergamon Press.

— 1969. (ed.) *Stimulation in Early Infancy*. London, Academic Press.

Beach, F.A. 1955. "The Descent of Instinct", *Psychological Review* 62:401-410.

Beach, F.A./Jaynes, J. 1954. "Effects of Early Experience upon the Behaviour of Animals", *Psychological Bulletin* 51:239-263.

Bliss, E.L. 1962. (ed.) *Roots of Behavior*. New York, Harper.

Blurton Jones, N. 1972. "Comparative Aspects of Mother-Child Contact", pp. 305-328 in: N. Blurton Jones (ed.) *Ethological Studies of Child Behaviour*. Cambridge University Press.

Bowlby, J. 1973. *Separation, Anxiety and Anger*. Vol. 2. *Attachment and Loss*. London, Hogarth.

Bronfenbrenner, U. 1968. "Early Deprivation in Mammals: a Cross-Species Analysis", pp. 627-764 in: G. Newton/S. Levine (eds.) *Early Experience and Behavior*. Springfield, Thomas.

Caldwell, B.M. 1962. "The Usefulness of the Critical Period Hypothesis in the Study of Filiative Behavior", *Merrill-Palmer Quarterly* 8 (4) 229-242.

Campbell, B.A. 1967. "Developmental Studies of Learning and Motivation in Infra-Primate Mammals", pp. 43-71 in: H.W. Stevenson/E.H. Hess/H.L. Rheingold (eds.) *Early Behavior, Comparative and Developmental Approaches*. New York, Wiley.

Clark, Le Gros. 1960. *The Antecedents of Man*. Chicago, Quadrangle Books.

Coghill, G.E. 1929. *Anatomy and the Problem of Behavior*. New York, Macmillan.

Connolly, K. 1970. (ed.) *Mechanisms of Motor Skill Development*. London, Academic Press.

Connolly, K./Elliott, J. 1972. "The Evolution and Ontogeny of Hand Function", pp. 329-383 in: N. Blurton Jones (ed.) *Ethological Studies of Child Behaviour*. Cambridge University Press.

Crook, J.H. 1972. "Social Organization and the Environment: Aspects of Contemporary Social Ethology" pp. 101-115 in: C.L. Brace/J. Metress (eds.) *Man in Evolutionary Perspective*. New York, Wiley.

Darwin, C. 1859. *On the Origin of Species by Means of Natural Selection*, 1st ed., London, John Murray. (Reprinted New York, Philosophical Library, 1951.)

De Beer, G.R. 1928. *Vertebrate Zoology*. London, Sidgwick & Jackson.

Delius, J.D. 1970. "The Ontogeny of Behavior", pp. 188-196 in: F.O. Schmitt (ed.) *The Neurosciences Second Study Program*. New York, Rockefeller University Press.

Denny, M.R./Ratner, S.C. 1970. *Comparative Psychology*. Homewood, Dorsey.

Eckensburger, L.H. 1973. "Methodological Issues of Cross-Cultural Research in Development Psychology", pp. 43-64 in: J.R. Nesselroade/H.W. Reese (eds.) *Life-Span Developmental Psychology*. New York, Academic Press.

Emlen, J.T. 1958. "The Art of Making Field Notes", *Jack-Pine Warbler* 36:178-181.

Erlenmeyer-Kimling, L. 1972. "Gene Environment Interactions and the Variability of Behavior" pp. 181-208 in: L. Ehrman/G.S. Omenn/E. Caspari (eds.) *Genetics, Environment and Behavior*. New York, Academic Press.

Fantz, R.L. 1967. "Visual Perception and Experience in Early Infancy: a Look at the Hidden Side of Behavior Development", pp. 181-224 in: H.W. Stevenson/E.H. Hess/H.L. Rheingold (eds.) *Early Behavior, Comparative and Developmental Approaches*. New York, Wiley.

Foss, B.M. 1961. (ed.) *Determinants of Infant Behaviour*, vol. 1. London, Methuen.

Freud, S. 1950. "Instincts and Their Vicissitudes", (1915) pp. 60-83 in: E. Jones (ed.) *Collected Papers* 4. London, Hogarth.

Fuller, J.L./Thompson, W.R. 1960. *Behavior Genetics*. New York, Wiley.

Gesell, A. 1954. "The Ontogenesis of Infant Behavior", pp. 335-373 in: L. Carmichael (ed.) *Manual of Child Psychology*, 2nd ed. New York, Wiley.

Glass, D.C. 1968. *Environmental Influences*. New York, Rockefeller University Press.

Hall, K.R.L. 1963. "Some Problems in the Analysis and Comparison of Monkey and Ape Behavior", pp. 273-300 in: S.L. Washburn (ed.) *Classification and Human Evolution*. Chicago, Aldine.

Hamburg, D. 1969. "A Combined Biological and Psychosocial Approach to the Study of Behavioural Development", pp. 269-277 in: J.A. Ambrose (ed.) *Stimulation in Early Infancy*. London, Academic Press.

Harlow, H.F. 1958a. "The Nature of Love", *American Psychologist* 13:673.

— 1958b. "The Evolution of Learning", pp. 269-290 in A. Roe/G.G. Simpson (eds.) *Behavior and Evolution*. New Haven, Yale University Press.

— 1965. "The Affectional Systems", pp. 287-334 in: A.M. Schrier/H.F. Harlow/G. Stollnitz (eds.) *Behavior of Nonhuman Primates*, vol. 2. New York, Academic Press.

— 1972. "Generalization of Behavioral Data Between Nonhuman and Human Animals", *American Psychologist* 27 (8) 709-716.

Hebb, D.A. 1953. "Heredity and Environment in Mammalian Behaviour", *British Journal of Animal Behaviour* 1:43-47.

Hess, E.H./Petrovich, S.B. 1973. "The Early Development of Parent-Young Interaction in Nature", pp. 25-42 in: J.R. Nesselroade/H.W. Reese (eds.) *Life-Span Developmental Psychology*. New York, Academic Press.

Hinde, R.A. 1959. "Some Recent Trends in Ethology", pp. 561-610 in: S. Koch (ed.) *Psychology: a Study of a Science*. Vol. 2, *General Systematic Formulations, Learning, and Special Processes*. New York, McGraw-Hill.

— 1961. "The Establishment of the Parent-Offspring Relation in Birds, with Some Mammalian Analogies", pp. 175-193 in: W.H. Thorpe/O.L. Zangwill (eds.) *Current Problems in Animal Behaviour*. Cambridge University Press.

— 1970. *Animal Behavior*. New York, McGraw-Hill.

Hooff, J.A.R.A.M. van. 1972. "A Comparative Approach to the Phylogeny of Laughter and Smiling", pp. 209-238 in: R.A. Hinde (ed.) *Non-verbal Communication*. Cambridge University Press.

Hutt, C. 1966. "Exploration and Play in Children", pp. 61-81 in: P.A. Jewell/C. Loizos (eds.) *Play, Exploration and Territory in Mammals*. Symposia of the Zoological Society of London 18. London/New York, Academic Press.

Inhelder, B. 1956. "Criteria of the Stages of Mental Development", pp. 75-107 in: J.M. Tanner/B. Inhelder (eds.) *Discussions on Child Development*, vol. 1. London, Tavistock.

Jacobsen, C.F./Jacobsen, M.M./Yoshioka, J.G. 1932. "Development of an Infant Chimpanzee During Her First Year", *Comparative Psychology Monographs* 9:1-94.

Jay, P. 1963. "Mother-Infant Relations in Langurs", pp. 282-304 in: H.L. Rheingold (ed.) *Maternal Behavior in Mammals*. New York, Wiley.

Kaplan, J./Schusterman, R.J. 1972. "Social Preferences of Mother and Infant Squirrel Monkeys Following Different Rearing Experiences", *Developmental Psychobiology* 5 (1) 53-60.

Kellogg, W.N./Kellogg, L.A. 1933. *The Ape and the Child*. New York, McGraw-Hill.

King, J.A. 1968. "Species Specificity and Early Experience", pp. 42-64 in: G. Newton/ S. Levine (eds.) *Early Experience and Behavior*. Springfield, Thomas.

Knox, W.E. 1955. "The Metabolism of Phenylalanine and Tyrosine", pp. 836-866 in: W.D. McElroy/H.B. Glass (eds.) *A Symposium on Amino Acid Metabolism*. Baltimore, Johns Hopkins.

Kohts, N. 1935. "Infant Ape and Human Child: Instincts, Emotions, Play, Habits", *Scientific Memoirs of the Darwinian Museum, Moscow* 3:

Kortlandt, A. 1962. "Chimpanzees in the Wild", *Scientific American* 206 (5) 128-138.

Kuo, Z.Y. 1967. *The Dynamics of Behavior Development: an Epigenetic View*. New York, Random House.

Langer, J.A. 1970. "Werner's Comparative Organismic Theory", pp. 733-771 in: P.H. Mussen (ed.) *Carmichael's Manual of Child Psychology*, vol. 1. New York, Wiley.

Lehrman, D.S. 1953. "A Critique of Konrad Lorenz's Theory of Instinctive Behavior", *Quarterly Review of Biology* 28:337-363.

— 1962. "Interaction of Hormonal and Experiential Influences on Development of Behavior", pp. 142-156 in: E.L. Bliss (ed.) *Roots of Behavior*. New York, Hoeber.

Lorenz, K. 1941. "Vergleichende Bewegungsstudien an Anatinen", *Journal für Ornithologie* 89 (Sonderheft) 19-29.

Mason, W.A. 1965. "The Social Development of Monkeys and Apes", pp. 514-543 in: I. DeVore (ed.) *Primate Behavior*. New York, Holt, Rinehart and Winston.

— 1968. "Early Social Deprivation in the Nonhuman Primates: Implications for Human Behavior", pp. 70-101 in: D.C. Glass (ed.) *Environmental Influences*. New York, Rockefeller University Press.

McGraw, M.B. 1963. *The Neuromuscular Maturation of the Human Infant*. New York, Hafner.

Mussen, P.H. 1970. (ed.) *Carmichael's Manual of Child Psychology*, 3rd ed. New York, Wiley.

Newton, G./Levine, S. 1968. (eds.) *Early Experience and Behavior*. Springfield, Thomas.

Nissen, H.W. 1951. "Social Behaviour in Primates", pp. 423-457 in: C.P. Stone (ed.) *Comparative Psychology*. Englewood Cliffs, NJ, Prentice-Hall.

Nottebohm, F. 1970. "Ontogeny of Bird Song", *Science* 167:950-956.

Nunnally, J.C. 1973. "Research Strategies and Measurement Methods for Investigating Human Development", pp. 87-109 in: J.R. Nesselroade/H.W. Reese (eds.) *Life-Span Developmental Psychology*. New York, Academic Press.

Piaget, J. 1953. *The Origin of Intelligence in the Child*. London, Routledge and Kegan Paul.

Preston, D.G./Baker, R.P./Seay, B. 1970. "Mother-Infant Separation in the Patas Monkey", *Developmental Psychology* 3 (3) 298-306.

Rheingold, H.L. 1963. (ed.) *Maternal Behavior in Mammals*. New York, Wiley.

— 1967. "A Comparative Psychology of Development", pp. 279-293 in: H.W. Stevenson/E.H. Hess/H.L. Rheingold (eds.) *Early Behavior, Comparative and Developmental Approaches*. New York, Wiley.

Riesen, A.H. 1961. "Studying Perceptual Development Using the Technique of Sensory Deprivation", *Journal of Nervous and Mental Disease* 132:21-25.

Riopelle, A.J./Rogers, C.M. 1965. "Age Changes in Chimpanzees", pp. 449-462 in: A.M. Schrier/H.F. Harlow/F. Stollnitz (eds.) *Behavior of Nonhuman Primates*, vol. 2. New York, Academic Press.

Rosenblatt, J.S./Turkewitz, G./Schneirla, T.C. 1961. "Early Socialization in the Domestic Cat as Based on Feeding and Other Relationships Between Female and Young", pp. 51-74 in: B. Foss (ed.) *Determinants of Infant Behaviour*. London, Methuen.

Rosenblum, A.R. 1971. "The Categorizing of Behaviour. General Discussion", pp. 129-133 in: H.R. Schaffer (ed.) *The Origins of Human Relations*. London, Academic Press.

Rosenblum, A.R./Kaufman, I.C. 1967. "Laboratory Observations of Early Mother-Infant Relations in Pigtail and Bonnet Macaques", pp. 33-41 in: S.A. Altmann (ed.) *Social Communication Among Primates*. Chicago University Press.

Schneirla, T.C. 1950. "The Relationship Between Observation and Experimentation in the Field Study of Behavior", *Annals of the New York Academy of Sciences* 51:1022-1044.

— 1952. "A Consideration of Some Conceptual Trends in Comparative Psychology", *Psychological Bulletin* 49:559-597.

Scott, J.P. 1950. "The Social Behavior of Dogs and Wolves: an Illustration of Sociobiological Systematics", *Annals of the New York Academy of Sciences* 51:1009-1021.

— 1963. "The Process of Primary Socialization in Canine and Human Infants", *Monographs of the Society for Research in Child Development* 28 (1) 1-47.

— 1968. "The Process of Primary Socialization in the Dog", pp. 412-439 in: G. Newton/S. Levine (eds.) *Early Experience and Behavior*. Springfield, Thomas.

Scott, J.P./Fuller, J.L. 1965. *Genetics and the Social Behavior of the Dog*. University of Chicago Press.

Spencer Booth, Y. 1970. "The Relationships Between Mammalian Young and Conspecifics Other Than Mothers and Peers: a Review", pp. 119-194 in: D.S. Lehrman/R.A. Hinde/E. Shaw (eds.) *Advances in the Study of Behavior*, vol. 3. New York, Academic Press.

Stevenson, H.W./Hesse, E.H./Rheingold, H.L. (eds.) 1967. *Early Behavior, Comparative and Developmental Approaches*. New York, Wiley.

Struhsaker, T.T. 1971. "Social Behavior of Mother and Infant Vervet Monkeys", *Animal Behaviour* 19 (2) 233-250.

Tanner, J.M. 1970. "Physical Growth", pp. 77-155 in: P.H. Mussen (ed.) *Carmichael's Manual of Child Psychology*, vol. 1. New York, Wiley.

Thompson, W.R. 1960. "Early Environmental Influences on Behavioral Development", *American Journal of Orthopsychiatry* 30:306-314.

Thompson, W.R./Grusec, J.E. 1970. "Studies of Early Experience", pp. 565-654 in: P.H. Mussen (ed.) *Carmichael's Manual of Child Psychology*, vol. 1. New York, Wiley.

Thompson, W.R./Schaefer, T. 1961. "Early Environmental Stimulation", in: D.W. Fiske/S. Maddi (eds.) *The Functions of Varied Experience*. Chicago, Dorsey.

Tinbergen, N. 1963. "On Aims and Methods of Ethology", *Zeitschrift für Tierpsychologie* 20:410-433.

Uexküll, J. von. 1934. *Streifzüge durch die Umwelten von Tieren und Menschen*. Berlin, Springer. Translated in: C.H. Schiller (ed.) *Instinctive Behavior*. New York, Universities Press, 1957.

Vince, M.A. 1961. "Developmental Changes in Learning Capacity", pp. 225-247 in: W.H. Thorpe/O.L. Zangwill (eds.) *Current Problems in Animal Behaviour*. Cambridge University Press.

Walk, R.D./Gibson, E.J. 1961. *A Comparative and Analytical Study of Visual Depth Perception*. Psychological Monographs 75. Washington, American Psychological Association.

Warren, J.M. 1965. "Primate Learning in Comparative Perspective", pp. 249-281 in: A.M. Schrier/H.F. Harlow/F. Stollnitz (eds.) *Behavior of Nonhuman Primates*, vol. 1. New York, Academic Press.

Werner, H. 1957. "The Concept of Development from a Comparative and Organismic Point of View", pp. 125-148 in: D.B. Harris (ed.) *The Concept of Development*. Minneapolis, University of Minnesota Press.

Zimmerman, R.R./Torrey, C.C. 1965. "Ontogeny of Learning", pp. 405-447 in: A.M. Schrier/H.F. Harlow/F. Stollnitz (eds.) *Behavior of Nonhuman Primates*, vol. 2. New York, Academic Press.

FRANÇOIS BRESSON

Inferences from Animals to Man: Identifying Behaviour and Identifying Functions

Making inferences from animals to man raises two kinds of problems, epistemological problems and problems that can be related to the sociology of science.

The epistemological problems mainly include those encountered in the biological study of systems that change over time and evolve. What is at issue is the choice of those invariants which make it possible to describe and explain changes in systems over time. Such problems are common to both phylogenetic and ontogenetic studies. In addition, these epistemological difficulties are often encountered in comparative studies, whether they involve making comparisons between species, between moments in the development of an organism, between the ecological environments of a species, or between cultures, in the case of human societies.

Problems related to the sociology of science are occasioned by the current polemics with respect to *certain* inferences from animals to man, whereas other inferences, apparently analogous from an epistemological point of view, do not lead to controversy. Physiologists and biologists have long considered the human species as one among others. Consequently, conclusions drawn from other species are considered applicable to man as long as homologous systems are involved. It thus becomes economical to study systems wherever it is relatively easy to do so. One may therefore study vision *in general,* assuming that, *mutatis mutandis,* the findings based on research on pigeons or on cats will provide information about the mechanism of vision in all eyes, provided these are similar enough. Behaviour itself has been studied from this perspective: Thorndike and Pavlov attempted to isolate specific features of behaviour that would hold for the whole of the animal kingdom. This endeavour to identify common animal behavioural features was nonetheless accompanied by demonstrations of specific features or distinctions between different specific features of one or another species (Schneirla, 1949; Bitterman, 1965;

Hodos/Campbell, 1969; Seligman, 1970). We must therefore ask why certain inferences are singled out for rejection by a part of the scientific community. Alongside a general problem raised by comparative inferences or by the identification of homologies and analogies, there would seem to be a more specific problem insofar as some of the inferences are concerned. For instance, the von Frisch data concerning communication in bees and his interpretation of these findings have long been opposed. This is also the case for Lorenz' work on aggression. Similar instances may be cited for ontogenetical studies of man as in the resistance to Freud or Erikson's efforts to establish infant sexuality. The problem is thus one of interpreting these social phenomena whereby certain inferences are unacceptable to a part of the scientific community. Is this the result of resistance to a change in paradigm (as defined by Kuhn)? Or is there a problem related to inferences in domains in which it is difficult to establish adequate proof so that some find the evidence sufficient and others do not? Or are certain theses rejected because they go against well-established value systems as in the case of the long-standing opposition to the theory of evolution in the name of religious beliefs? It remains to be seen whether the sociological fact of the polemics surrounding certain inferences from animals to man or from adults to infants is symptomatic of epistemological difficulties, of difficulties related to science, or those related to the scientist, or both at the same time.

BEHAVIOUR IDENTIFICATION AND EVOLUTION: ONTOGENETIC EXAMPLES

One dilemma arises in identifying the invariants which make it possible to understand the phylogenetic as well as the ontogenetic evolution of a system. A problem encountered by the various evolutionist theories derives from the necessity of reconciling the stability of animal species (as evidenced in the difficulties of hybridization) with the transformation of species as it is described by the phylogenetic tree. There is in fact no experimental evidence for the transformation of one species into another and, as attempts to manipulate lineage have produced no new genetically stable species, the possibility of going from intra-species variability to species differentiation is thus still conjectural. The phylogenetic tree which biologists consider a scientific fact does not have the same status as other biological concepts and the kind of evidence advanced for it accounts for why some epistemologists call it a metatheoretic concept. Some new data have however been supplied by studies on the genetic code (Ohno, 1970; Markert, 1975). These data support morphological findings at the biomolecular level and provide a model of the evolutive mechanism by specifying the homological relations between proteins. A double mechanism appears at the biomolecular level, one controlling the formation of new genetic information and the other regulating variations in the way this genetic information is used or expressed. This last mechanism also appears in ontogenetic development through the regulation of gene functions and it is thus that metabolic variations are explained indirectly.

The phylogenetic as well as ontogenetic filiation of behavioural responses raises problems still more complex. Morphological and especially biochemical research has made it possible to specify objects that are relatively stable and identifiable so that homologies can be determined in a relatively reliable manner. The definition of behavioural responses is, however, a more complicated task. Four kinds of criteria can in fact be used conjointly to identify behaviour: 1) the segments concerned, or the physical form of the response; 2) the stimuli eliciting the response; 3) the effects of the response; 4) the functional characteristics of the response. This is not the case in morphological or biological studies where physically stable structures can be defined independently of either their function or the way in which they function. The complexity of identifying behaviours entails difficulties in identifying evolutive series and filiations whether these be phylogenetic or ontogenetic.

Grasping

Scientists agree for instance that the grasping reflex of human infants is an "ancestor" of prehension. However, the grasping reflex differs from prehension by the form of the hand movement since there is no finger-thumb opposition; it also differs by the eliciting stimulus as palmar stimulation is necessary; and finally it differs by the function, for its adaptive value is not known. The grasping reflex can be related to prehension only by the effect and the organ concerned, which is to say by the closure of the hand around the object and the resistance to withdrawal of the object. The current thesis of a filiation between the grasping reflex and prehension thus infers an ontogenetic homology from a very limited analogy.

Walking

An identical analysis can be made if we consider the question of the neonate walking reflex: is this "primary walking" and thus an ancestor of "secondary walking"? In this case, however, André-Thomas and Sainte-Anne Dargassies (1952) have shown that systematic training lengthens the duration of primary walking and later accelerates the development of secondary walking, thus confirming the conjecture of there being a homology (André-Thomas/Autgarden, 1953; Zelazo/Kolb, 1972). But here again primary walking differs from secondary walking in its function, its effects, and the conditions for its prolongation. Previous to the findings provided by André-Thomas and Saint-Anne Dargassies, only the identity of the body segments involved and the apparent similarity of movements justified the "walking" denomination.

Babbling

We may also ask whether infant babbling is an ancestor or a "precursor" of language. Here again the form and the effect as well as the eliciting stimuli differ. The only common feature of both babbling and language is the vocalization of sound involving the apparatus of phonation. In its initial phase, between the fifth and sixth months, babbling is present in deaf as well as in hearing infants (even if there are some quantitative differences as shown by Lenneberg, 1967). This fact leads to the conclusion that auditory feedback is not necessary to the organization and regulation of babbling, whereas we know that auditory feedback is essential where language is concerned. The second phase of babbling, in the second half of the first year, is however absent in deaf infants. Moreover, certain features of this second phase depend on features of the language spoken by adults in the infant's environment. At this stage there is an adjustment of pitch to that of the adult's voice and thus a differentiation between babbling in the presence of the father and babbling in the presence of the mother (Lieberman, 1967) and the appearance of differences between the babbling of infants raised in tone language environments (Chinese) and those raised in an English-speaking environment (Weir, 1966). It is possible to infer that the first phase of babbling is independent of language since an essential feature of the second phase is lacking. On the other hand, it can also be inferred that the similarity in utterance systems and in the functioning of the phonation system constitute an adequate basis for ascribing this behaviour to the language lineage. It may be added that when deafness occurs at the time of the second babbling phase (and even up to three years) the regression observed is comparable to the behaviour of deaf-born infants. This shows that auditory feedback is not only essential to the regulation of the form of babbling, but even to the subsistence of babbling itself. In the absence of experimental control, as in training at the first babbling phase, we are confronted with a situation similar to the one described with respect to grasping.

Reaching

It appears that manual reaching, elicited and controlled by the vision of an object, can be shown from birth (Bower *et al.*, 1970; Bower, 1974) although this behaviour subsequently disappears and then reappears at about four or five months. The conditions for eliciting visually guided reaching behaviour do not remain identical over time, however, and autonomous prehension of an object, regardless of the conditions of presentation, appears relatively late.

In this case we have behavioural responses which differ partially by the eliciting conditions. Initially quite specific, they subsequently disappear and then reappear, becoming less and less specific. Is it the same behaviour? Most psychologists are in agreement that it belongs to the same evolutive lineage, although its functional conditions evolve (Humphrey, 1969; White *et al.*, 1964; White, 1971). However, they do not necessarily agree about the causes and the mechanisms of this evolution.

Imitation

The development of studies on the ontogeny of human behaviour makes it possible to multiply the examples of such problems. We can cite precocious imitation in infants as in the gestural-echo response: opening and closing the mouth, tongue protrusion, etc., on seeing these movements made by an adult. Studied some time ago by Zazzo (1957), this behaviour has been taken up again in more recent research. Here again, the behaviour disappears for a time. When similar behaviour later appears it differs from the first by some of its features and particularly by its functional effects.

Variability of Criteria

While the criteria are stable in the evolution of molecular forms or structures, the criteria used to identify behavioural responses which are to be united in the same ontogenetic sequence are variable. Moreover, their functional signification may vary or may be specified for only one stage of evolution, just as their eliciting conditions may change. In morphological phylogeny, the carriers of indexes or of criteria, that is, the individuals representing species, can be isolated and identified. The problem is one of describing the mechanism of transformation from one individualized form to another: the difficulty does not lie in the discontinuity, but rather in establishing a continuity in spite of the separation between species. In morphological ontogeny, the continuity is assured by the unity of the index conveyor, that is, the evolving individual. The discontinuities are less clear-cut, except when, depending on the level of observation, they appear through sufficiently rapid change and through the appearance or disappearance of elements such as those connected with birth or puberty, or the transformation from the larval state to the imago. In many instances, however, there are quantitative variations that do not constitute real discontinuities. In the case of the ontogeny of behaviour, the problem seems much more complex, since it is necessary to establish the identity of the elements of the evolutive lineage, and since the form and the function of the behaviour pattern may evolve simultaneously. And yet, in spite of the difficulties just described, scientists are generally in agreement as to the identification of ontogenetic behavioural lineages.

HOMOLOGIC INFERENCES AND BEHAVIOURAL PLASTICITY

The problem of behavioural inferences from animals to man or, more generally, from one species to another, like the problem of inferences from one ontogenetic stage to another, corresponds to the broader problem of behavioural plasticity, whether this be from the point of view of its morphological or functional evolution. The status of a behaviour pattern has to be specified not in isolation, but with respect to the general system of adaptation of the organism to its environment. More specifically, every time a comparison is made, we have to ask what remains stable and what changes; the interpretation must take into account both stability and variation. The morphological identity of two behavioural responses is no guarantee of functional identity just as, conversely, functional identity (however defined, since a function is not observable) is no guarantee of morphological identity. The problem is rather one of understanding how the transformations of the organism and the transformations of the adaptive conditions either preserve or transform the function of the behaviour pattern.

General vs. Specific Elicitation Conditions

The examples of ontogenetic behavioural evolution cited above are not equally significant. Babbling is a "spontaneous" behaviour in that it is observed in all infants and elicited under a very wide range of endogenous and exogenous conditions. The walking reflex is, on the other hand, only elicited in special situations where the infant is handled by an adult. A possible interpretation of the difference between the two observable responses is that babbling is a functional behaviour as it appears independently of the variations in environmental conditions and is a general manifestation of a specific developmental stage. Conversely, the singularity of the conditions under which the primary walking response is elicited has suggested the notion of "segmentary" behaviour to some authors because it involves a system in which the regulatory relations with other systems of the organism are not yet established. For this reason it has received both a "maturational" explanation for its disappearance (according to Gesell or Humphrey's definition) and an "integrationist" explanation (according to Hughling Jackson's definition).

Intermediate Examples

Were we faced only with such divergent instances of behaviour as primary walking and babbling, we could limit the functional interpretation to those examples in which the eliciting conditions are necessarily encountered, or at least encountered quite frequently in the developmental environment. However, the other instances cited above fall between these two extremes.

The visually guided gestural echo-response or precocious "imitation" observed in infants is obviously dependent on the presence of the adult, but beyond this general condition, the adult must also behave in a certain manner to elicit the echo-response (*e.g.* repetitive and slow closing and opening of the mouth or of the hand, protrusion and withdrawal of the tongue). In the absence of these specific eliciting conditions, are there no similar but less directly observable interactions? A number of recent studies have dealt with other kinds of visually regulated interactions between the mother and the infant, raising the question of whether the gestural echo-response is a member of this functional class, or whether it constitutes a particular "segmentary" case as its disappearance at about two months would suggest.

The case of reaching behaviour is similar. The eliciting conditions are quite specific, involving the presentation of a small, slowly moving object within a certain distance of the infant, himself in a position that allows him to make certain arm movements. This behaviour also disappears about two months after birth and, as we observed above, it later reappears, but with an autonomy in prehension (no longer confined to grasping) which confers a functional value. It is in terms of the conditions for its initial appearance comparable to "segmentary" behaviour such as primary walking, but however uncommon the eliciting conditions may be, they are not totally artificial in the case of reaching.

Evolution of Behavioural Constraints

The underlying difficulty comes from an oversimplification of the nature of the problem in opposing the innate and the learned, or internal and external conditions. Behaviour is in fact always subject to these constraints. The problem is rather to determine how in each case the constraints are combined. Certain behavioural responses, such as babbling in its first phase, are elicited under a very wide range of situational conditions and seem to conserve their form throughout the variations that take place in situations. It would almost seem that these are determined by internal conditions as far as both the elicitation and regulation of the behaviour are concerned. Indeed the same could be said of breathing. But what is characteristic of babbling is the evolution of the internal and external constraints. In its first phase, babbling appears to be almost independent of external variables, there being no specificity in the conditions. In its second phase, however, the form of behaviour seems to be regulated by external conditions, its maintenance being dependent on auditory feedback, that is, on its "external" effects. It becomes impossible thereafter to determine the course of this behaviour when language-like utterances appear. Moreover, some effects of this behaviour, such as arousing the adult's interest or attention, seem to be constant in both phases. It has even been possible to distinguish some distinct and stable differentiations in sound that have communicatory value (Halliday, 1973). The extent to which this behaviour is increasingly regulated by external constraints is not, however, incompatible with the functional character of babbling as pre-language training, but as likely as this role seems to be, it has not been proven.

An Exception

When a behaviour pattern appears in only a few individuals or even in only one, the singularity may be due either to exceptional environmental conditions to which the individuals have to adapt, or to a modification of the genetic pool (in which case the new pool may be subject to different conditions of adaptation to the environment). Inter-individual variation in the adaptive conditions of different behavioural responses reveals the varying degrees to which these are free from internal constraints, just as the dependence of a behaviour on universal factors, such as the field of gravity, shows the unavoidable character of certain external constraints.

Plasticity and Modes of Adaptation

Thus plasticity is a problem of variability in modes of adaptation. Indeed, if we consider phylogenetic evolution, we can distinguish two diametrically opposed modes of adaptation to variations in ecologic conditions, the time scale being the same for all species living in one ecologic niche. The first mode is bound to variations in genetic conditions and is related to the greater reproductive capacity of the best-adapted individuals. But since individuals are rigid, it is the species (or the genome) which adapts. This implies a short infancy if the rate of reproduction is to be high. The counter-examples that can be cited such as lengthy larval states in some invertebrates are in fact generally found in conjunction with highly restricted conditions of environmental variation (as, for instance, the chrysalis enclosed in the cocoon or hidden in the ground, etc.). Conversely, if the rate of reproduction falls, intra species variability may be partly compensated by intra-individual variability, the plasticity of individuals taking over to some extent for the plasticity of the species. It seems that in phylogenetic evolution there has been a progressive substitution of the second mechanism for the first.

It therefore appears that plasticity is a necessary characteristic of certain behavioural responses. Its functional character is probably bound to the adaptive necessity of adjusting to variations in environmental conditions (at the level of the organism), as for example in food-seeking behaviour. However, a consequence of this plasticity is that it is sometimes possible to obtain "arbitrary" behavioural responses by creating exceptional conditions. This is what happens in many psychological experiments involving learning and conditioning where the system is observed in its extreme functioning conditions. All too often the error is made of taking these as representative of a general case or, *a fortiori,* as "elements" of behaviour.

AN EXAMPLE: LANGUAGE

All these problems are clearly seen in the case of the phylogeny and the ontogeny of human language. Studies carried out by the Gardners as well as Premack or Rumbaugh have shown that the chimpanzee can be trained to handle some aspects of a man-made communication system involving combinative features (Gardner/Gardner, 1969; Premack, 1972; Rumbaugh *et al.,* 1974; Gill/Rumbaugh, 1974). On the other hand, research on nonverbal communication in humans has developed in parallel (Blurton Jones, 1972; Eibl-Eibesfeldt, 1972; Hinde, 1972; Cranach/Vine, 1973). All of these studies have drawn attention to the problem of the phylogeny of language taken as one particular system of communication. They have also tended to reintroduce some continuity between man and animals insofar as the behaviour concerned was considered to be specifically human. At the same time, Chomsky's views have posed the problem in quite different terms, since he contends that human language is innate and is included in the genome while maintaining at the same time that language is uniquely human (Chomsky, 1968).

Everyone agrees that language is specific to at least *Homo sapiens,* since little or nothing is known about extinct species of the genus *Homo* (Lieberman, 1973). Man is the only animal to have developed this type of communication, as he is the only animal to have used fire. However, saying that this is specifically human behaviour does not imply that it appeared abruptly at some time during phylogenetic development, complete with all the features of adult language in our human societies. It thus seems worthwhile to raise the question of the phylogeny and the ontogeny of language.

Phylogeny of Language

Studies on the phylogeny of language have been carried out in different ways, either starting from behaviour observed in man and seeking its ancestral traces in other animal species, or, inversely, attempting to find behavioural responses in man related to those observed in animals. We therefore have a double process of inferences, one from man to animals and the other from animals to man. The same aspects of language behaviour are not necessarily considered, however, and research is sometimes based on analogies, sometimes on homologies. This complex picture serves to bring out some of the essential aspects of the problem of inference.

Phylogeny of the Syntactico-Semantic System

An initial particularity of human language is the infant's ability to modulate utterances in accordance with an external auditory model provided either by the utterances of another person or by his own utterances. We know that the nervous structures necessary to the acoustic regulations of utterances are found only in humans and in various bird species. It is furthermore possible to liken some features of babbling to the acquisition of singing in certain birds. In the chaffinch, for example, a pre-singing phase is found which is comparable to the first babbling phase because it appears even in deaf individuals, and the same extinction of singing as of vocalization takes place if deafness occurs before stabilization (Konishi/Nottebohm, 1969). The phylogenetic distance here excludes any homology, as the difference between the ecological niches excludes, at first sight at least, a simple convergence. Some authors have evoked analogies, although the arguments are not very convincing (Mowrer, 1960). And yet the parrot's imitation of the human voice was already puzzling to Descartes. The mutation which has in this regard separated the genus *Homo,* or at least some of the *Homo* species, from other primates constitutes a fundamental phylogenetic event, for it is not thereby necessary to confine vocalization to a fixed pool of the species, however large it may be. The example of birds nevertheless shows that this mutation is not a sufficient condition for assuring the production of a language with combinative features.

A second particularity of human language is its syntactic combinative structure. This was long held to be indicative of a specifically human capacity, even if experiments on sign-like substitutions could be undertaken with chimpanzees. A number of previously cited studies have shown that it is possible for a chimpanzee to learn to produce and to understand sequences of signs based on syntactical combinative rules, sequences that have semantic value. The success of these experiments demonstrates that the chimpanzee has some of the cognitive capacities necessary to acquire language and that its cognitive performance is qualitatively similar to that of the human infant, if not quantitatively identical to that of the human adult.[1] The syntactical combinative ability and to a certain degree the ability to make referential use of signs are present in the communication elicited in these experiments, but it must be added that some of the features of human language which progressively appear in the human child after three years seem to be absent from chimpanzee language. All the sentences produced by Sarah, Washoe and Lana are so constituted that the semantic value of an element remains constant.[2] This feature is known as context-free in the theory of syntax. Moreover, the sentences do not exhibit elements with *referential* value which depend on the context. For instance, there are no referential marks with variable value (Jakobson's "shifter") such as the pronouns "I" and "you", nor are there any marks for the modification of the declarative value of sentences such as modal or aspectual relations, certain propositional negations, "yes or no" questions, or hypothetical propositions. However, if utterances presenting these features have not yet been observed, it does not follow that it is impossible for the chimpanzee to produce them.

It should be noted that these language features have two important characteristics: 1) the possibility of making sentences whose referential value is independent of the immediate situation; and 2) the possibility of modifying the value of a part of a sentence by using a modifier that changes the nature of the speech act. Moreover, these two characteristics are found in all known languages. The first characteristic is obviously difficult to elicit under experimental conditions with a chimpanzee insofar as learning would require a preestablished language. Furthermore, it would be difficult to envisage what the conditions for control might be. This property of language which makes it possible to build references that are completely independent of an actual situation requires a much broader capacity of representation than that observed in the chimpanzee: delayed reactions, construction of tools—attesting to the capacity of relating means and ends, and conservation of tools—showing that they have taken on a character of generality.

Phylogeny of Representation

That which constitutes the third particularity of human language then is the fact that this particular communication system includes representational properties. This raises the question of whether the phylogeny and the ontogeny of language have to be studied only through communication behaviour. On the one hand, research on nonverbal human communication shows that communication behaviour is not limited to language. This poses the problem of language specificity with respect to the whole set of communication systems that insure interpersonal regulations of behaviour. On the other hand, if communication is not limited to language, language is not limited to a system of communication either, for it is also a system of representation. The study of the philogeny and the ontogeny of language must not be restricted to *only one* of these two functions. The known systems of nonverbal communication in animals as well as in humans (except, of course, the gestural language of the deaf-mute) are always bound to interpersonal relations in the immediate communication situation and do not seem to include representational elements. But this does not exclude the possibility of demonstrating these representational elements in other animal behavioural responses.

The study of the phylogeny of representation must not be limited to language, but one must nevertheless determine which properties of language make possible certain specific aspects of the functions of representation. In humans, the capacity to envisage possible alternatives or of referring to references that do not exist *in situ* is manifested in language, visual imagery and visuo-gestural expression. What is interesting about this latter representational capacity is that it develops, similarly to language, in children after the age of two in accordance with ordered stages of increasing complexity and that it functions independently of language. This autonomy indicates that the capacity is not a simple and direct consequence of language acquisition. Exhibited in such rudimentary forms as making a simple knot (an ability found in three-year-olds), this cognitive capacity is likewise found in highly complex forms which, because *forecasting rules* are involved, suppose the level of cognitive development which Piaget [3] has defined as the operatory stage. The work of tapestry or silk-brocade craftsmen is an example of this cognitive ability to execute the error-free reorganization of complex warp arrangements necessary for changing the design. Such activities have two important characteristics. Firstly, they cannot be taught by verbal means alone (which does not exclude verbal commentary in guiding attention or scanning the operations), but since they can be taught to the blind, they cannot be considered to be purely visually guided activities either. Secondly, these activities have been found in all known human societies and they most likely date from at least the Neolithic period. On the other hand, they have never been observed in animals, even in primates.[4] We do not know whether such activities could be taught to the chimpanzee in the absence of experiments similar to those carried out on language.

This nonverbal capacity to envisage possible alternatives and to refer to references that do not exist *in situ* is also found in another kind of representational activity, at least at the conclusion of the child's ontogenetic development. It consists of visual imagery such as it is found in the computing procedures of champion chess players which are unrelated to language (de Groot, 1965; Chase/Simon, 1973).

With regard to the phylogeny of the kinds of activities just mentioned, we have almost no data concerning visual imagery (except for capacities such as delayed response, making and conservation of tools). The precedents for tracings or drawings executed by primates are rare and unconvincing

(Holmes, 1911; Morris, 1962; Premack, 1975). The symbolic games that Piaget considers the source of systems of representation raise, as do the games observed in monkeys and apes, difficult problems of interpretation. It seems almost certain that these games belong to the ontogeny of representational capacity, but it is not so evident that animal games belong to the phylogeny of representation.

Phylogeny of Speech

The phylogeny of human language appears to be the outcome of an encounter between two types of transformations: a transformation in cognitive and representational capacities (which did not have to be linked to the communication system), and a transformation in the vocal communication system. The latter consisted of a mutation, as already stated, which enlarged the vocal repertory by making the auditory-laryngeal regulation by feedback possible. It also consisted of a transformation of the laryngeal tract, completing the transformation of the nervous system (Lieberman, 1973). In humans, unlike other primates, the supra-laryngeal tract opens into the pharynx and not into the oral cavity. This new disposition makes it possible, at the cost of a respiratory disadvantage, for the capacity of phonetic production to develop. The new apparatus is able to produce well-differentiated contrasts with the least effort in attaining articulatory precision. As a consequence, the number of acoustically differentiated productions is increased. This extension of the phonetic system permits making full use of the opportunities provided by the auditory-motor feedback and allows for the phonetic variety necessary to establish a large number of lexico-syntactic compounds. It should be noted here that if contemporary spoken languages differ greatly from one another in their *phonemic* systems, there is considerable overlap in the way in which their *phonetic* systems make use of the possible compounds of the oppositions of features.

It is possible, as some data seem to suggest, that the mutation of the vocal system of production has been accompanied by a transformation in the system of reception. The human infant of one to four months already appears to be able to discriminate the acoustical features corresponding to the opposition /b/ – /p/ (Eimas *et al.*, 1971). This opposition between voiced and voiceless forms of stopped consonants is both quite advantageous and very widespread in a number of languages. Yet this acoustical opposition is related to the differences in intervals between the onset of the first formant relative to the second and third formants. Given the nature of this opposition, an important problem in the phylogeny and the ontogeny of language is raised by the fact that it is detected in infants before any production of it has taken place, that is, before any differentiation can be made by the contrast between the articulatory modes of these consonants (Lieberman *et al.,* 1967).

Analogical and Digital Systems

These transformations of the vocal system, registered in the genetic pool of the species, have, by enlarging the capacity of the system, made possible its convergence with a representational system, itself having an enlarged capacity. However, we can only take note of this encounter, without being able to say whether these two transformations took place simultaneously and independently or whether one resulted from the other. It is nevertheless worth observing that the specificity of the representational system, such as it is realized in language, is the result of this convergence of systems. The visual and visuo-gestural representational systems mentioned above are *analogical-type* systems in the sense of analogue computers, being based on what Pierce calls iconic signs. Conversely, human language is a *digital-type* system (the linguist's sign-arbitrariness). This last feature makes possible the specialization of certain elements of language that function as orders or signals to effectuate an action or an operation (as in arithmetic or in algebra where the signs of operations—or operators—function as orders to carry out an operation). In language, marks such as "not", "if", "and", "or", etc. work in this way. Representing such orders is practically impossible in analogical systems. And yet having signals that convey orders is essential to the functioning of a representational communication system. Systems of cries and vocalizations, because they are unable to designate objects even if they can function as instrumental communication, seem to be particularly apt as conveyors of orders. The linearity of sonorous sequences is also a necessary factor in this functioning; it is in fact the only way in which the scope of an operator can be defined. Such precision is not possible in a system with more than one dimension such as images, pictures, schemata, or gestures.

The representational specificity of language resides in its double capacity: firstly, that of marking, when constructing references to the non-present (possible alternatives or past events), the "computing" operations to be effectuated; and secondly, that of being able to communicate to others the results of these representational operations. Other systems of representation do not have the same capacities. Marking the operations to be effectuated is not possible in images or in pictures where only the representation of a state resulting from the operations is provided. And visuo-gestural representations are not directly communicable: they are transmitted by imitative examples.

Lateralization

The inneist thesis has been supported not only by advancing the data concerning auditory-laryngeal feedback and phonetic discrimination in infants, but also by data concerning cerebral lateralization. The lateralization problem raises several questions that should be distinguished. Firstly, there is the question of whether the organization of the nervous centres dealing with one class of behavioural responses is unilateral or bilateral. Secondly, there is the question of whether the associations between nervous centres that coordinate two classes of behaviour are such that the classes are always coordinated by centres belonging to the same hemisphere or always by centres belonging to opposite hemispheres. Thirdly, there is the question of why in the majority of cases the nervous centres controlling language are in the *left* hemisphere, a fact that suggests determination by conditions common to the whole species and renders individual chance determination improbable.

Findings concerning lateral predominance in the organization of certain activities are still far from clear in animals (other than man), even though the notion of cerebral functional symmetry is now rejected. Unilateral motor preferences in animal behaviour demonstrates the possible lateralization of motor organization in an individual, but this organization is not stable in the species (Cole, 1957; Lehman, 1970; Trevarthen, 1971). Furthermore, studies on information processing do not always show that each category of information processed corresponds to a lateral organization.[5] On the contrary, the dissymmetry of the cerebral organization has been established for humans, at least for the human adult. Studies in pathology (split-brain and focal lesions) and in normals (unilateral presentation of information) have led to reliable attributions of certain behavioural functions to either the left or the right hemispheres. For example, in cases of acalculia following left cerebral damage, subjects seem to lose algorithmic ability in computations. In cases involving right cerebral lesions, the algorithmic ability seems to be intact, but subjects no longer know how to apply the procedure. It would appear to be a matter of loss in these subjects of "addressing" capacity, to use computer language (Bresson *et al.,* 1972).

Insofar as manual organization is concerned, the problem is not so much to determine whether one of the hands is "preferred" or "more skillful", but to know, as in the example of acalculia, the ways in which the skills of each hand are complementary. The invariable lateralization of the functions of the two hands seems to be the main way in which manual organization distinguishes humans from other primates. It is only in humans that we find the capacity of controlling the activity of one hand by the activity of the other; for instance, holding the apple with one hand and the knife with the other and *coordinating* the complementary and reciprocally guided movements of both hands. This reciprocal guiding activity is not compatible with the idea of two distinct and quasi-independent brains as certain studies of split-brain patients have suggested. Because it involves a double regulation or a modification of one activity by another—the second acting on the input information of the first, this manipulative ability is qualitatively distinct from nonhuman primate activity. This dissymmetry in the role of the two hands appears early in the ontogeny of infant behaviour. In a study on infant reaching, we have found that when infants are presented with an object held by the tip of an adult's fingers, they most often reach for the object with their right hand. This is observed as early as twenty-six weeks. When at the same age the same object is presented on a small wooden tray, the two hands cooperate as follows: one hand, usually the left one, is first placed on the tray, then the right hand reaches for the object, even if at this age there is still no real prehension but only reaching. This behaviour is maintained for several weeks, the left hand acting it would seem as a preliminary locator.

Do behavioural responses showing complementarity in the activity of the two hands have something to do with the right cerebral lateralization of "spatial functions" and left cerebral lateralization of language? Do they have something to do with the evolution of cognitive capacities? The case of the acalculia patients cited above would seem to support such a conjecture. In a theory of behavioural ontogeny of the Piagetian type where cognitive schemes are derived from action schemes, it is reasonable to think that bilateral coordination, where actions act upon actions, may play a role in the development of cognitive activity similar to the one we find in language.

Ontogeny of Language

Turning now to language ontogeny, we are again faced with some of the difficulties raised by phylogenetic studies. Following Chomsky's initial publications, a number of studies tried to replace the thesis that reduced language acquisition and functioning to classical learning and conditioning systems by a new, but equally excessive, nativist thesis. Washoe, Sarah and Lana's exploits have since had the advantage of serving as a reminder that phylogenetic discontinuities are relative. They have also been a reminder that, once mutations have taken place, it is possible, by investigating the series of the species, to find *post hoc* continuity: *natura non facit saltus*. But once the jumps have been made, we can at least try to find out where they started from. We in fact know that before language appears in the infant, a form of communication exists which is already a system of meaning (Halliday, 1973). More importantly, we also know that the more complex functions of language appear only after an intricate preliminary edifice has been constructed. The development of informatory language which involves the capacity of imagining "possible alternatives" is not completed before seven to nine years of age. Lastly, we know that the process of acquisition does not follow the consecutive subdivisions of a linguistic description. There are no distinct phases of phonetic, syntactic, lexical, semantic, and pragmatic learning. Rather, there is a progressive construction of systems of meaningful communication, each adapted to a corresponding level of cognitive capacity and fully fitted to function meaningfully at that level. The prelinguistic system of oral and gestural communication is very similar to what we observe in the chimpanzee. But what is more important is that the cognitive operations effectuated before language is established are not directly transferable in linguistic form. As an example, the child at about eighteen months is capable of grouping objects in a collection. Presented with cups, sticks, and marbles, he will gather all the sticks (or all the marbles) in a cup, even if the objects are spread out at some distance from where he is sitting. Between twenty months and two years, the same child, presented with a collection of marbles, will answer the question, "What is this? " pointing to each object, "marble, marble, marble...". A little later he answers the same question using an article "*a* marble, *a* marble, *a* marble..." *(une bille, une bille, une bille),* and at about twenty-five to twenty-eight months he answers, "*a* marble, *another* marble..." *(une bille, une autre bille).* Finally at about thirty months he will respond with "*some* marbles" *(des billes).* Thus while capable of actively grouping similar objects together in a collection, he was only able to *represent* the collection one year later. This décalage shows that the process of naming does not involve a simple association of objects and words, but consists rather of constructing a new system of representation.

One last point concerning the ontogeny of language should be made. There is a feature of language that has attracted little attention among linguists in spite of the importance of its role, and it is one that appears early in infant development. This is the relation between the linguistic and discursive structure of a sentence and the breath group (Lieberman, 1967). In the second babbling phase, vocalizations are regulated as we have already seen by both adult language and communication functions, but they are also organized into unities which are subject to the rhythm of respiration. Later on, the parts of the primitive sentence, that is, a sentence of the "theme-predicate" or "topic-comment" type (Gruber, 1967), are also governed by this rhythm which is that of intonative melody. Lieberman's analysis of this phenomenon is quite persuasive. The syntactic and semantic organization would accordingly remain dependent on a scanning system constituted by a basic rhythm which may not be necessary to the production of meaning, but which is probably essential for the perceptual découpage, even if semantically neutral. Vocal production, whether it be vocalization or speech, is thus embedded in a more basic gestural behaviour. Furthermore, the scansion which introduces a basic regularity from which modulating variations emerge, is not confined to language; we find it in dancing, singing, and in still other activities.

CONCLUSION

The foregoing examples taken from phylogenetic and ontogenetic studies of language reveal a diversity of problems concerning behavioural inferences from animals to man.

Behavioural Homologies

Behavioural homologies are founded on diverse criteria of identification: the form of behavioural responses, stimulus situations, effects of behavioural responses, functions, etc. While their identification may not be unequivocal, homologies are often deemed satisfying and convincing. This use of similitudes, supported by a minimal number of criteria, undoubtedly has heuristic value, but to go further requires techniques for establishing proofs. In the first place, a change of criterion should not be made arbitrarily, but justified theoretically. For example, if we take the case of behavioural responses such as primary walking found in behavioural ontogeny, the interpretation in terms of a segmentary system is based on one ontogenetic theory. The interpretation of the findings will vary depending on whether it is made within the framework of a purely maturational theory or within the framework of a more complex interactionist theory such as the epigenetic model proposed by Waddington.

The same could be said of the studies of the ontogeny of vision carried out since Hubel and Wiesel's findings. There can be no straightforward interpretation of experiments that involve the effects on the visual system as a result of deprivation of visual experience at birth or at various intervals after birth. What is the meaning of the effects on the specialization of neuronic systems as a result of non-functioning? Does the deprivation, during a critical period, result in the attrition of a system which is genetically determined and consequently established at birth (Hubel/Wiesel, 1970)? Or does this deprivation disturb the establishment, during a critical period, of the functionality of a system which is thus not complete at birth (Held, 1968; Pettigrew, 1972)? Advancing sufficient proof is a difficult task at this point, although data are more numerous in support of the thesis of a perturbation during the completion of a development that depends on postnatal experience. Whereas in this instance the species' most constant environment is a condition for accomplishing a developmental transformation, the form of the developmental transformations seems to be determined by the genetic pool of the species: the normal environmental conditions to which the species is adapted must be radically changed if a variation is to take place.

Behavioural Analogies

Analogies too have considerable heuristic value. If it were confirmed for instance that the auditorily regulated acquisition of singing found in certain birds is always accompanied by a left cerebral lateralization of this behaviour, then it would be plausible to assume that cerebral lateralization is a condition of this regulation and that the left hemisphere has something to do with this function. It would still be necessary however to find experimental data to support this assumption. Analogy involves a general theory of functions. The identity or the similitude of two systems of relations found in different species suggests the hypothesis of the functional necessity of these systems of relations or at least of their optimalization.

Adaptation and Functional Stability

The third point and the most delicate concerns the function of a biological system or of a behaviour pattern. Indeed, a distinction should be made between: *a)* the necessity of postulating, in a general manner, the adaptation of behaviour and of organisms (Bresson, 1971; Rosen, 1967; Sommerhoff, 1969), and *b)* the question of the function of a behavioural response, an organ, or, more generally, a sub-system. Nothing in fact authorizes going directly from the equivalence between survival and adaptation to the logical implication that if there is a behavioural response, then there is a function (and similarly for an organ or any sub-system). Thus three questions must be asked concerning behavioural ontogeny:

1. Is the observed behavioural response functional or not at the ontogenetical stage of evolution (from birth to death) reached by the organism under study?
2. Has the observed behavioural response been previously functional or will it become functional at some point in this evolution?
3. If the observed behavioural response has been functional at different moments of this evolution, has it always had the same function or has it had different functions at different times?

It is the third question that is capital in phylogeny. The two others are related to the problem of identifying behavioural responses. Behaviourists define these by their form and by the stimulus used to elicit them; functional problems are secondary, if they are considered at all. Ethologists on the other hand are primarily interested in behavioural responses that recur in the natural environment and appear to have functional value. Within the evolutionist perspective, studies on phylogenetical determinations tend to emphasize the relation between species membership and functionality. Although this point of view is legitimate, this relation does not at all imply postulating either that behavioural responses are always functional, or that a behavioural response, if functional, always has the same function.

We can speak of the *function* of a system (an organism or a functioning organ) when it acts in such a way as to produce effects that establish or maintain a state or a property in the system itself or in a larger system of which it is a component. Furthermore, the resulting state or property must assure an equilibrium or, in some way, an optimum. It is then evident that the functional properties of such an organization can change for several different reasons.

First, the effect of the functioning of the sub-system (the function of which is under study) does not necessarily involve a modification in the overall system because the sub-system is either not yet or is no longer related to the overall system. In this case there is indeed an observable behavioural effect, but it is one that is not yet or is no longer functional, for example, primary walking.

Second, there may be a modification of the overall system such that this state or functional property of the sub-system no longer assures an equilibrium or an optimum.

Third, a new sub-system may appear which is more effective in maintaining an optimal state in the system as for instance a sub-system that extends the conditions that assure the system's stability.

This inventory is not exhaustive, but it serves to indicate how functional variations in behavioural responses can be taken into account by contemporary models (Bresson, 1967; Sommerhoff, 1969). Establishing homologies between different forms of behaviour (in the above examples, between different effects of a sub-system) does not necessarily mean that they are functionally homologous. We have only to consider language and communication behaviour which we have already discussed at some length. Nonverbal communication behaviour is essential in sexed species to assure mating, essential between mothers and offspring when the young have no autonomy at birth, and essential also in intra- and inter-species social relations in the ecological niche. These behavioural responses must therefore belong to the species' genetic pool. But is their function identical when they constitute all or the better part of the communication system and when they are associated with open systems such as human language? And in the genus *Homo*, have they had the same function throughout evolution? Our knowledge of man is limited to a small part of this evolution: even for *Homo sapiens* the duration of the Upper Palaeolithic period (third Wurm period) represents three times the evolutionary span dating from the Neolithic Age. The interpretation of the functional signification of behaviour is indeed difficult under these circumstances.

Even in ontogeny, what we have said concerning the evolution of the relationship between vocal utterances and the breath group shows that one behavioural regulation may have a function that evolves concomitantly with its integration in a different system.

Finally, the functional interpretation can change when some aspects of the observed system are isolated. In humans and in animals that have a retina with fovea, two visual functions can be distinguished: 1) the primary, foveal, system which is adapted (or adaptable) to exploration or analysis of complex patterns; and 2) the phylogenetically more recent secondary system, incapable of fine discriminations but sensitive to moving objects; it functions as a warning system, entailing, when stimulated, the rotation of the eyes which brings the image of the stimulating source into the fovea (Trevarthen, 1968). This peripheral vision has long been described as "inferior", "lacking acuity", etc., that is, as *deficient* as compared to foveal vision. However, if the warning system is located in the *same* organ as the refined discriminating system, if it is a differentiated region of the retina, then it is *necessary* for this region, first of all to be peripheral, and second, to be insensitive to the cristallin accomodation variations in order to function as a warning signal. Foveal vision is indeed linked to the cristallin accomodation which, with convergence capacity, provides both a clear image and distance cues. If the warning mechanism has to function for any visual distance independently of the distance to which, at any given moment, the cristallin is accommodated, then clearness of the peripheral image is incompatible with this function. In other words, the optimization of isolated sub-systems is not necessarily compatible with the optimization of the overall system in which they are integrated: the composition that maximizes the overall functioning rarely corresponds to the composition of the maxima.

A second example is provided by the fact that whereas the function of human language is to communicate meaning and references, an infinite set of sentences that are neither meaningful nor refer to anything can be produced. This is due to the fact that the capacity of language to construct possible alternatives, rules out semantic rules that would prohibit composing syntactically correct but semantically empty sentences. The resulting secondary functions that consequently appear in human language would then be assonant games and poetry. This fact may be compared to the existence of non-alimentary mastication in alimentary behaviour.

Given the difficulties inherent in the specification of behavioural functions, it can happen that, after having identified a human behavioural pattern that is common to several animal species, one is tempted to generalize the signification of this behaviour, giving it the *same* meaning in the different species. While this functional identity has heuristic value, its assertion also requires proof. One of the difficulties is due to the fact that the rapid evolution in the modes of social utilization of certain human capacities does not imply a corresponding evolution in the biological conditions of behaviour: man's biological adaptive structure is most probably still very close to that of the Palaeolithic hunter. When common behavioural patterns are identified, their functional signification in animals may appear incompatible either with what we know concerning the way contemporary human societies function or with well-established systems of values or with both. It is thus that we are confronted with the sociological problem of science mentioned earlier.

Establishing behavioural homologic or analogic relations does not constitute an explanation any more than does establishing behavioural phylogenetic or ontogenetic filiations. It is the functions of behavioural responses that have to be explained by taking both differences and similarities into account.[6]

NOTES

1. The data concerning these cognitive capacities are not limited to the acquisition of the capacity to compound such combinative sequences. We should also mention findings showing the capacity of the chimpanzee to identify its mirror image as its own image, while the same ability is not observed in the Macaca (Gallup, 1970). We should also mention studies concerning the use of tools in wild or restricted environments (van Lawick-Goodall, 1970).
2. Washoe, Sarah and Lana are the three chimpanzees raised in human context by the Gardners, Premack and the Rumbaugh groups respectively (see Gardner/Gardner, 1969; Premack, 1972; Rumbaugh *et al.*, 1974).
3. It has only been relatively recently that the mathematical theory of knots and braids was developed (Artin, 1974; Crowell/Fox, 1963). This research demonstrates the level of algebraic complexity necessary for correctly and precisely translating such activities into discrete language.
4. The analogy with a behaviour pattern observed in certain birds is however worth noting. The weaver-bird performs a kind of weaving in building its nest, as its name indicates (Collias/Collias, 1962).
5. This has a certain importance with regard to the problem of language lateralization. Nottebohm (1971, 1972) has shown that in the adult chaffinch, the "full song" is mediated by the left hypoglossal nerve. However, when this nerve is sectioned in the young chaffinch before singing has developed, it nevertheless develops normally under right nerve control. These data call to mind what takes place in the case of human language acquisition.
6. Translated from the French by Barbara Lulich.

REFERENCES

Artin, E. 1947. "Theory of Braids", *Annals of Mathematics* 48:643-649.

André-Thomas/Autgarden, S. 1953. "Les deux marches", *La Presse médicale* 61:582-584.

André-Thomas/Saint-Anne Dargassies. 1952. *Études neurologiques sur le nouveau-né et le jeune nourrisson.* Paris, Masson.

Bitterman, M.E. 1965. "Phyletic Differences in Learning", *American Psychologist* 20:396-410.

Blurton Jones, N. (éd.). 1972. *Ethological Studies of Child Behaviour.* Cambridge University Press.

Bower, T.G.R. 1974. *Development in Infancy.* San Francisco, W.H. Freeman and Co.

Bower, T.G.R./Broughton, J.M./Moore, M.K. 1970. "Demonstration of Intention in the Reaching Behavior of Neonate Humans", *Nature* 228 (5272) 679-681.

Bresson, F. 1967. "Adaptation, équilibre et modèles tirés de la théorie des jeux", pp. 105-126 in: Bresson *et al.* (eds). *Les Processus d'adaptation.* Paris, P.U.F.

Bresson, F./Schonen, S. de/Tzortzis, C. 1972. "Etude des perturbations dans des performances logico-arithmétiques chez des sujets atteints de diverses lésions cérébrales", *Langages* 7 (25) :108-122.

Chase, W.G./Simon, H.A. 1973. "The Mind's Eye in Chess", pp. 215-281 in: W. Chase (ed.). *Visual Information Processing.* New York/London, Academic Press.

Chomsky, N. 1968. *Langage and Mind.* New York, Harcourt, Brace and World Inc.

Cole, J. 1957. "Laterality in the Use of the Hand, Foot, and Eye in Monkeys", *Journal of Comparative and Physiological Psychology* 50:296-299.

Collias, N.E./Collias, E.C. 1962. "An Experimental Study of the Mechanisms of Nest Building in a Weaverbird", *Auk* 79:568-595.

Cranach, M. von/Vine, I. (eds.) 1973. *Social Communication and Movement. Studies of Interaction and Expression in Man and Chimpanzee.* London/New York, Academic Press.

Crowell, R.M./Fox, R.H. 1963. *Introduction to Knot Theory.* Boston, Ginn.

De Groot, A. 1965. *Thought and Choice in Chess.* The Hague, Mouton.

Eibl-Eibesfeldt, I. 1972. *Love and Hate: The Natural History of Behavior Patterns.* New York, Holt, Rinehart and Winston.

Eimas, P.D./Siqueland, E.R./Jusczyk, P./Vigorito, J. 1971. "Speech Perception in Infants", *Science* 171:303-306.

Gallup, G.G., Jr. 1970. "Chimpanzees: Self-Recognition", *Science* 167:86-87.

Gardner, R.A./Gardner, B.T. 1969. "Teaching Sign Language to a Chimpanzee", *Science* 165:644-672.

Gill, T.V./Rumbaugh, D.M. 1974. "Mastery of Naming Skills by a Chimpanzee", *Journal of Human Evolution* 3:483-492.

Gruber, J.S. 1967. "Topicalization in Child Language", *Foundations of Language* 3:37-65.

Halliday, M.A.K. 1973. "Early Language Learning: a Sociolinguistic approach", in: W.C. McCormack/S. Wurm (eds). *Language and Man.* The Hague, Mouton.

Held, R. 1968. "Dissociation of Visual Functions by Deprivation and Rearrangement", *Psychologische Forschung* 31:338-348.

Hinde, R.A. (éd.). 1972. *Non-verbal Communication.* Cambridge University Press.

Hodos, W./Campbell, C.B.G. 1969. "*Scala Naturae:* Why There is No Theory in Comparative Psychology", *Psychological Review* 76:337-350.

Holmes, S.J. 1911. *The Evolution of Animal Intelligence.* New York, Holt.

Hubel, D.H./Wiesel, T.N. 1970. "The Period of Susceptibility to the Physiological Effects of Unilateral Eye Closure in Kittens", *Journal of Physiology* 206:419-436.

Humphrey, T. 1969. "Postnatal Repetition of Human Pre-Natal Activity Sequences with some Suggestions of their Neuroanatomical Basis", in: R.J. Robinson (ed.). *Brain and Early Behavior.* New York, Academic Press.

Inhelder, B./Lézine, I./Sinclair, H./Stambak, M. 1972. "Les débuts de la fonction symbolique", *Archives de Psychologie* 41:187-243.

Konishi, M./Nottebohm, F. 1969. "Experimental Studies in the Ontogeny of Avian Vocalizations", in: R.A. Hinde (ed.). *Bird Vocalizations in Relation to Current Problems in Biology and Psychology*. Cambridge University Press.

Lehman, R.A.W. 1970. "Hand Preference and Cerebral Predominance in 24 Rhesus Monkeys", *Journal of Neurological Sciences* 10:185-192.

Lenneberg, E.H. 1967. *Biological Foundations of Language*. New York, Wiley.

Liberman, A.M./Cooper, F.S./Shankweiler, D.P./Studdert-Kennedy, M. 1967. "Perception of the Speech Code", *Psychological Review* 74:431-461.

Lieberman, P. 1967. *Intonation, Perception and Language*. Cambridge, Mass., M.I.T. Press.

— 1973. "On the Evolution of Language: a Unified View", *Cognition* 2:59-94.

Markert, C.L. (ed.). 1975. *Isozymes. III. Developmental Biology, IV. Genetics and Evolution*. New York, Academic Press.

Morris, D. 1962. *The Biology of Art*. London, Methuen.

Mowrer, O.H. 1960. *Learning Theory and the Symbolic Process*. New York, Wiley.

Nottebohm, F. 1971. "Neural Lateralization of Vocal Control in a Passerine Bird: I. Song", *Journal of Experimental Zoology* 177:229-262.

— 1972. "Neural Lateralization of Vocal Control in a Passerine Bird: 2. Subsong, Calls and a Theory of Vocal Learning", *Journal of Experimental Zoology* 179:35-50.

Ohno, S. 1970. *Evolution by Gene Duplication*. Heidelberg, Springer-Verlag.

Pettigrew, J.D. 1972. "The Importance of Early Visual Experience for Neurons of the Developing Geniculostriate System", *Investigative Ophthalmology* 11:386-394.

Piaget, J. 1946. *La Formation du symbole chez l'enfant*. Neuchâtel/Paris, Delachaux et Niestlé.

Premack, D. 1972. "Language in Chimpanzees", *Science* 172:808-822.

— 1975. "Putting a Face Together", *Science* 188:228-236.

Rosen, R. 1967. *Optimality Principles in Biology*. London, Butterworth.

Rumbaugh, D.M./Glasersfeld, E. von/Warner, H./Pisani, P./Gill, T.V. 1974. "Lana (Chimpanzee) Learning Language: a Progress Report", *Brain and Language* 1:205-212.

Schneirla, T.C. 1949. "Levels in the Psychological Capacities of Animals", in: R.W. Sellars/V.H. McGill/M. Farber (eds.). *Philosophy for the Future*. New York, Macmillan.

Seligman, M.E.P. 1970. "On the Generality of the Laws of Learning", *Psychological Review* 77:406-418.

Sommerhoff, G. 1969. "The Abstract Characteristics of Living Systems", in: F.E. Emery (ed.). *Systems Thinking*. Harmondsworth, Penguin.

Trevarthen, C.B. 1968. "Two Mechanisms of Vision in Primates", *Psychologische Forschung* 31:299-337.

— 1971. "Manipulative Strategies of Baboons and the Origin of Cerebral Asymmetry" in: M. Kinsbourne (ed.). *Hemispheric Asymmetry of Function*. London, Tavistock.

Van Lawick-Goodall, J. 1970. "Tool-Using in Primates and Other Vertebrates", pp. 195-249 in: D.S. Lehrman/R.A. Hinde/E. Shaw (eds.). *Advances in the Study of Behavior*. New York/London, Academic Press.

Weir, R.H. 1966. "Some Questions on the Child's Learning of Phonology", pp. 153-172 in: F. Smith/G.A. Miller (eds.). *The Genesis of Language: a Psycholinguistic Approach*. Cambridge, Mass., M.I.T. Press.

White, B.L. 1971. *Human Infant — Experience and Psychological Development*. Englewood-Cliffs, N.J., Prentice Hall.

White, B.L./Castle, P./Held, R. 1964. "Observations on the Development of Visually-Directed Reaching", *Child Development* 35:349-364.

Zazzo, R. 1957. "Le probleme de l'imitation chez le nouveau-né", *Enfance* 1:135-142.

Zelazo, N.A./Kolb, S. 1972. " 'Walking' in the Newborn", *Science* 176:314-315.

KLAUS FOPPA

Comparative Implications of Learning Psychology

Learning psychologists have never been much concerned with methodological problems of infrahuman-human comparisons. From the very beginning of learning experiments most of them seemed to have been more or less convinced that the identification and analysis of elementary learning functions in lower organisms would almost automatically lead to the understanding of more complex forms of learning. The obvious success of animal experimentation, the possibility of formulating functional relationships, and the applicability of these "laws" to human learning (as in the case of behaviour therapy), rendered superfluous the question of whether inferences from animal to human learning processes are feasible. Furthermore if inferential problems of some kind existed, their solution should be possible by the well-known means of inferential statistics.

Despite these firmly rooted convictions, comparative learning psychology has its serious methodological problems. And so does any general learning psychology, at least insofar as it is based on animal data. It is by no means self-evident that, for example, the classical conditioning of leg flexion in the sheep implies the "same" learning process as the conditioning of finger bending with human subjects. The logic of comparison in this rather simple case rests, as far as I can see, on one of two possible strategies. Type I comparisons or inferences consider the problem as one of *generalization*. Observations can be generalized as far as the observed sample of organisms (or learning mechanisms) is representative for the statistical population in question. Type II comparisons or inferences are based on *structural identities* of the observed organisms. I am not quite sure whether these two types of inferential thinking are really independent or not. Since each of them leads to rather different practical consequences, the question may stay open.

THE RELATION BETWEEN INFRAHUMAN AND HUMAN LEARNING AS A PROBLEM OF GENERALIZATION

As mentioned above, Type I comparisons presuppose that the observed sample of animals is representative for that population for which general (learning) laws are formulated. Therefore, all the members of the sample (and the respective population) must have at least one feature in common. Consequently, for Thorndike (1911, p. 280): "Formally the crab, fish, dog, cat, monkey, and baby have very similar intellects and characters. All are systems of connections subject to change by the law of exercise and effect." Even more explicit is Hull's statement, years later (1952, p. 4). "It may be added that most of these postulates have been based on the behavior of lower organisms, particularly the rat, in the belief that the behavior of all mammals operates to the same primary laws." With respect to certain general learning functions all organisms are thought to be identical. Therefore, it is of no importance what animals are taken as subjects of learning experiments. Since these learning functions are supposed to be characteristic for a wide range of adaptive behaviour, they should be valid irrespective of the behaviour segment for which they originally were formulated. Or as Skinner put it: "I suggest that the dynamic properties of operant behavior may be studied with a single reflex (or at least with only as many as are needed to assure the general applicability of the results) (1938, pp. 45 f.). Note that neither the identity of organisms with respect to central learning mechanisms nor the generality of these mechanisms is an object of empirical investigation. It is rather a question of belief. Consequently, the problem of generalizability of results has only formally been solved: if a sample of individuals is observed, the observations obtained are—according to the canon of inferential statistics—generalized for all unobserved members of the population. But nobody can be quite sure what this population is like. Does it include crab, fish, dog, cat, monkey, and baby; all mammals or only higher mammals or primates?

These difficulties of generalization are not overcome by restricting oneself to species-specific learning behaviour (Bitterman, 1960, 1965; Bolles, 1970; Seligman, 1970). Neither do they lose importance by the fact that classical learning theory is somewhat outdated (despite Spiker, 1970; Weiss/Miller, 1971); so are the extreme versions of Hull's or Skinner's early positions. But despite Skinner's much more cautious formulation of today (1971, 1974) the central aspects of the generalization problem remain unsolved:

> Anything like an adequate sampling of the species of the world is out of the question, even for the field ethologist, but the experimental analysis of behavior has been extended to a fairly large number of species, among

> them *Homo sapiens*. There are excellent reasons for beginning with simple cases and moving on only as the power of the analysis permits. If this means, as it seems to mean, that one begins with animals, the emphasis is no doubt upon the features which animals and people have in common. Something is gained, however, since only in this way can one be sure of what is uniquely human. (1974, pp. 226 f.)

One is not only left in uncertainty about which features "animals and people have in common," but the question remains: "what are the adequate methods for demonstrating these common features?" Take, for example, the case of a simple discrimination experiment (of the type sketched below). Whatever species one is experimenting with, one is almost certainly not interested in this special type of learning situation alone. Consequently one prefers to speak about discrimination learning or spatial habit reversal in general. Without knowing how representative the specific experimental set-up and the observed behaviour with respect to the field of interest are, one is generalizing to an unknown domain. Since this seems to be quite a usual procedure, one obviously relies upon the power of intention: whenever an experiment is *intended* to be representative for a certain type of situation, it is treated as being, in fact, representative.

Confronted with this twofold generalization problem (generalization over species and situations) learning psychologists behave as if they could predict the entries of an n-species x m-situations matrix, given the observation of members of *one* species in *one* experimental situation. It should be quite clear that this amounts to a sheer guessing game.

THE RELATION BETWEEN INFRAHUMAN AND HUMAN LEARNING AS A PROBLEM OF STRUCTURAL OR FUNCTIONAL IDENTITY

As mentioned above, the attempt to generalize specific learning data to different species and different situations presupposes at least one common property (learning mechanism) of the population's members. Some of the difficulties of generalization may arise from the fact that this commonality of properties is not empirically checked. The demonstration of structural or functional identities demands a different methodological approach. Whereas the comparison of two different species could be a possible *consequence* of Type I inferences, it is the (trivial) *prerequisite* for Type II inferential procedures: without direct comparison neither structural nor functional identities can be detected.

Before I discuss the application of this comparative method in learning psychology, let me say a word about the method itself. Structural identities in the behaviour of two species, under certain circumstances (which so cogently have been presented in the previous papers), *may* indicate that they are homologues. Although there exists a catalogue of special criteria of homology, I am not quite sure that it is a comparison method *sui generis*. Without going into any detail, let me briefly refer to Wickler's comparison (1961) of "showing the nest entrance" *(Nestzeigen)* with 3- and 10-spined sticklebacks. The problem can be put in the following way: Is "showing the nest entrance" of species A structurally identical to "showing the nest entrance" of species B? Even if you have no information about the behaviour in question, you can infer something about the answer. Probably it will depend on how specific and precise the descriptions of the compared behaviour segments are. The more detailed the definitions, the higher will be the probability that the two examples have to be judged as non-identical. Note that the result primarily depends on the kind of definition one is using. Moreover, the criteria of homology (*e.g.* Remane, 1952) are of importance only insofar as they determine the aspects which necessarily have to be part of the definition. Wickler's question therefore could be reformulated: Does the behaviour of species B satisfy the definition of "showing the nest entrance" of species A? There are not many comparative learning studies of Type II. But to illustrate the specific difficulties of demonstrating structural identities in the learning process of different species a single example should be sufficient. The following discussion should therefore be taken neither primarily as the exposition of the learning psychologist's standard position nor as its critique. It is only intended to clear up some of the problems of Type II inferences in learning psychology.

In his paper on "Phyletic Differences in Learning" (1965), M.E. Bitterman summarizes the results of several learning experiments with "a variety of animals" in the following table.

Animal	Spatial problems: Reversal	Spatial problems: Probability	Visual problems: Reversal	Visual problems: Probability
Monkey	R	R	R	R
Rat	R	R	R	R
Pigeon	R	R	R	F
Turtle	R	R	F	F
Decorticated rat	R	R	F	F
Fish	F	F	F	F
Cockroach	F	F		
Earthworm	F			

Behaviour of a variety of animals in four classes of problem which differentiate rat (R) *and fish* (F) *expressed in terms of similarity to the behaviour of one or the other of these two reference animals (Bitterman, 1965, p. 408. Copyright by the American Psychological Association. Reprinted by permission.)*

The results of four types of experiments (spatial and visual reversal training; spatial and visual probability learning) are classified as either "ratlike" (R) or "fishlike" (F). This means that the respective learning function (*e.g.* of the turtle for spatial habit reversal) resembles the mean learning function of rats in an analogous experimental learning situation. It is immediately obvious that this kind of classification involves several difficulties. First of all there is the question of whether two analogous experimental situations are, in fact, equivalent. Or, as Bitterman puts it: "Who can say, for example, whether the sensory and the motor demands made upon the two animals in these experiments were exactly the same?... It would, I must admit, be a rare coincidence indeed if the conditions employed for the two animals were exactly equal in all of these potentially important respects" (1965, p. 399). Whether Bitterman's proposal to compare two situations by the means of *control by systematic variation* could be a feasible solution, will not be discussed, since other aspects of the problem seem to be more important. The following question is of particular interest to us: which are the aspects of the learning function underlying the designations "ratlike" or "fishlike"? Let us take as an example the *spatial habit reversal.* The experimental situation is rather simple: The animal is trained to choose one of two possible sides (left or right) in a maze. After a certain criterion is reached, the former correct choice is made incorrect and vice versa. This alternation procedure is continued for a certain number of trials. The figure shows the learning curves of the rat, the turtle, the decorticated rat, the fish, the cockroach and the earthworm for spatial reversal problems (see Bitterman, 1965). With respect to these problems, Bitterman states "that improvement has appeared in the rat under a wide variety of conditions—it is difficult, in fact, to find a set of conditions under which the rat does not show improvement. In the fish, by contrast, reliable evidence of improvement has failed to appear under a variety of conditions" (Bitterman, 1965, p. 400). Even though we know that the initial performance of the fish (namely, an African mouth breeder) does not differ from its performance at the end of the experiment and therefore no reversal learning has occurred (the same is true for the cockroach and the earthworm), it is for several reasons difficult to draw clearcut conclusions.

First it has been known for a long time that the interpretation and comparison of average learning curves is quite problematic (*e.g.* the summary of arguments on this topic by Foppa, 1965). This is due mainly to the fact that individual learning curves generally are of rather different shapes, which do not correspond to the smooth, "typical" learning curve. Averaging different functions does not lead to some "true" function lying beneath these individual functions. In fact, it produces an artificial function. Of course nothing changes, unless these functions are depicted graphically. The problem arises

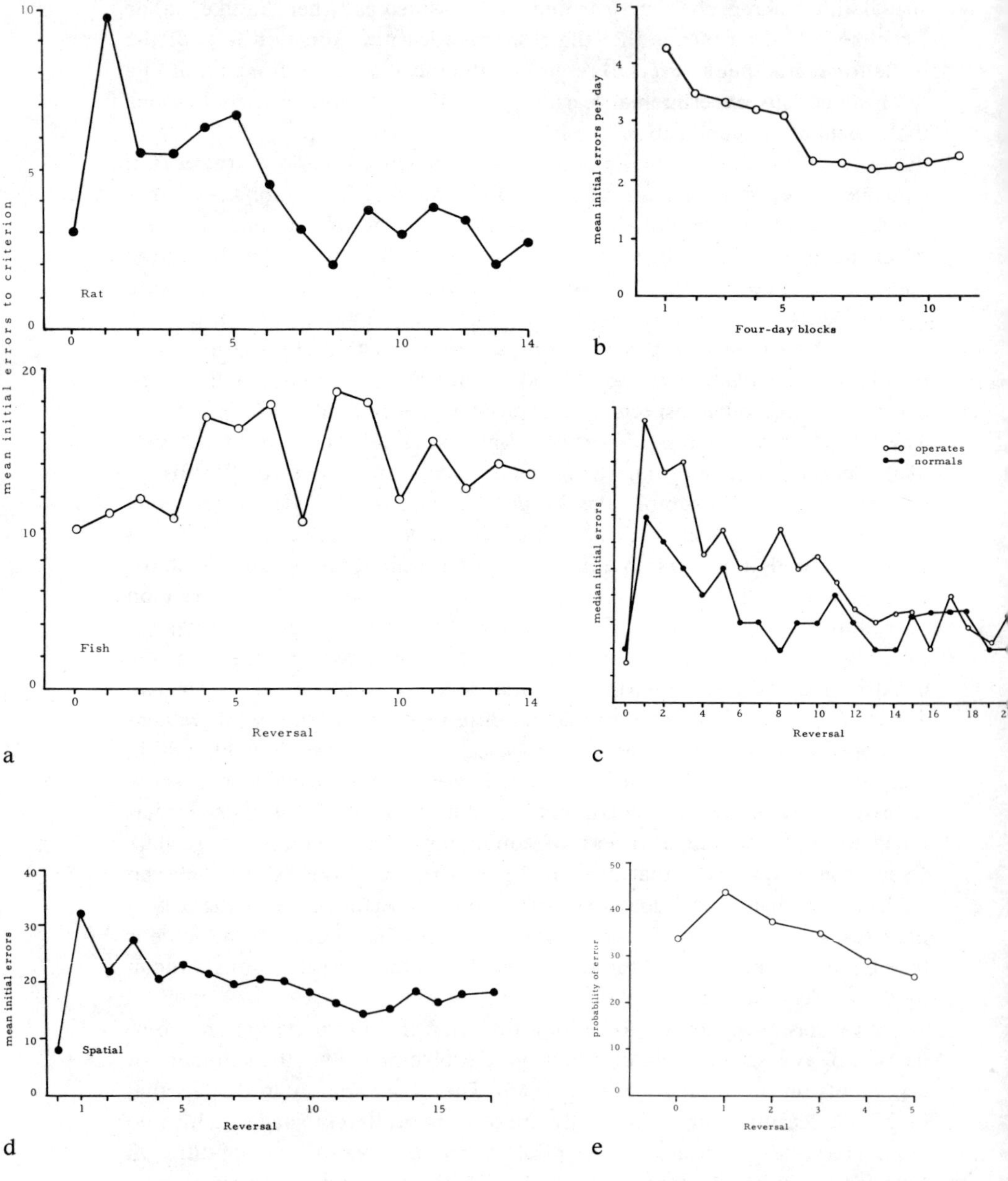

Spatial habit reversal in different animals: a) fish and rat; b) cockroach; c) normal rats and rats extensively decorticated in infancy; d) turtle; e) earthworm. (After Bitterman, 1965.)

as soon as individual scores (*e.g.* number initial errors to criterion) are averaged over single segments of the learning process (*e.g.* reversals). Let us add here that it would not bring about any change if a great number of individuals whose individual learning functions were of the same type, was tested.

Second, a comparison would still be difficult, even if the curves in the graphs represented individual learning functions. A glance at the figure shows that four different learning scores were used in Bitterman's example, namely: mean initial error per day, mean initial error to criterion, median initial errors, and probability of errors. It is impossible to decide on a theoretical basis whether these are equivalent learning measures. Quite probably the different scores are not perfectly correlated (different learning measures almost never are), which means that the single measures do not reflect the same aspects of the learning process. Therefore, in comparing two of these processes, uncomparable things possibly are related.

Third, with data like those of Bitterman, we are faced with an additional problem. He considers "fishlike" behaviour the fact that spatial reversal problems are not mastered by the members of a particular species. In other words, the subject makes more or less the same number of errors in the first reversals as in the later ones. It is impossible to prove by the usual procedures of inferential statistics that there is really no difference between two statistics. If one finds—while comparing two distributions—, no differences of statistical significance, then the conclusion that no such difference exists is not justified: H_0 cannot be proved to be true. What Bitterman classifies as "fishlike" spatial behaviour therefore is not to be characterized as "reversal behaviour in which no learning can be observed". His classification rather refers to individuals or species which do not show any statistically significant improvement in a certain type of situation. The possibility of finding such an improvement may, however, depend on very different factors: given the well known variance of learning achievement between individuals it will be easy to render differences non-significant by examining just a few cases. On the other hand, even small absolute differences can be "made" significant by considering large samples.

Fourth and finally, this leads to a central problem, which arises not only with respect to the comparison of learning data. It is the question under what conditions two compared phenomena have to be considered "equal" or "similar". However numerous the discussions about the concept of similarity and the methods for assessing it have been, little attention has been paid to the problem of equality or sameness. This might be due to the fact that the equality of two objects, processes, etc., seems much more directly perceptible than their similarity.

Despite the divergent opinions about the types of factor on which the similarity of two phenomena depends, it seems beyond question that *relative* indications are meant here. Two unrelated white persons may seem similar when compared to a Negro. But they will seem quite dissimilar when compared with the twin brother of one of them. The meaning of the statement that two things are similar always depends on the *tertium comparationis*. The assessment of equality seems to be independent of such reference points. Therefore similarity is considered a "weak" form of equality, an imperfect realization of full identity.

The fact that neither the assessment of equality or sameness can occur in an absolute sense is overlooked quite easily. If one says, for example, that two species adapt in the same manner to spatial reversal problems, then this assertion refers, of course, to certain experimental situations and to a certain level of analysis, on which the data are processed. Even though this seems to be quite trivial, one cannot help feeling that the assessment of equality itself will lead to an *absolute* assertion, which needs no further explanation.

This does not apply only to those cases in which equality is assessed more or less globally (*e.g.* in case of coins of a certain nominal value, which, only by this token, are treated as equal). Even with respect to more analytical procedures, for example the measurement of body size, equality is no absolute relation. Two individuals might be "equally tall" if measured in centimetres while differences of size may be observed when measured, for example, in millimetres. The degree of correspondence of any two scores depends on the *frame of analysis*. Standard measurement scales (like the cm/g/sec-system) do not change this relativity completely; they only make possible more explicit and precise assertions about the level of analysis.

What does this mean for the problem of interspecific comparison of learning data? It seems obvious that comparisons of learning phenomena of different species are not very meaningful, unless at least one of the following conditions is met: *a) the determination of the aspects of analysis* on the basis of more general biological considerations on species-specific mechanisms, which determine the respective learning process; *b) the determination of the level of analysis* on the basis of a measurement scale. While the first condition is the prerequisite for the *denotation* of the comparative dimensions, the measurement scale renders the *determination* of the level on which equality or sameness are judged. The higher the type of measurement scale, the more exact this determination can be. Therefore, absolute or ratio scales should be sought.

Despite the fact that during the last few years some advances have been made towards a better understanding of species-specific mechanisms of adaptation (*e.g.* Bolles, 1970, 1974), we are still far from formulating biologically established homology criteria for learning processes (similar to those of Remane for morphological structures). The attempt to develop universal and fundamental learning measures, which would allow for a comparison where structural criteria of comparison are absent seems therefore more promising. To conclude, I therefore would like briefly to consider an approach, which still leaves us with some difficulties, but, which despite these, could give rise to some interesting considerations.

The example of Bitterman's experiments shows clearly the situation, in which learning psychology—at least as far as the problem of adequate measuring procedures is concerned—finds itself. There are many possible learning measures and a rational choice among them seems hardly possible. The experimenter is free to choose what measure he finally wants to select for a particular experiment. Apart from considerations of practicality, the change in latency is as "correct" a learning measure as the decrease in errors or the increase of the response strength. This would not be so impor tant if the correlations between the different measures were all high. But this is not precisely the case. If someone still wants to measure the "learning process," he finds himself in a similar situation to someone who wants to measure the growth of different children with tape measures of different elasticity.

Quite apart from interspecies comparisons, the formulation of one or several universal learning measures therefore seems to be indispensable for learning psychology. But how could such a learning measure be arrived at as long as no satisfactory theory of learning exists? I do not, of course, believe that measurement in its broader sense can be established independently of all theoretical considerations. On the other hand, it seems obvious that measurement—at least in some cases—is only a preliminary but not a sufficient condition for the explanation of a phenomenon; hence the presupposition for the formulation of a theory. In this sense, measuring the size of a body is not a sufficient condition for explaining its growth during development, but it is, in any case, a preliminary condition for it. Neither will a universal learning measure be sufficient for explaining the learning process. The formulation of a learning theory is quite another thing.

Since more formal considerations on measurement theory would be out of place, I shall only sketch some possible general properties of such a learning measure.

Let us take as an example any learning experiment. Usually one starts from the assumption that the learning process takes place "between" the behavioural baseline and the criterion behaviour. Although some authors believe that discrete changes occur in the elementary "parts" of this process (the all-or-none models of learning, *e.g.* Estes, 1960), it is generally assumed that the process of behavioural adaptation consists of a more or less continuous transition from the initial to the criterion behaviour. But the assumption of continuous change does not seem to be justified, not even on the molar level of behaviour. If the criterion behaviour is unequivocally defined (for example, 17 "correct" out of 20 trials), this performance level will either have been reached or not. If only 16 out of 20 trials were correct, then the criterion is not "almost" achieved, it is simply not achieved. Otherwise it would make no sense to call the criterion unequivocally defined. Consequently there is no continuous, but a steplike transition from one behavioural state to the other. The choice of the respective criterion, of course, is arbitrary. But once the two behavioural states are defined, the difficulty of the specific learning task can be completely characterized by the time needed for the transition from one to the other. Instead of a usual learning curve, a single time score (given by the sum of the time scores for every trial) would result. To put it a little more precisely, the difficulty of any learning problem for any specific individual behaviour could be determined as the transition time between two behavioural states. This does not mean that no learning curve could be generated, but this function would not be based on performance scores in consecutive trials. Rather it would be defined for the transition time scores over increasingly difficult learning tasks.

Extreme cases, at least, can demonstrate the plausibility of such an approach. If the criterion behaviour lies beyond the possibilities of an individual, transition time would be infinite, for example, practically undeterminable. If, on the other hand, initial and criterion behaviour coincide, no additional time is needed to reach this criterion.

It seems quite evident what the advantages of such a procedure (or of a similar one) would be. First of all, the extent to which the transition times for physically equal criteria correspond to each other could be assessed. Differences in the amount of time needed could inform us about differences in "subjective" difficulty of "physically equal" demands. As a consequence, learning tasks could be ranked for different individuals and different species. Secondly, we could determine which learning tasks are of the same difficulty for different individuals and species. Last but not least, the fact that the influence of experimentally induced conditions on the difficulty of the task could also be analysed without difficulties makes this approach interesting.

It should be mentioned that the first attempts to apply this measuring procedure to paired associate learning met with some technical problems, for example, the construction of tasks with continuously increasing difficulties for single individuals. Nevertheless, attempts to define more universal learning measures are necessary. The success of comparative learning psychology will in the end depend on the success of such attempts.

NOTE

The essential points of the discussion have been incorporated into the body of the paper.

REFERENCES

Bitterman, M.E. 1960. "Toward a Comparative Psychology of Learning", *American Psychologist* 15:704-712.

— 1965. "Phyletic Differences in Learning", *American Psychologist* 20:396-410.

Bolles, R.C. 1970. "Species-Specific Defense Reactions and Avoidance Learning", *Psychological Review* 77:32-48.

Estes, W.K. 1960. "Learning Theory and the New 'Mental Chemistry' ", *Psychological Review* 67:207-223.

Foppa, K. 1965. *Lernen, Gedächtnis, Verhalten*. Köln, Kiepenheuer & Witsch.

Hull, C.L. 1952. *A Behavior System*. New Haven, Yale University Press.

Remane, A. 1952. *Die Grundlagen des natürlichen Systems, der vergleichenden Anatomie und der Phylogenetik*. Leipzig, Geest & Portig.

Seligman, M.E.P. 1970. "On the Generality of the Laws of Learning", *Psychological Review* 77:406-418.

Skinner, B.F. 1938. *The Behavior of Organisms*. New York, Appleton-Century-Crofts.

— 1971. *Beyond Freedom and Dignity*. New York, Knopf.

— 1974. *About Behaviorism*. New York, Knopf.

Spiker, C.C. 1970. An Extension of Hull-Spence Discrimination Learning Theory", *Psychological Review* 77:496-515.

Thorndike, E.L. 1911. *Animal Intelligence*. New York, Macmillan.

Weiss, R.F./Miller, F.G. 1971. "The Drive Theory of Social Facilitation", *Psychological Review* 78:44-57.

Wickler, W. 1963. "Ökologie und Stammesgeschichte von Verhaltensweisen", *Fortschritte der Zoologie* 13:305-369.

MARIO VON CRANACH

Inference from Animal to Human Behaviour: Conclusions

In this chapter, I outline the present state of methods for the comparison of animal and human behaviour. I also intend to find answers to the questions posed in the introduction; not all of them can be answered, however, since some were inappropriately posed. I am trying to systematize what I learned from the conference. In doing so, I am drawing on the conference papers and discussions as well as on additional literature, without accepting all the viewpoints contained in these sources. This chapter, however, is not a summary of the conference proceedings; the discussions and the papers were at once too rich and too divergent for such an enterprise. I am not repeating, except where necessary, what has been said, and the reader cannot make up for reading the other papers by looking up this final chapter.

In contrast to many of the conference papers which deal with factual problems, using them as examples or for their own weight and interest, in this chapter I am intentionally concentrating on the purely methodological point of view. "The logic of inference from animal to human behaviour" was the original name and intent of the enterprise, and I am sticking to that purpose. Thus, I give considerable attention to questions of definition and terminology. I further try to develop some rules for a basic comparative methodology and I investigate, from the viewpoint of concepts and operations, the comparative methods used in ethology. Finally, I attempt an evaluation, establish some rules of application, and arrive at some conclusions concerning further methodological development.

It is possible that some readers will consider my approach too petty and my conclusions too scrupulous, and will feel that the emphasis on clarity of concepts and neatness of operations is exaggerated. In methodological issues, however, it is desirable to try to achieve as much precision as possible. In the long run, even a certain formalization will prove necessary.

I have to confess that I myself am surprised by the difficulty of the matter and the meagreness of methodical possibilities. My hope, therefore, is that at least I should succeed in making some topics objects for rational discussions rather than for taste or dogmatic assertion, as they have often been up to now.

TWO BASIC PROBLEMS OF HUMAN ETHOLOGY

The task of establishing a somewhat unitary science of human behaviour will demand the permanent cooperative efforts of all behavioural and social sciences. The contribution of human ethology should be to explain the behaviour of man as a species that has evolved phylogenetically under specific circumstances, as did any other species. The development of such a theory of human ethology depends, as far as I can see, on the solution of two problems, both referring to relations between animal and human behaviour, namely, the *problem of comparison* and the *problem of integration.*[1] These problems are not completely independent, nevertheless they can be treated separately to a certain degree. This book aims at the methodology of the comparison problem; the integration problem will however be touched, and a final conclusion of this chapter will be that an ultimate solution of the comparison problem presupposes at least a partial solution of the integration problem.

The *problem of comparison* is concerned with the basic question of how the behaviour of animal and man can be compared. It often takes the form of the question of whether behaviours in the two organisms correspond to each other, and what conclusions may be drawn from such a correspondence. In a more general way the comparison may concern complex behavioural systems (*e.g.* "instincts"), or the question may aim at establishing commonalities in the behaviour of species that make use of a certain general behavioural strategy, for example, maternal transport of the baby ("*Tragling*"), the specific question being whether man also uses that strategy. The comparison normally aims at inference from animal to man, although in social ethology an increasing use of concepts derived from human social psychology can be observed in the study of other species.

The importance of the comparison problem may be judged by the many futile polemics, accusations and debates concerning, for instance, Lorenz's book *On Aggression;* the deviant treatment of the problem in various disciplines, such as human ethology and learning theory; and by examples found in this book. The problem is one of special difficulty for two reasons: first is the fact that the complicated principles of comparative methodology have never been properly codified; second, are the differences both between different animals and between a given animal and man, in the complexity of form, function and history of the behaviour to be compared.

Even if the comparison problem were solved, the *problem of integration* would remain. This concerns the different functions and meanings of the identified and compared behavioural elements or subsystems with respect to their total behavioural context. Let us assume that the chimpanzee's "silent bared-teeth face" were to correspond to the human smile[2]: it will immediately become evident that this supposedly phylogenetically old element has a new function in this new context, accompanied by subjective interpretations in man which are unexplorable in the chimpanzee. There are friendly, shy, sardonic and cruel smiles; a smile can be voluntarily produced without appropriate affect; smiling is associated with other communicative behaviours, orienting behaviour and speech; and it is regulated by cultural norms, particularly noticeable from one viewpoint in peoples of the Far East. Generally speaking, any specific human behaviour will be related to a rich context of other human behaviours which partly determine its nature, and any attempt at explanation will have to relate at one and the same time to different information sources, such as genetic information, individual experience and social rules and norms (see Masters, 1975). The more important the behaviour to be studied, the heavier the impact of this contextual integration will be.

There are, however, practically no studies of the integration problem. No rules have been established, no generalizations have been drawn and few examples will be found in the literature. Worse still, the problem is hardly acknowledged as a scientific problem worth theoretical and empirical efforts: few human ethologists take the trouble to add more than a casual remark to their comparative findings that, of course, human behaviour was far more complex. Probably the problem will have to await further development of interdisciplinary research. For the time being, however, when speaking about integration we are literally moving in unknown terrain.

We shall now turn to the comparison problem; but in doing so, we shall continually be reminded of the importance of the integration problem.

BASIC CONSIDERATION ON COMPARATIVE METHODOLOGY

A Classification of Comparative Operations

In this section we are proposing a conceptual framework for a classification of some operational steps that we consider generally indispensable in comparative investigations. This classification should correspond to the logic of the procedure and therefore contain specifications regarding at least two questions: 1) What is the function of the comparative operation in the entire procedure? 2) In view of this function, what information is to be considered indispensable in order to execute the operational step, and what information is generated by this step?

In our classification we distinguish the following operations: comparative investigations are executed in the service of *comparative questions* concerning similarities or differences between the compared systems[3] at a specific *level of comparison*. As major steps in such investigations we distinguish *comparative assessment* and *comparative analysis*. Let us discuss these concepts in more detail.

The Comparative Question

The Concept of the Comparative Question

Comparative research, like all other research, should be performed in the service of a particular question that constitutes the beginning of the investigation. The question does not, however, constitute the beginning of thinking, but will in turn be based on an *"erkenntnisleitendes Interesse"* and have its particular place in a wider context, theoretical or other; it is important to state clearly the nature of this relationship so that the question can be phrased correctly and in detail. Furthermore, this wider context will supply the meaning of the question and help us to ask it in a way that permits us to deal with the "so what" argument that may be raised in the light of the results.

We thus require that the question be appropriately situated in its wider context, especially if the answer is needed in order to arrive at a decision within this same context. There is however one exception to this requirement, namely the aim of pure description.

How much detail is needed can be clarified by considering the function of a question, and this is that it should lead to a clear answer. Because the answer must be obtained by means of empirical investigation, the question must contain the information necessary to perform the empirical steps, that is to determine a level of comparison and to perform comparative assessment and analysis. To the extent that the question remains unspecific, the answer also will be either irrelevant or unspecific.

Similarities and Differences

Similarities are studied if the comparative question concerns characteristics that are common to the compared systems; differences are studied if it concerns what makes them unequal.

On the assumption that both the compared systems possess a limited number of characters and must be similar or different in respect to every one of them, similar and different characters constitute complementary sets which add up to the total set of characters.

Whether we define a given character of both systems as similar or different depends on use of a criterion, that is we can only compare two given actualities in relation to a third one, the criterion. (A test of the hypothesis: "Boys are more aggressive than girls" requires an operational definition of aggressiveness, *e.g.* "physical assault" as a criterion). The criterion used may take the form of a scale which is a set of quantitative criteria ordered along one dimension. A change in the criterion may lead to a change in the classification: similarity may become difference.

The choice of the criterion depends both directly upon the level of comparison (to be discussed in the following paragraph) and indirectly on the nature of the comparative question.

Level of Comparison

Complex systems can be compared in different ways. Behaviours are particularly complex systems, and there are many ways with respect to which even a well-defined and limited piece of behaviour can be subjected to comparison. (For example, the smile of an infant and of an adult person may be compared in respect of the number, kind and combination of the muscles involved; frequency of appearance; temporal course and development; intensity; context, especially other seemingly connected behaviours or antecedents and consequences, to name only a few possibilities). The determination of a "level of comparison" enables us to choose between different modalities of comparison. In order to serve this function, it should contain prescriptions for the comparative analysis and assessment that are sufficiently detailed to provide a basis for these methodical steps.

The problem of choice among different concepts and operations is not new in empirical research. Frequently the concept of a "level of analysis" (or level of integration) is used to designate the choice of particular "units of analysis" (concepts) which subsequently determine the definition of "units of measurement" (operations) (von Cranach, 1964). The level of comparison serves a similar, although superordinated function. It determines the choice of the mode of comparative assessment, specifying operations applied to the compared systems. Also it is itself determined by the comparative question and by a priori assumptions about the compared systems—theory being one of the most prominent forms of the said assumptions.

Comparative Assessment and Analysis

The concepts of these two methodical steps follow from our previous considerations:

Comparative assessment comprises all those methodical operations that aim at gaining data[4] on the compared systems. The main aspects of these procedures will be those of *description* or measurement (pertaining to the question: how are data assessed?); and those of sampling (pertaining to the question: how are the instances to be assessed selected?).

All instances and details of operations executed for description and sampling belong to the category of comparative assessment. Comparative assessment thus constitutes the operational definition of that part of the comparative question that aims at properties of the compared systems on a certain level of comparison. The operations should be well defined as to the degree of reliability that is intended by the investigator; they will, to the same degree, specify the concrete operational meaning of the comparative question. Thus, if a method or instrument is chosen or constructed prior to, and independently of, the comparative question, it will determine how that comparative question is posed.

Our consideration of questions regarding similarities and differences suggests that their respective methodical treatment may be related. In comparative assessment, however, the two cases may differ in regard to the degree of correspondence of comparative operations to be applied to both the compared systems, which depends on assumptions of likeness in the underlying comparative question.

Having achieved comparative assessment, we then possess descriptions of the compared systems with respect to properties suggested in the comparative question at a particular level of comparison. In order to answer the comparative question, however, we have to analyse the data. The concept of *comparative analysis* includes all the methodical operations by which an answer to the comparative question is derived from the set of data. It is therefore based on information contained in the comparative question, and on factual information provided by comparative assessment. The outcome of comparative analysis must constitute an answer to the comparative question. If it does not, then the investigation will have failed.

An obvious and apparently simple instance of such an analysis would be to put the data of both systems side by side and to compare them by inspection. This might give rise to a judgment such as: "look, A is different from B". Comparative analysis, however, is not always restricted to "simple" inspection, it may consist of other forms of tests, especially statistical ones. In the latter case, the comparative question must be phrased as a statistically testable hypothesis. Obviously, the more precisely the operation of comparative analysis is designed, the more specific will it be in yielding answers to questions either of similarity or of difference.

A Research Example

Having distinguished a number of concepts referring to methodical steps in the comparative process, it may now be helpful to illustrate these concepts by analysing a research example. The reader should keep two things in mind: First, we are here interested in the nature and connection of the research operations performed, not in their correctness or in the justification of the conclusions the author may have drawn. Second, the author has not in fact proceeded according to our propositions, nevertheless we are applying our conceptual classification to his work.

A Comparative Study of Facial Expression of Emotion

For our purpose we have chosen a comparative study of facial expression.[5] From a comparative viewpoint this example is relatively simple since no inter-species comparison is involved and the comparison refers to simple and overt rather than more complex and covert behaviour. In its procedures, however, it is more sophisticated than many other comparative studies in this area. The study is part of a research program whose general purpose is illustrated by the following paragraph:

Quotation 1

> Does a particular facial expression signify the same emotion for all peoples? Or does the meaning of any facial expression depend upon the culture of the expressor and the observer? Does an expression composed, for example, of a raised brow, widely opened eyes, and a dropped-open mouth always signify surprise, or only for Americans, or only for certain Americans? Might the same facial expression signify sadness for Japanese and anger for Chileans, and have no emotional connotations whatsoever in some other culture? A long-continuing dispute has been waged over the question of whether there are universal facial expressions of emotion or whether facial expression of emotion is specific to each culture. The argument stems from differing theoretical explanations of how facial muscular movements become related to emotional states... (Ekman, 1971, p. 207.)

Ekman's empirical investigation of this problem is based on his "neuro-cultural theory of facial expression of emotion" (pp. 212-235).

Although the reader is referred to the source, a short quotation will give an impression of the theory's general trend:

Quotation 2

> We have called our theory *neuro-cultural* because it emphasizes two different sets of determinants of facial expressions, one of which is

> responsible for universals and the other for cultural differences. *Neuro* refers to the facial affect program—the relationships between particular emotions and the firing of a particular pattern of facial muscles. This program, as we will explain, is at least partly innate, and can sometimes be activated with relatively little prior cognitive processing or evaluation. *Cultural* refers to the other set of determinants—most of the events which elicit emotion, the rules about controlling the appearance of emotion, and most of the consequences of emotion. These, we hold, are learned and vary with culture (*op. cit.,* p. 212).

Ekman and his co-workers performed several studies to investigate hypotheses derived from this theory. The study we are concerned with here, the second experiment of his series, follows a research type called by the author, "components approach", which is characterized as follows:

Quotation 3

> In a components approach, facial behavior is treated as a response and is directly measured. There are two fundamental questions which can be answered through a components approach. Do the measurements of facial behavior differ depending upon *when* the behavior occured, or *who* showed it? An example of the first type of question is whether the brow is more elevated when people watch stress films than pleasant films. An example of the second type of question is whether there is a difference between Japanese and American subjects in the frequency of brow elevations when watching stress film (*op. cit.* p. 236).

The purpose of the experiment is stated as follows:

Quotation 4

> This experiment, like the first, was designed to determine whether facial expressions of emotion shown by members of two different cultures in a particular exciting circumstance are the same or different. However, rather than a judgment approach, a components approach was used. The facial behavior of the Japanese and American subjects was measured directly to eliminate the possibility that the high correlations found in the first experiment were achieved because observers could interpret culturally differing facial behavior correctly, and to establish that the facial behavior (and not only the judgments) was similar across cultures (*op. cit.* p. 246).

The subjects employed were 25 Japanese students from Waseda University in Japan and 25 American students from the University of California. The selection of the eliciting situation and stimulus material was guided by the following considerations:

Quotation 5

Our theoretical framework suggested that it would be difficult to obtain evidence of the universal aspects of facial expression unless the facial expressions were drawn from a situation which met three criteria: by and large, the same emotions must be elicited in both cultures; display rules to disguise the facial behavior must be inoperative or similar in their operation in the two cultures; and behavioral consequences which might obscure the facial expressions of emotion must be unlikely (*op. cit.* p. 240).

In consequence, both groups were exposed to the same stimulus material which had the following characteristics:

Quotation 6

A set of stress-inducing films, shown to both Japanese and American subjects, was the elicitor selected. This elicitor was chosen because Lazarus (Lazarus/Opton/Tomita/Kodama, 1966) had obtained evidence that these films elicited a comparable verbal report of emotional arousal from Japanese and American subjects, and although there was some ambiguity in the psychophysiological measures of arousal, those data were interpreted (Averill/Opton/Lazarus, 1969) as allowing the inference that the stress films have common eliciting properties in these two cultures (*op. cit.* pp. 240-241).

The subjects viewed a 23-minute film containing both neutral material (a travelogue) and a series of Lazarus's stress-inducing films (*op. cit.*, p. 242).

The behaviour to be compared in the two groups was sampled in the following way:

Quotation 7

The sample of facial behavior of the Japanese and American subjects consisted of the last 3 minutes of each subject's facial behavior videotaped during the neutral film, and the entire 3 minutes recorded during the last stress film (*op. cit.* p. 248).

The facial behaviour was then measured by means of an observation system which is described as follows:

Quotation 8

Measurement utilized a new procedure developed by Ekman, Friesen and Tomkins, the Facial Affect Scoring Technique (FAST). We can here provide only a brief description of these scoring procedures. Details about

its derivation, use and reliability are reported elsewhere (Ekman/Friesen/Tomkins, 1971).
There are two separate steps in FAST measurements, location and classification. In location, scorers determine the beginning and end (and thereby the duration) of each observable movement of the face. In classification, each of these located facial events is compared with a list of items and assigned the score which represents the item it most closely resembles. Location and classification are done separately for the brows-forehead, the eyes-lids, and the lower face. When one of these facial areas is scored, the other two areas are blocked from the view of the FAST scorer. Location requires slowed- and stop-motion viewing. Classification requires consulting a list of facial items for each of the three facial areas. There are 8 items for the brows-forehead, 17 for the eyes-lids, and 59 for the lower face. Each item represents a distinctive appearance of the face. The items do not include or mention emotion, but instead describe a behavioral appearance, for example, a lip press. Most of the items are depicted visually by photographic examples: a few of the items for the lower face are instead described in words (*op. cit.* pp. 248-249).

Most of the items were based on theory, although some were empirically derived (*op. cit.* p. 249).

Results were then calculated for the three facial areas (Brows-Forehead, Eyes-Lids and Lower Face) as well as for single items (movements) as for so called "emotions". The latter term is misleading here; what is meant are particular expressions consisting of combinations of single items and belonging, according to prior research, to specific emotions. These measures were then correlated for both groups:

Quotation 9

The most direct way to test our hypothesis was to correlate the facial measurements of the American and Japanese subjects to determine whether their facial behavior was similar during the stress film (*op. cit.* p. 253).

The item correlations were based on the actual scoring items utilized in the classification of facial behavior. The emotion correlations were based on combining particular items which our theory specified as variants of the same emotion. Let us take as an example the item and emotion correlations for the eyes-lids. There were 17 eyes-lids items. The frequency of occurence of each of these 17 items across all Japanese subjects was correlated with that for the American subjects, separately for the neutral and stress conditions. When the correlation was calculated for

> emotions, the 2 eyes-lids items which our theory specifies as variants of anger were combined to yield a single anger frequency, the 3 items for happiness were combined to yield a single frequency for happiness, etc. Thus, combined item frequency scores for each emotion were obtained, and these emotion frequencies were correlated. Items then describe actual behavior, all of the observable facial behavior. Emotions are groups of items categorized in accordance with our theory (*op. cit.* pp. 253-254).

The computation yielded correlation coefficients of varying size that corresponded in general to the experimenter's expectations:

Quotation 10

> With one exception, the emotion correlations are higher than the item correlations. This is because differences between cultures on specific items were eliminated or reduced when the items were grouped according to emotion. The largest such difference was in the eyes-lids area correlations during the neutral condition. The Japanese and American subjects differed in the frequency of occurence of those eyes-lids items which by our theory are variants on sadness. When the items were grouped into emotions, this difference in frequency was eliminated and the correlation increased from .39 on an item level to .86.
>
> Except for this instance, the correlations on facial behavior in the three separate areas of the face for both items and emotions were high. As we had expected (and explained in introducing this experiment), these correlations were higher for the facial behavior during the stress than the neutral condition, but even for the latter the correlations were substantial (*op. cit.* p. 254).

Still another procedure, comparison of whole face areas, was applied and yielded similar results.[6]

Analysis of the Example

Since we have now obtained an overall view of the study, let us re-examine what Ekman has done in terms of our classification of methodical steps.

Statement of the Comparative Question

First we find that the study is done in the context of a broad comparative question as stated in a general way in Quotation 1. This general question is attacked within the framework of a rather explicit theory, outlined in Quotation 2, and transformed into the *specific comparative question* cited in Quotation 4. It consists of four components which we have further analysed with respect to four viewpoints: where was the component *derived* from? Which *specific aspect* of the comparative question is contained in the component? What operational prescription is given for the treatment of this

	Component	*Derived from*	*Guiding aspect of comparative question*	*Operational prescription*	*Function of component*
1.	Members of two different cultures	General comparative question	What is to be compared?	Statement of comparative objects and general sampling prescription	Specification of comparative question
2.	Facial expression of emotion	General comparative question	In what respect	Determination of kind and site (locus) of movement	Preliminary definition of level of comparison
3.	Same or different	General comparative question	How should question be answered?	Determination of outcome of comparison	General statement of goal of analysis
4.	Particular exciting circumstance	Theoretical framework	How should be compared?	Description of experimental situation	Rough determination of aspects of comparative assessment

Components of the specific comparative question: "This experiment . . . was designed to determine whether facial expression of emotion shown by members of two different cultures in a particular exciting circumstance are the same or different."

aspect? What is the function of the component in the total comparative investigation? For details, the reader is referred to the Table.

Two results of our analysis seem especially noteworthy: 1) The specific comparative question is derived from the general comparative question and the underlying theoretical framework (which is an a priori assumption). 2) The specific comparative question contains only preliminary and rough determinations of the further steps, namely, the level of comparison, the comparative analysis and the comparative assessment. Thus the question arises as to how these parts of the comparative investigation are further specified.

Level of Comparison

Above, we have maintained that the level of comparison should be determined by the comparative question and by a priori assumptions about the compared systems. In our example, we do not find any explicit statement of a comparison level. The statement implicitly contained in the research question is only a preliminary one, and generally phrased: "facial expression of emotion". Since there are many ways of comparing facial expression of emotion—consider only the possibility of varying the degree of fineness by varying the temporal solution of movement by the recording instrument—, where do we find a detailed definition of level of comparison underlying the operations in this study? Almost entirely in the applied method of comparative assessment, FAST.[7]

In view of our requirements on the level of comparison stated on page 359, how does FAST determine the comparison level?

First, it contains information on details of the comparative assessment, for example, how to break up the total facial expression into components, what categories to apply to these, and what degree of fineness to consider in the sampling procedure. In these details, it is based on a variety of assumptions and decisions, most of them of a theoretical nature (Quotation 8).

Second, the details of FAST permit comparison, since it has been constructed with a view to this very purpose. This is achieved by basing FAST on a theory that postulates a certain amount of cross-cultural invariance of facial expression and then develops this principle in detail. It is the same theory that already partly determines the comparative question. On this basis, it seems unnecessary to provide for comparability by other constructive means (*e.g.* using models and judges from both cultures in the construction of the observation system), and we cannot find a clear statement about such means in Ekman, Friesen and Tomkins (1971).

Comparative Assessment

The method of comparative assessment is documented and discussed in some detail in Quotation 8 and in our remarks above, and we have nothing to add here. The reader is also referred to Ekman, Friesen and Tomkins (1971). It should be mentioned that FAST was neither constructed specifically for this particular investigation nor independently of it; it was, however, constructed on the same theoretical basis. This may frequently happen in practice.

Comparative Analysis
For an illustration of this methodical step in our example, the reader is referred to Quotations 9 and 10. It may be noted that, in order to make comparative analysis possible, the author transforms his comparative question for "sameness or difference" into a hypothesis of similarity (Quotation 9). Obviously, this directed hypothesis gives a better picture of the experimenter's true intentions than the relatively neutrally formulated, bi-directed comparative question.

The method for testing the hypothesis statistically, namely, correlational statistics, is chosen on the basis of the hypothesis and of knowledge of the method of comparative assessment that had already been established when the method of comparative assessment was chosen.

The Connection between the Comparative Operations: Logic and Sequential Order.

Since we have now established the function of the single methodical steps in the total comparative investigation, we can define in general a requirement they should meet: each step should be executed for the purpose of providing sufficiently detailed information so that the next step can be performed. Hence, they cannot be performed in random order. On the contrary, our theoretical considerations (pp. 357-360) would suggest the following sequence:

1. Specify the *comparative question*
2. Determine the *level of comparison*
3. Adopt a method of *comparative assessment* for both of the compared systems.
4. Define the method of *comparative analysis*
5. Execute the operations and evaluate the outcome in the light of the comparative question.

The example we have analysed in the preceding section does not, however, follow this sequence; the picture is far more complex.

Obviously, its sequential order is determined by historical factors as well as logical considerations.—We are well aware that this will be a frequent, if not the normal case.—The historical development of a particular piece of research will not necessarily follow our logically established order. As many important findings have been achieved (contrary to methodological rules) before the fitting hypothesis was formulated, it may also happen, for example, that fruitful comparisons may take place before a level of comparison is explicitly defined. In particular, the existence of a potent measurement instrument (*i.e.* a method of comparative assessment) may very often instigate the formulation and investigation of comparative questions. We still think, however, that consideration of the logical order of methodical steps may help to prevent confusion. For the same reason we suggest that in the case of a deviant sequence the requirements of information on which each step is to be based be observed and the source of this information inspected.

Subsequent Comparison

The comparison of data originally collected for other purposes presents a different case. In biology, however, this is no exception since a lot of data comes from purely descriptive studies. Comparisons of this kind should in general follow the strategy of investigating whether the data to be compared could have been collected in the course of a unitary comparative study complying with the methodical requirements just outlined. Special attention may be devoted to the determination of a hypothetical level of comparison.

ETHOLOGICAL METHODS IN BEHAVIOUR COMPARISON

Essential Components of Comparative Questions and Explanatory Concepts

We are claiming that questions concerning correspondences of animal and human behaviour hold a limited number of essential components (the reader may test this claim). These are: structure and function; similarity, sameness and difference; relationship; adaptation to environment and convergence. Let us now inspect the meaning of these terms more closely.

Structure and Function

In an earlier article, I have defined these concepts as follows (von Cranach, 1972, p. 372 f): "The *structure* of a system can be defined as the totality of relations between its subsystems... The structure is thus a map or diagram of relations, and to that extent it has a static quality. The concept of *function,* in everyday parlance, has two uses and can signify: 1) The action of one unit on another unit,[8] whether the relationship between these units is that of element and system or of system and environment, and 2) the behaviour of a structured whole."

Both concepts of function accordingly denote a dynamic aspect of the system. The two are fundamentally similar, for, in the second case, the concept of "behaviour patterns" in open systems must be taken to include interaction with the environment of the system. The second definition is more narrow, for it stipulates some kind of behaviour, whereas the first requires no more than existence.[9] In the following paragraphs and throughout the rest of this paper, the concept of function will be used in the first sense.[10]

As can be seen, the concepts of structure and function relate comparative research to the so-called "general theory of systems" (see von Cranach, 1972, p. 371 ff). This is justified, since the consideration of behaviour as a system yields rich insights.

Of course, this relationship is by no means always clearly stated in comparative questions concerning human and animal behaviour, nor is there general agreement on the definition of the terms structure and function. We hold however that any such question will contain components that may be labelled structure and/or function. That is to say, the description of a piece of behaviour in its elements (on any level of analysis) will describe what will be labelled its structure, and the description of the effect of that behaviour on other behaviours or on the environment will represent what we label its function.

Similarity, Sameness and Difference

These concepts carry two functions in comparative investigations: they are the object of the comparative question and therefore the goal of the comparative operations (assessment and analysis). Although we have already treated similarity and difference, we have to take up the issue again because of the somewhat confusing use of the concepts in comparative studies of behaviour.

Similarity, as we have put it (and also "sameness"), is "sought if the comparative question concerns common characteristics of the compared systems". By closer inspection it turns out that the search may aim at different classes of common characteristics covered by partly identical, partly different terms. We can distinguish between several forms of similarity:

— *Structural similarity* refers to the instance in which two systems show similar configurations of elements and their relations (*i.e.* similarity of structure). If similarity is of a dimension that affects the observer's sense organs (*e.g.* by Gestalt perception), which is not always the case, we may talk about *phenomenal similarity* as a subclass of structural similarity.

— *Contextual similarity* refers to correspondence of two elements with respect to their location in a system, by their relation to other elements; this relationship may refer to both structural and functional aspects. Thus two elements—within two systems—may be similar because they possess a similar position, or because they carry a similar function (= *functional similarity*) even if the systems themselves lack structural similarity.[11]

Besides similarity, the term *sameness* is used, and in a twofold sense. It designates either an extreme form of similarity or, more important, it refers to the *identity* of two systems due to a historical process. In this second sense, the two compared systems may possess a quality of correspondence without obvious similarity, as in the relation of an egg to the chicken into which it later develops. In the following discussions, the term sameness will be used in this second sense.

Difference can refer to a lack of similarity in any of the quoted respects, or to a lack of sameness; therefore in any given case it will be necessary to further specify whether we are dealing with structural, phenomenal, or contextual difference, or difference of historical source.

Relationship designates a historical connection of two systems, in which one system originates from the other or both from a third one. It therefore refers to the explanatory part of the concept of sameness, since it is assumed that two systems show a certain quality of identity because they are related. The relationship is supposed to operate in the "same" systems because of determining common factors, such as "common information" or "common genes".

Adaptation to environment refers to all changes of a system's structure (including behavioural structures) which improve, with respect to a given criterion, the relation to its environment (*i.e.* its function). Thus, the concept is, in its logic and content, related to the concept of function. The necessity to adapt is considered to be an essential cause of any change in a system.

Convergence is the process of independent adaptation of two systems to (similar) environments, resulting in increasing similarity.

The various components outlined above, in different combinations, not only represent the logically essential elements of ethological questions about behaviour comparison, they also constitute the two basic paradigms to which these questions refer according to their specific combinations of components, namely homology and functional analogy. Thus, we may preliminarily define *homology* as sameness of structure based on relationship, and *functional analogy* as similarity of structure resulting from adaptation to environment; thus, analogy can be seen as the result of convergence.[12] Since both paradigms refer to forms of similarity, a paradigm for the *comparative analysis of differences* has to be developed; most probably, it should be a concept derived from that of functional analogy.

Homology and analogy are both explanatory concepts, since they are aiming at an explanation of sameness or similarity; but they also constitute methodological concepts, consisting of rules for the application of criteria making it possible to choose between the alternative explanations in a given empirical case.

In terms of our basic comparative methodology, the concepts of homology and functional analogy refer to specific types of comparative questions, namely those concerning sameness or those concerning the correspondence on a certain level, between structural similarities (homology) and functional similarities (functional analogy). The homology and analogy criteria constitute specific methods of comparative analysis, which in turn determine the methods of comparative assessment.

The Concept of Homology

Concept and method of homology are treated in the papers by Cain, Senn, and Wickler, and the discussions which follow, and the reader is referred to these papers. In this section, I shall try to discuss some open questions, to summarize and to draw conclusions.

Definition and Criteria of Homology

Owen (1848) defined homology as "the same organ under every variety of form and function". Later, the concept underwent considerable change, and at times several competing definitions were in use. Today, Simpson's definition (1950) of homology as "sameness of characters based on genetical relationship" which specifies the sameness by adding the term of genetical relationship, seems to be widely accepted.

From the double function of the concept as an explanatory principle and a methodical tool, it follows that only those parts of the definition are

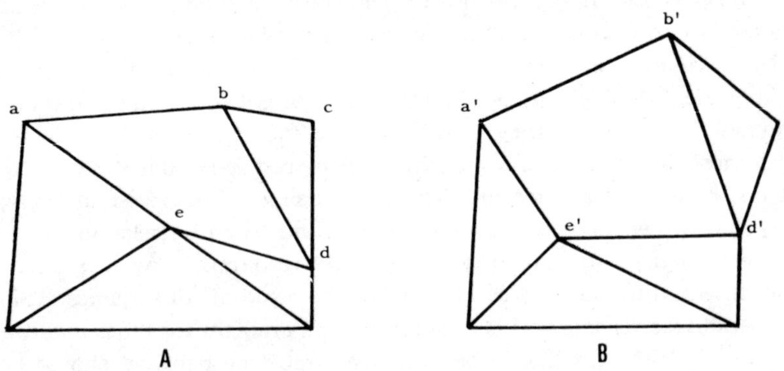

FIGURE 1. *Schematic representation of homology criterion* I
a = a', c = c', *etc. (structural similarity).*

meaningful which correspond to empirically useful criteria. Remane (1952) has systematized these criteria as summarized in the following paragraphs:[13]

The criterion of location[14] (criterion I). Elements in two separate systems are identified by their links with other elements (similarity of location within the structure; see Fig. 1). Since behaviour patterns are structures in time, the placing of their elements must also be defined in relation to a temporal structure. In the case of behaviour patterns, additional confirmation can be obtained by identifying the organs involved in the temporal structure (Wickler, 1961).

The criterion of situation can be applied only if two conditions are met: *the number of components in the two systems must correspond,* otherwise it will not be possible to identify the separate elements accurately, and the *connections between them must be constant in time*. This latter condition means that the criterion of situation cannot be applied to structures whose elements have continually changing connections.

The criterion of the specific quality of structures (criterion II). This criterion is fulfilled when the structures share a maximum number of peculiar characteristics (see Fig. 2). It might therefore more aptly be called "the criterion of the specific quality of elements".

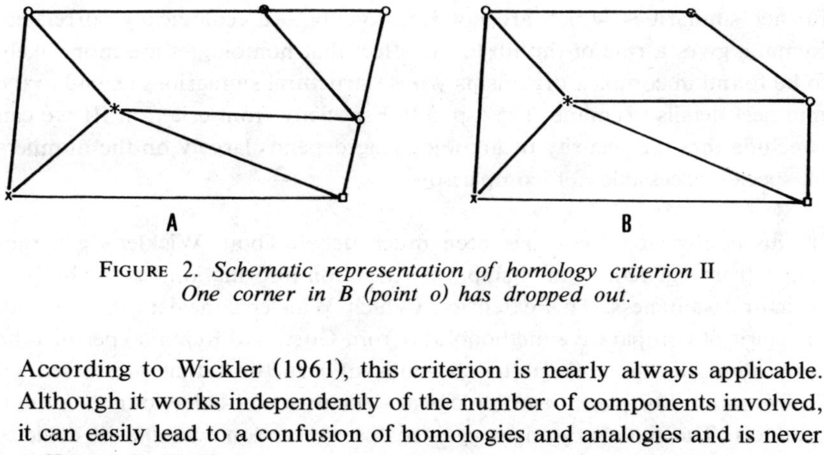

FIGURE 2. *Schematic representation of homology criterion* II
One corner in B (point o) has dropped out.

According to Wickler (1961), this criterion is nearly always applicable. Although it works independently of the number of components involved, it can easily lead to a confusion of homologies and analogies and is never sufficient in itself.

The criterion of relatedness through intermediate forms (criterion III). This criterion can only be applied in a second operation after the first two criteria have been tried. Two distinct structures can be regarded as homologous if they are linked by sufficient intermediate forms satisfying the above homology criteria. These intermediate forms must be descended from organisms whose overall organization lies between that of the organisms to be compared. It is necessary to ensure that the first and last characteristics in the series represented by the two structures are not simultaneously present in one of the intermediate forms (see Fig. 3).

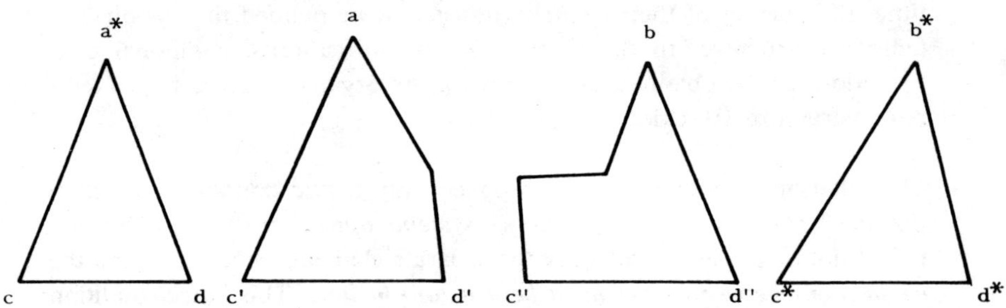

FIGURE 3. *Schematic representation of homology criterion* III
a* *and* b* *are not homologous (as the intermediate forms indicate).*
c *and* c* *are homologous.*

Along with these main criteria there are other auxiliary ones; for example, simple structures are likely to be homologous if they occur in a large number of closely similar species, or if the species in which they occur exhibits further similarities which are not functionally and ecologically correlated. Remane gives a rule of thumb to the effect that homologies are more likely to be found in complex organisms whose structural similarities extend to the minutest details (Remane, 1952, p. 59). Especially from criterion III we can conclude that the security of homologizing depends largely on the numbers of species accessible for comparison.

In this conference, there has been much debate about Wickler's generalization from "genetic relationship" to "common information" as the basis of structural sameness. This extension (which Wickler considers in line with the spirit of comparative methodology from Goethe to Remane) permits the application of the concept to all kinds of structures that are based on "information" or plan, from biological structures and behaviour to cultural products. The main argument against this conception was that it deviates from the established use in biology, giving new meaning to an old term and thus adding to confusion. In favour of Wickler's position, it has been stated that the term genetic relationship promises more than it appears to mean since no practicable criterion for the genetical transfer of the homologized character is used in past and present research. Let us now analyse the issue more profoundly.

Above, we have distinguished three main classes of resemblance of systems, namely, structural (with the sub-class of phenomenal) similarity, contextual similarity, and sameness (based on historical connection). Wickler argues that Remane's homology-criteria do, in fact, prove sameness, and that the next step, the choice between the two competing explanations of sameness, will be more or less arbitrary until it can be based either on positive proof, or on exclusion of one of the two possibilities (which implies inventing a criterion). Actually, in inspecting Remane's first two criteria, I come to the conclusion that the first one refers to contextual similarity, the second one to structural similarity. Sameness (defined as identity in the historical process) can only be inferred from these criteria, and this inference is only justified 1) if no competing explanation for the established similarity exists, and 2) if a historic relationship *in general* (not just with regard to the compared characters) can be assumed. (This is why a proved relationship of the two systems is a precondition of homologizing its parts.) It is only the third criterion that tests the identity in a historical process by taking into account intermediate forms, and thus refers to sameness.

We have depicted these logical steps in Figure 4 (p. 376). We are concluding that whether Simpson's or Wickler's definition is used, the transition from similarity to sameness involves, at least in the absence of criterion III, a guess; while Simpson's definition involves an additional guesslike choice between alternatives (without criteria). In morphology, of course, tradition is ruled out and no other competing explanations exist so that inference to heredity seems rather safe. However, in the case of behaviour tradition is in principle always possible. We therefore tend to prefer, as it stands, Wickler's position.

Application to Behaviour

Homology criteria were originally introduced for application to morphological structures. The expectation that their application to behaviour would be an issue in the discussion of our conference failed, since nobody questioned this possibility.[15]

My previous proposition that behaviour may be homologized *instead* of the central nervous structures that produce it, is to be abandoned in the light of Mr. Hécaen's paper, since the proposition contains an untenable assumption that there is a somewhat firm relationship between brain and behaviour. Instead, we should consider that overt behaviour produces environmental consequences and thus constitutes the point where selection operates. It is therefore conceivable that a behavioural structure remains relatively stable, while its underlying neuronal basis undergoes shifts and changes.

This line of thought directs our attention towards another important

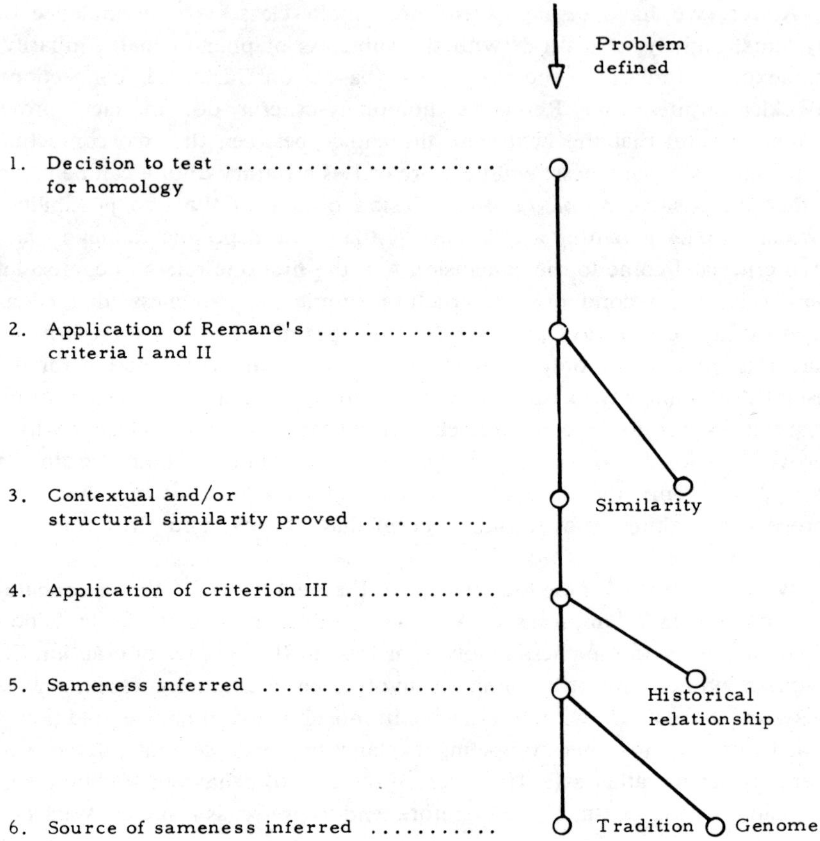

FIGURE 4. *Methodical steps in homologizing.*

point: the movement itself as a hereditary unit is the character on which selection operates, and only characters can be objects of homologizing. Even if character in this sense refers to hereditary complexes as units, it means restriction of the concept's use to rather narrow and well described motor patterns, such as facial expressions, orienting movements, sucking, certain greeting movements and the like.

The confinement to heredity results in still another restriction: concept and method, as stated above, can only be applied to sameness (or similarity), not to difference, for the reason that *common* genes only transmit *common* but not different characteristics. The difference between man and woman, for example, can not be homologized.[16] Another feature of behaviour to be dealt with is its transient nature, the fact that is does not fossilize. Thus we are confined to comparison among extant forms; this will in most cases lead to cladistic homology (Senn, this vol.), and will be the more fruitful the more related species can be investigated.

Relationship and Circularity

Relationship, as we have seen in "the Comparative Question" (p. 358), is a precondition to the inference of sameness. If now, as it often happens, homology methods are applied in the service of a comparative question which aims at an explanation of the observed sameness by relationship, the suspicion of circularity arises. Our discussions concluded, that there is no logical circle involved, if the required statement of a relationship concerns the phylogenetic connection of the compared systems in general the homology question, however specific characters. This reasoning is based on the (generally confirmed) expectation that parts of the compared systems' structures will be determined by common information, parts by adaptation to environment. This is sound logic, since if we assumed that *all the compared structures* are based on common information, the question about the source of any sameness would implicitly be answered.

Another question concerns the term "relationship". Obviously, its specific meaning cannot be generally determined, but has to be redefined in any given investigation depending on the comparative question. In the ideal case in morphology, where comparison takes place in the context of Linné's established systematic order, relationship is always established, as far as I can see, on a somewhat higher branch of the phylogenetic tree than is determined by the comparative question.

In behaviour comparison, in order to gain firmer ground, it might be considered worthwhile to ask about the relationship between the organs that subserve the compared movements. Certainly, a more detailed establishment of rules for the required relationship would be desirable.

Homology of Intra-Individual Behaviours

There are two instances of intra-individual research that imply homology problems, which are normally overlooked.

Serial homology (see Senn, this vol.) comprises, in application of the distinctions developed in comparative morphology, those cases in which behaviour executed by different but corresponding organs is to be compared. An example is the comparison of right and left hand movements.

The so-called *"homonomy"* seems to be the more important case. I do not intend to go into the problems of developmental comparison which Ambrose has so profoundly discussed (this vol.). To state the problem of homonomy, however, I should like to point out that the interspecific comparison of development involves, in general, two steps, namely, the establishment of the developmental sequence within the compared species, and their comparison between the species. It is the first of these that demands our attention here. As far as the underlying hypothesis involves the notion of structural change, it will be necessary to identify the same behaviours at different age levels, which is a homology problem. Sameness of ontogenetically successive states is called homonomy by taxonomists, but there is, to my understanding, no difference from homology that justifies a terminological distinction.

A further application of this reasoning is widely overlooked especially in experimental psychology: wherever a hypothesis involving the notion of sameness is tested in an experiment by comparing behaviours in successive trials, the test of homology is logically required.

Homology of Complex Behaviour Systems

The application of the homology concept to simple movements is perhaps not too controversial; it is mainly the comparison of more complex and comprehensive behaviours and behaviour systems that has been a hot iron. We feel that the question whether these can be treated as cases of homology reasoning can be decided on the basis of the definitions and rules that we have established in this chapter. Let us now have a short look at three examples.

Releasing Mechanisms. The question of whether releasing mechanisms are homologous arises if we compare the *"Kindchen-Schema"* (Lorenz, 1943)[17] in various species. Even the most simple model we could adopt consists of two distinct parts: a perceptual mechanism that results in selective responsiveness to releasing stimuli, and a (released) movement of more or less specific character. Even brief consideration will show that the first part, the perceptual mechanism, can at present only be described as to its function, while a structural description has to await further progress in neurophysiology. Thus homology is out of the question. It is only the motor part of the mechanism that may be considered a case for the homology method.

Motivational Systems. The comparison of motivational systems has been treated by Leyhausen (this vol.) and in the discussion following his paper. Although it is my feeling that in our efforts we have not achieved ultimate clarity, I should like to advise the reader to study these pages carefully. Obviously the matter is, in regard to comparative methodology in general and to homology in particular, fairly complex; let us now try to disentangle it at least as far as it is possible in a short overview.

According to the variety of theories in this field, a motivational system can be defined in many different ways; here we shall concentrate on some common aspects of these definitions. Most of them are based on the assumption that *a set of movements is caused by,* among other factors, *a system of states of readiness.* Let us look at these components in more detail.

a) This set of movements is normally conceived as characterized by a systemic structure organized by related and normally hierarchically ordered functions. To homologize pairs of single movements separately does not constitute a fundamental difficulty; it is the systemic character that is problematic. Comparing a system of characters obviously is a case of application of Remane's second criterion; its restriction as to number and constant connections will, especially in view of the relative independence of the motivational systems stated by Leyhausen, present some difficulty.

b) The system of states of readiness is also of a complex nature; it may include energetic, affective and cognitive components. More important, however, is the fact that it is not constituted by characters at all, but is of a hypothetical nature; in psychology it would be called a "hypothetical construct". There is no justification for applying homology criteria to such a structure.

c) The causal relationship linking movements and states of readiness, finally, is an epistemological principle of a functional nature, that cannot be subject to the homology method. Therefore we come to the conclusion that the motoric portions might, with considerable difficulty, be subject to homologizations; but not the motivational system as a whole.

Social Systems. A social system as a whole does not constitute a hereditary unit and hence is not genetically transferred. It does not appear to constitute a suitable object for homology comparison.

We can conceive of a borderline case under three conditions: 1) *if* two given social systems are exclusively determined by individual behaviour so that there is no social phenomenon that cannot be reduced to individual behaviour; and 2) *if* the relations between the individual behaviour constituting the two systems are invariant; and 3) *if* homology can be proved for all individual behaviours corresponding in the two systems, *then* something equivalent to a homology of the two systems will have been established.[18] These conditions, however, do not normally apply in vertebrates (see Crook, this vol.; Masters, 1975). Social systems, at least in animals like geese or primates, can be characterized as nested sets of relations on hierarchically ordered levels of integration, and partly determined by the social systems' specific histories and environments. Thus they constitute considerably more than a mere set of their members' individual behaviours. We may therefore conclude that in cases that are relevant for human ethology, homology reasoning cannot be applied to social systems.

The Concept of Functional Analogy

Like the term homology, "functional analogy" designates both an explanatory concept and a method. It is treated in the papers by Cain, Masters and Rössler, and the reader is referred to these. In the following section, I shall point out the conclusions in relation to the afore-mentioned methodical principles.

Definition and Properties of the Concept

Owen (1848) defined an "analogue" as "a part or organ in one animal which has the same function as another part or organ in a different animal" (after Cain, this vol.). It has been pointed out (*e.g.* Remane, 1952, p. 88 ff), that in this form the concept is not in contradiction to homology; it is, however, of small explanatory value. The definition has subsequently been changed, mainly by restricting it to the case of the structural similarity of two organisms which is generated by the same function, that is adaptational similarity in Darwin's sense. Simpson's definition (after Masters, this vol.) is nowadays widely accepted: "The development of the same sort of evolutionary opportunity by different groups of organisms produces the evolutionary phenomena known as parallelism and convergence (...) when the element of an ultimate ancestral structure in common is lacking (...) the structures developed for the function in question are not homologous but only analogous." In short, *analogy is structural similarity resulting from adaptation to environment.*

As pointed out, adaptation to environment implies the concept of function. For this reason, the concept of analogy as just defined is better labelled *"functional analogy"* (compare, however, our considerations in the section "functional analogy and the problem of integration".)

The concept of "functional analogy" in this sense is different from the concepts of "analogy" and of "model" as used in the philosophy of science (although the latter two are obviously differently used). Analogy seems to refer to a specific and restricted similarity between a given object and a theoretical object; Achinstein (1964, pp. 332-333) points out: "(...) an analogy is drawn between certain objects or phenomena described in a model (a model of an X), or theory, and other objects or phenomena which may be more familiar; (...) some important analogies involve a comparison between two theoretical objects, or phenomena, themselves (...). When an object cited in an analogy is shown to have certain properties similar in particular respects to properties of objects in a given theory, it is frequently referred to as an analogue of the theoretical objects." The term "model", according to the same author, refers to a much stricter relationship, namely (p. 329, after Braithwaite) : " . . . a model for a theory T is another theory M which corresponds to the theory T in respect of deductive structure . . . (the) model is another interpretation of the theories calculus." From Achinstein's discussion it can be seen that the matter, especially with regard to the logical and the physical correspondence, may be very complex. For our purpose it is sufficient, but also necessary to keep in mind that "functional analogy" goes beyond "analogy" and "model", in that it involves an explanation for an assessed similarity, namely, the function of the similar properties.

Remane (pp. 97, 98) names several types of functional analogies, distinguished by their different origin. In our discussion, we are always referring to analogy based on functional adaptation.

The Relation of Functional Analogy and Homology

It is sometimes argued that any assessed similarity between systems in biology is based on homology, analogy or on chance. The notion of chance as an explanation of similarity implies either the assumption that the development of the structures in question has no adaptive significance; or else the operation of as yet unknown selective forces. The first case seems highly improbable, at least as far as conspicuous characters are concerned. The second case implies the assumption of an analogy with an unknown identical adaptive function. We therefore see homology and analogy as the only explanations of similarity that are theoretically meaningful and have been validated by empirical proof.

Can a certain organ be homologous and analogous at once? To discuss this question, we have to remember the concept of level of comparison. It is generally accepted that similar structures can be homologous at a certain level of comparison and analogous at another one. In the literature, the term "*homeology*" designates analogies that have developed on the basis of homologous organs. In our discussions (see p. 29-30), it has been pointed out that homeology cannot be considered an independent mechanism, if comparison levels are clearly separated.

The question whether homology and analogy can be assumed to exist at the same level of comparison is much more difficult. Researchers working in the field of comparative morphology have stated that homology and analogy exclude each other so that only one of them can be assumed (*e.g.* Remane, p. 92). As far as behaviour is concerned, Wickler (1961) has adopted the same opinion. By the restriction to cases of non-homology, the definition of analogy has assumed, as it were, an additional character.

It is easy to understand why morphologists adopted the rule of exclusion. The purpose of their studies was to establish or complete the phylogenetic tree; therefore, they had to assess genetical relationships. Homological similarities were important for them, while analogies were mainly considered a nuisance, to be eliminated if possible. It is important to see that in comparing animal to human behaviour we are not following the same purpose. It is not our aim to establish a phylogenetic tree, but to better understand the functioning of the human organism by analysing the descendance, development and function of its parts. Our decision whether homology and analogy exclude each other should be based on an evaluation of the mechanisms underlying those concepts. The explanatory power of homology rests on heredity, that of analogy on adaptation. Adaptation, as Masters (this vol.) points out, is the key concept of evolutionary theory and highly important for the function of all kinds of characters, independently of their hereditary connections. There is no logical reason why a given character should not be shaped at once by heredity *and* adaptation. Neither do I know of any theoretical reason or factual finding that speaks against this possibility. I therefore cannot see why, in our context, homology and analogy should, at the same level of comparison, exclude each other. If a certain similarity has been proved homologous, we should go ahead and study its adaptational meaning, and vice versa.

Analogy Criteria and Knowledge of Function

According to Remane (p. 92), these criteria are based on "the following two characteristics of analogy: 1) their relation to function and way of life, and 2) their contradiction to homology in the sense of 'either/or'." This second characteristic is ruled out in consequence of our decision against the mutual exclusion of homology and analogy. The first leads to the following generalization (von Cranach, 1972, p. 385): "Structures are most likely to be analogous if they are found in large numbers in species sharing a particular way of life or a particular environment, independently of any relationship (Wickler, 1961). The same will be true if they occur in species with the same environment or way of life, whose relatives in other environments or with a different way of life lack the structures in question. Neither criterion can rule out homologies altogether."

These criteria are somewhat insufficient, and for obvious reasons: as pointed out by many authors, the notion of function inherent in the concept refers to specific functions in specific environments; thus "function" cannot easily be generalized. In every case, it is necessary to establish the specific system-environment relationships that justify the assumption of the adaptedness of a given character that is instrumental for its development. For both systems, the functions are to be assessed *before* the comparison can be done, or in terms of Rössler's , the two legs have to be established before the bar that connects them can be constructed. Depending on the specific functional analogy to be drawn, the criteria *may* be quite specific.

In any case, however, the analogy conclusion will be the safer the greater the number of species that can be studied in regard to the function in question.

Adaptation to Environment

Adaptation to environment constitutes the key concept of functional analogy, since it represents the mechanism on which the explanation of similarity is based. We have to consider it more thoroughly, particularly since the process in question is based on the assumption of a similarity of environment. Obviously, this leads to a comparison of environments, which again requires a method. We cannot go into details here. Let us note, however, that it is not the total environment that has to be similar, but only that dimension to which the hypothetical function of the structure in question relates. Thus the concept of comparison level applies also to the comparison of environments. For a given physical environment, many different comparison levels may be stated.

Obviously, the hypothetical function has also to be proved. This seems to require two steps: first, the demonstration of a coincidence of environmental similarity (on a dimension defined by the chosen comparison level) with a general behavioural strategy, a "way of life" on the part of the compared organisms (*e.g.* "swimming" or "nectar sucking"). Second, the demonstration of the interaction between the environmental aspect and the "way of life" in the production of the specific behaviour to be explained. (The coincidence of hanging flowers and the strategy of nectar sucking may explain similar flying behaviour in colibris and hawk moths.) Let us emphasize that the concept of adaptation to environment does not require that the adaptation is to be achieved the same way in the two organisms. Very different anatomical or behavioural structures may be transformed in a convergent process in order to serve a similar function.

In view of our methodical claims concerning the concept of function, behavioural comparison from animal to man presents some particular difficulties. Man's natural environment, in any known state of his existence, is culture; this lends unique qualities to the environment's physical properties. Therefore in every case the decision has to be made whether a given aspect of human environment is not unique and fundamentally different from that of his animal counterpart. In this decision, one of the difficulties will consist in the ascription of a particular quality of an environmental object to the environment or to man's adaptive behaviour; in the latter case, this aspect would constitute an instance of how the adaptation is achieved, and correspond to the concept of functional analogy. Is the symbolic quality of a given physical object to be ascribed to the environment or to human (symbolic) adaptive behaviour? This question certainly goes beyond the scope of this paper.

Functional Analogy and the Problem of Integration

In examining analogy criteria and knowledge of function we have located a major difficulty in the use of functional analogies: the necessity to know function in advance. As Remane puts it: "A necessary precondition for judging on analogies as functional similarities seems to consist in the exact knowledge of the total functional connections within the organism, and between the organism and its environment, since only this knowledge permits assessment of the functional apparatus of a system." (p. 92, our translation.)

Remane here points to "functional connections within the organisms", an aspect that we have so far neglected. In fact it is a matter of course that any system is functionally connected as much to its own subsystems as to its environment. Looking at this fact from the point of view of evolution, we have to distinguish between *external* and *internal adaptation*. In a functional analysis of a system, any structure has to be studied from these viewpoints. The horns of a deer, for example, have to be studied in view of the animal's adaptation to its ecological and social environment as well as from that of what it means for the organism to produce several kilogrammes of bones every year.

Consequently, we have to complete our definition: Analogy is structural similarity resulting from an organism's adaptation to his environment (external adaptation) and/or to conditions of his own organization (internal adaptation).

In general, we consider internal to be as important as external adaptation. Internal adaptation becomes more important the more complexly the system under study is organized. We agree with Remane that the study of internal adaptation is a prerequisite for the judgment of functional analogy. At the beginning of this paper, we defined the *problem of integration* as concerning "the different functions . . . of behavioural elements . . . in their total behavioural context". In other words: the problem of integration largely refers to functional adaptation. We therefore conclude that the judgment of functional analogy ultimately presupposes the solution of the problem of integration.

Functional Analogy of Complex Behaviour Systems

The weakness of homology lies in the fact that it refers to characters and can be applied to complex behaviour systems only with difficulty. The advantage of functional analogy is often seen in its greater tolerance in this respect. Is this assumption justified?

Functional analogy is based on the assumption of an evolutionary process, namely, adaptation to environment or internal conditions. Evolution is a combination of three types of processes: variation, selection and retention (Campbell, 1965). All three of them are bound to genetic characters: variation is a product of mutation, selection operates on the character, and retention takes place by transmission of the character's variation norm. Thus, in functional analogy we are dealing with characters as well.

There is, however, a remarkable difference between homology and analogy in this respect. In classifying comparative operations, we have distinguished between the operations of comparative assessment and comparative analysis. In homology, the sameness of characters is the essence of comparative analysis; in functional analogy, characters are objects of comparative assessment. To use Rössler's (this vol.) picture of the letter π in homology the concept of genetic characters is part of the bar, in functional analogy it forms the basis of the two legs. In the latter case, we are therefore not restricted to the comparison of relatives. Also, it might be easier to develop rules and criteria for the analogical comparison of complex behaviour systems; such rules should concern the relation of characters within the two compared systems, and thus confirm the consistency of the combinations of characters. Without the development and application of such rules, the analogical comparison of complex behaviour systems remains a mere analogy in the ordinary sense, without the functional connotations which constitute the specific value of the concept and the method of functional analogy.

CONCLUSIONS

At the end of this chapter, I should like to review my results, putting the emphasis on method rather than content.

At the conference, in our efforts to find at least a preliminary solution to the comparison problem, we suffered from the lack of a general comparative methodology. In order to establish a minimal framework, I have tried to identify some basic methodical steps that should, to my understanding, be part of any comparative study: the statement of a comparative question, the definition of a level of comparison and the operations of comparative assessment and comparative analysis. Reference to this framework could be helpful when confusion arises (as it did so often in our discussions): in particular, the clear definition of a comparison level seems to be helpful in behaviour comparison.

I have then tried to clarify some of the terms used in ethological behaviour comparison. The confusion in regard to the general concept of similarity was a special obstacle for my own thinking. Of course, an agreement on a common terminology would be a necessary requirement for any methodological development.

Next, I have reviewed concepts and methods of homology and functional analogy as used in ethology. Since these are, up to now, the only methods of behaviour comparison that have been at all developed, a short discussion of their utility and limitations for the problem of inference from animal to human behaviour may be appropriate.

By means of the method of homology, we can answer questions about, whether a particular human behaviour is "the same", that is, phylogenetically related to a particular behaviour of a particular animal. The answers would be of considerable value for the construction of the detailed human phylogenetic tree. This would be an important scientific goal if there were enough closely related species. The method's usefulness for the interpretation of human behaviour is limited since it refers only to the sameness of structure of behaviour, that is the physical configuration of the movement; it has to be noted however that the behaviour's function may vary independently (von Cranach, 1972).

Further, we have to take into account the limitations of the method that are inherent in the nature of its basic concept, due to factual circumstances: the methodical restriction to comparison of relatively simple movements (characters) in closely related species, and the difficulty of dealing with complex behaviour systems.

For these reasons, the usefulness of homology for the comparison of animal and human behaviour should not be overrated. The method might yield very interesting results, however, for intraspecific comparison of behaviours in different cultures.

Functional analogy explains similarity through adaptation of the behaviour to the environment; behavioural scientists are likely to be more interested than systematic morphologists in this aspect. The method therefore seems to pertain to the more important questions in our context. Again, conceptual and factual limitations have to be considered. The restriction to the comparison of characters may perhaps be overcome. There is no restriction as to comparison with relatives. The comparison of environmental adaptation, however, presupposes species in similar environments, and the natural human environment, culture, is somewhat unique. Furthermore, we have tried to show that a precondition of functional analogy is the solution of the "problem of integration".

Working on our evaluation of homology and functional analogy, we are impressed by another finding: both these methods aim at forms of similarity; there is no corresponding method that evaluates observed differences. The evaluation of behaviour differences is of course extremely valuable for an understanding of the functions of observed behaviour systems. The development of a comparative methodology for behaviour differences should be possible, making use of the basic logic of (normal) analogy (*e.g.* Hesse, 1963, pp. 112-150), and the model of experimental methodology, which mainly aims at the evaluation of differences.

Thus, we see four main methodological goals:
1. The establishment of a general comparative methodology.
2. The elaboration of homology and functional analogy methods, mainly in order to permit the comparison of complex behaviour systems.
3. The development of methods for the comparison of environments.
4. The development of methods for the comparative evaluation of behaviour differences.

These methodological objectives reflect our emphasis on the function of behaviour systems, which leads to comparative questions different from those of comparative morphologists. Animal ethologists, too, in the long run, will not content themselves with comparative methods that were developed mainly for the purpose of systematic classification.

Stress on function leads to our final point: as we have seen, functional analogy requires previous knowledge of the studied organism's functions, and ultimately presupposes the solution of the integration problem. Comparative studies, in other words, do not free us from the study of the single species, they are quite on the contrary based on a more profound and integrated research on the single species' behaviour. These studies, however, should from the beginning consider the possibility of subsequent comparison. This seems especially true of the study of human behaviour, where the integration of hierarchical levels plays such a major role. Thus, the problem of integration seems a most important object of study in human ethology.

NOTES

1. To my knowledge, the distinction of these two problems goes back to the late E.A. Dölle (see von Cranach, 1974, p. 210). Both these problems are relevant for all behavioural comparison but they gain special significance in inference from animal to human behaviour.
2. See van Hooff, this volume.
3. In this chapter the term "system" as well as related terms will be used in the sense of the "general theory of systems", *e.g.* von Bertalanffy, 1956.
4. We are here using the term data in a wide sense, comprising all kinds of statements about the compared systems that are intended to be descriptive and objective.
5. For examination of the total study, the reader is referred to Ekman, 1971.
6. We did our best to report and cite the crucial parts of Ekman's study; the reader may himself feel the necessity of studying the original paper in its entirety.
7. In order to investigate this in detail, we have looked up the original source on FAST cited in Quotation 8 (Ekman/Friesen/Tomkins, 1971), to which the reader is referred.
8. We should add here: of elements on the same systemic level.
9. The first definition stresses the relational aspect.

10. Masters (this vol.) has pointed to the possible confusion between function and *causation*. While function, as defined, means the action (including all kinds of effects) of one unit on another, causation refers to the way in which such an effect is brought about. Causal analysis is therefore the analysis of function itself and is thus to be performed on a lower level than functional analysis. Certain functions can be based on many causes, and a cause may serve different functions.
11. The swimming-bladder of a fish and the diving-tank of a submarine, for example, are structurally different, but functionally similar elements in structurally different systems.
12. I should like to mention the diverging opinion of P. Leyhausen, who defines analogy as "similarity of service to the system" and convergence as "structural similarity resulting from adaptation to environment". Consequently, he considers convergence a special case of analogy, but not a process leading to analogy.
13. Our summary of the homology criteria is based on Remane (1952) and Wickler (1961). For detailed information, the reader is referred to these sources.
14. In an earlier publication (von Cranach, 1972) I have named criterion I "Criterion of situation", but the term "location" seems more appropriate.
15. A paper by Atz (1970), treats difficulties in the application of the homology concept to behaviour; I discovered this only after this chapter was finished.
16. R. Masters and P. Leyhausen have directed my attention to further problems that should have deserved more than a short note. Although sex-differences, as *differences* between men and women, cannot be homologized, what about the comparison of sex differences between species, *e.g.* sameness of sex difference between man and chimpanzee? The application of the traditional homology method depends on whether the difference is contained in the same genetic information. This leads to questions like whether a given sex difference is a character (or rather two characters of the two sexes); or, consequently, whether the difference is based on common information or, in the traditional formulation, on common genes. Certainly, this question makes a closer determination of what should, in a given case, be labelled as "common information" or what constitutes the organic basis for the inheritance of a given character even more urgent. Recent articles on "evolution on different levels" (King/Wilson, 1975) and on the evolution of genes (Markert/Shakter/Witt, 1975) throw some doubt upon the apparently innocent formulation of the "common gene".
17. The term "Kindchen-Schema" refers to an innate releasing schema in the human being, which, according to Lorenz, increases the attention and intensifies nursing behaviour which mothers show towards their babies.
18. With respect to some apparent relations between aspects of vertebrate social systems, these statements may seem unduly restrictive. The reader should bear in mind that my arguments try to reflect the present state of methodology; and if the latter cannot satisfy important scientific demands, the proper way is to develop methodology rather than to violate its inherent logic.

I am grateful to Anthony Ambrose, Roger Masters and Paul Leyhausen for their aid in correcting my English and for their helpful critical comments.

REFERENCES

Achinstein, P. 1964. "Models, Analogies and Theories", *Philosophy of Science* 31 (4) 328-350.

Atz, J.W. 1970. "The Application of the Idea of Homology to Behavior", pp. 53-74 in: L.R. Aronson *et al.* (eds.) *Development and Evolution of Behavior. Essays in Memory of T.C. Schneirla.* San Francisco, Freeman.

Averill, J.R./Opton, Jr., E.M./Lazarus, R.S. 1969. "Cross-cultural Studies of Psychophysiological Responses during Stress and Emotion", *International Journal of Psychology* 4:88-102.

Bertalanffy, L. von. 1956. "General Systems Theory", *General Systems* 1:1-10.

Campbell, D.T. 1965. "Variation and Selective Retention in Socio-Cultural Evolution", pp. 19-49 in: H.R. Barringer/G.T. Blanksten/R.W. Mack (eds.) *Social Change in Developing Areas.* Cambridge, Mass., Schenkmann.

Cranach, M. von. 1964. "Die Analyse sozialer Organisationen (Grossgruppen)", pp. 44-49 in: *Biologische und kulturelle Grundlagen des Verhaltens.* Bericht über den 24. Kongress der Deutscher Gesellschaft für Psychologie. Göttingen, Hogrefe.

— 1972. "Ethology and Human Behaviour", pp. 370-392 in: J. Israel/H. Tajfel (eds.) *The Context of Social Psychology.* London/New York, Academic Press.

— 1974. "Duplizität und Dubiosität", in: Th. W. Hermann (ed.) *Dichotomie und Duplizität: Grundlagen psychologischer Erkenntnis.* Bern/Wien/Stuttgart, Verlag Hans Huber.

Ekman, P. 1971. "Universals and Cultural Differences in Facial Expressions of Emotions", in: J.K. Cole (ed.) *Nebraska Symposium on Motivation.* Lincoln, University of Nebraska Press.

Ekman, P./Friesen, W.V./Tomkins, S.S. 1971. "Facial Affect Scoring Technique: A First Validity Study", *Semiotica* 3 (1) 37-58.

Hesse, Mary B. 1963. *Models and Analogies in Science.* London/New York, Sheed and Ward.

King, Marie-Claire/Wilson, A.C. 1975. "Evolution at Two Levels in Humans and Chimpanzees", *Science* 188:107-116.

Lazarus, R.S./Opton, Jr., E.M./Tomita, M./Kodama, M.A. 1966. "A Cross-Cultural Study of Stress-Reaction Patterns in Japan", *Journal of Personality and Social Psychology* 4 (6) 622-633.

Leyhausen, P. 1968. "Biologie von Ausdruck und Eindruck", pp. 297-406 in: K. Lorenz/ P. Leyhausen (eds.) *Antriebe tierischen und menschlichen Verhaltens.* München, Piper.

Lorenz, K. 1943. "Die angeborenen Formen möglicher Erfahrung", *Zeitschrift für Tierpsychologie* 5:235-409.

Markert, C.L./Shaklee, J.B./Whitt, G.S. 1975. "Evolution of a Gene", *Science* 189:102-114.

Masters, R.D. 1975. "Political Behavior as a Biological Phenomenon", *Social Science Information* 14 (2) 7-63.

Remane, A. 1952. *Die Grundlagen des natürlichen Systems, der vergleichenden Anatomie und der Phylogenetik.* Leipzig, Geest & Portig.

Simpson, G.G. 1950. *The Meaning of Evolution.* London, Oxford University Press.

Wickler, W. 1963. "Ökologie und Stammesgeschichte von Verhaltensweisen", *Fortschritte der Zoologie* 13:303-365.

IMPRIMERIE DARANTIERE · DIJON-QUETIGNY